REA's Test Prep Books Are The Best!
(a sample of the <u>hundreds of letters</u> REA receives each year)

" A guaranteed "5" on the AP Biology. The book's strength is its six full-length exams with multiple-choice questions. "

AP Biology Student, Stanford University, Stanford, CA

" REA's AP Biology test prep is a good reference source with good tests. It has an excellent review section and the tests are complete. "

AP Biology Student, Detroit, MI

" Your book was such a better value and was so much more complete than anything your competition has produced — and I have them all! "

Teacher, Virginia Beach, VA

" Compared to the other books that my fellow students had, your book was the most useful in helping me get a great score. "

Student, North Hollywood, CA

" Your book was responsible for my success on the exam, which helped me get into the college of my choice... I will look for REA the next time I need help. "

Student, Chesterfield, MO

" Just a short note to say thanks for the great support your book gave me in helping me pass the test... I'm on my way to a B.S. degree because of you! "

Student, Orlando, FL

(more on next page)

(continued from front page)

" I just wanted to thank you for helping me get a great score on the AP U.S. History exam... Thank you for making great test preps! "
Student, Los Angeles, CA

" Your *Fundamentals of Engineering Exam* book was the absolute best preparation I could have had for the exam, and it is one of the major reasons I did so well and passed the FE on my first try. "
Student, Sweetwater, TN

" I used your book to prepare for the test and found that the advice and the sample tests were highly relevant... Without using any other material, I earned very high scores and will be going to the graduate school of my choice. "
Student, New Orleans, LA

" What I found in your book was a wealth of information sufficient to shore up my basic skills in math and verbal... The practice tests were challenging and the answer explanations most helpful. It certainly is the *Best Test Prep for the GRE*! "
Student, Pullman, WA

" I really appreciate the help from your excellent book. Please keep up the great work. "
Student, Albuquerque, NM

" I am writing to thank you for your test preparation... your book helped me immeasurably and I have nothing but praise for your *GRE* preparation."
Student, Benton Harbor, MI

(more on back page)

THE BEST TEST PREPARATION FOR THE
ADVANCED PLACEMENT EXAMINATION

BIOLOGY

Joyce A. Blinn
Reading Specialist – Biology
Bowling Green State University
Bowling Green, Ohio

Shira Rohde
Instructor of Physiology
Santa Rosa Junior College
Santa Rosa, California

Jay Templin, Ed.D.
Assistant Professor of Biology
Widener University
Wilmington, Delaware

Research & Education Association
61 Ethel Road West • Piscataway, New Jersey 08854

The Best Test Preparation for the
ADVANCED PLACEMENT EXAMINATION
IN BIOLOGY

Printed in the United States of America

Library of Congress Control Number 2002113706

International Standard Book Number 0-87891-652-0

Research & Education Association
61 Ethel Road West
Piscataway, New Jersey 08854

REA supports the effort to conserve and
protect environmental resources by
printing on recycled papers.

CONTENTS

ABOUT RESEARCH & EDUCATION ASSOCIATION

Research & Education Association (REA) is an organization of educators, scientists, and engineers specializing in various academic fields. Founded in 1959 with the purpose of disseminating the most recently developed scientific information to groups in industry, government, high schools, and universities, REA has since become a successful and highly respected publisher of study aids, test preps, handbooks, and reference works.

REA's Test Preparation series includes study guides for all academic levels in almost all disciplines. Research & Education Association publishes test preps for students who have not yet completed high school, as well as high school students preparing to enter college. Students from countries around the world seeking to attend college in the United States will find the assistance they need in REA's publications. For college students seeking advanced degrees, REA publishes test preps for many major graduate school admission examinations in a wide variety of disciplines, including engineering, law, and medicine. Students at every level, in every field, with every ambition can find what they are looking for among REA's publications.

Unlike most test preparation books—which present only a few practice tests that bear little resemblance to the actual exams—REA's series presents tests that accurately depict the official exams in both degree of difficulty and types of questions. REA's practice tests are always based upon the most recently administered exams, and include every type of question that can be expected on the actual exams.

REA's publications and educational materials are highly regarded and continually receive an unprecedented amount of praise from professionals, instructors, librarians, parents, and students. Our authors are as diverse as the subject matter represented in the books we publish. They are well-known in their respective disciplines and serve on the faculties of prestigious high schools, colleges, and universities throughout the United States and Canada.

ACKNOWLEDGMENTS

In addition to our authors, we would like to thank Dr. Max Fogiel, President, for his overall guidance, which brought this publication to completion; Larry B. Kling, Quality Control Manager of Books in Print, and John Paul Cording, Manager of Educational Software Publishing, for their supervision of the production of this revised edition; Amy Jamison, Project Manager, for her coordination of revisions; Michael Tomolonis, Assistant Managing Editor for Production, and Amy Laurent for their editorial contributions; and Paula Musselman for typesetting revisions.

ABOUT THE BOOK

This book provides an accurate and complete representation of the Advanced Placement Examination in Biology. Our six practice tests are based on the format of the most recently administered Advanced Placement Biology Exam. Each test is three hours in length and includes every type of question that can be expected on the actual exam. Following each test is an answer key complete with detailed explanations designed to clarify the material for the student. Also included in this book is an AP Biology Course Review. It is provided for students to use as a quick and handy reference for topics that will be covered on the AP Biology Exam. By going over the review section, completing all six tests, and studying the explanations that follow, students will discover their strengths and weaknesses and become well prepared for the actual exam.

ABOUT THE TEST

The Advanced Placement Biology Examination is offered each May at participating schools and multi-school centers throughout the world.

The Advanced Placement Program is designed to allow high school students to pursue college-level studies while attending high school. The participating colleges, in turn, grant credit and/or advanced placement to students who do well on the examinations.

The Advanced Placement Biology Course is designed to be the equivalent of a college introductory biology course, often taken by biology majors in their first year of college. The AP Biology Exam covers material in three areas:

1. Molecules and Cells – 25%
2. Heredity and Evolution – 25%
3. Organisms and Populations – 50%

The exam is divided into two sections:

1. **Multiple-Choice:** Composed of 120 multiple-choice questions designed to test the student's ability to recall and understand various biological facts and concepts. This section of the exam is 90 minutes long, and is worth 60% of the final grade.

2. **Free-Response:** Composed of four mandatory questions designed to test the student's ability to think clearly and present ideas in a logical and coherent way. The answers must be in essay form. Outlines alone, or unlabeled and unexplained diagrams are not acceptable. Each of the four

questions is weighted equally, and topics covered are as follows:

>Molecules and Cells – one question
>Heredity and Evolution – one question
>Organisms and Populations – two questions

Any one of these four questions may require students to analyze or interpret data or information from their laboratory experience as well as classroom lectures. This section of the exam is 90 minutes long, and counts for 40% of the final grade.

Students may find the AP Biology Exam considerably more difficult than most classroom exams. In order to measure the full range of student ability in biology, the AP Exams are designed to produce average scores of approximately 50% of the maximum possible score for the multiple-choice and essay sections. Therefore, students should not expect to attain a perfect or near-perfect score.

ABOUT THE REVIEW SECTION

This book also contains an AP Biology Course Review that can be used by students as both a refresher for subject matter that will be encountered on the exam, and as a quick reference for facts while taking the practice exams.

Chapter 1 – **Molecular Basis of Life and Cells:** Includes sections on Biochemistry, Cells, and Transformation of Energy.

Chapter 2 – **Principles and Theories of Genetics and Evolution:** Includes sections on Molecular Genetics, Heredity, and Evolution.

Chapter 3 – **Organismal and Population Biology:** Includes sections on The Interrelationship of Living Things, Diversity and Characteristics of the Kingdom Monera, The Kingdom Plantae, The Kingdom Animalia, and Principles of Ecology.

SCORING THE TEST

The multiple-choice section of the exam is scored by crediting each correct answer with one point, and deducting only partial credit (one-fourth of a point) for each incorrect answer. Questions omitted do not receive a deduction or a credit. Use this formula to calculate a "raw score":

$$\underline{} - (\underline{} \times 1/4) = \underline{}$$

number
right   number
wrong*   raw score

* DO NOT INCLUDE UNANSWERED QUESTIONS

Then,

$$\underline{} \times .75 = \underline{}$$

raw
score   multiple-choice raw score

The four essays are given a maximum score of 15 points each. Higher scores are awarded to essays that demonstrate in-depth knowledge and understanding of the subject or topic. The score should be given by a person who is both knowledgeable in biology and is able to be impartial, such as a teacher. Add the four scores together to calculate a second "raw score." Use this formula:

$$\underline{} + \underline{} + \underline{} + \underline{} = \underline{}$$

essay #1   essay #2   essay #3   essay #4   essay raw score

Now you will need to determine the composite score. To do this, you will need to add the raw scores from both sections. Use this formula:

$$\underline{} + \underline{} = \underline{}$$

multiple-choice score   essay score   composite score
(round to nearest whole number)

You may then convert your composite score to an AP Biology Test grade by using the scale that follows. Please note that there is not a fixed composite score range that is consistent from year to year. The following scale should be used as an estimate only.

COMPOSITE SCORE	AP GRADE
101–150	5 (extremely well qualified)
82–100	4 (well qualified)
62–81	3 (qualified)
40–61	2 (possibly qualified)
0–39	1 (no recommendation)

Your grade will be used by your college of choice to determine placement in its biology program. This grade will vary from college to college, and is used with other academic information to determine placement. Normally colleges participating in the Advanced Placement Program will recognize grades of 3 or better. Contact your college admissions office for more information regarding its use of AP grades.

CONTACTING THE AP PROGRAM

For registration bulletins or more information about the AP Biology exam, contact:

AP Services
P.O. Box 6671
Princeton, NJ 08541-6671
Phone: (609) 771-7300
Website: www.collegeboard.org/ap

AP BIOLOGY
INDEPENDENT STUDY SCHEDULE

The following study schedule allows for thorough preparation for the AP Biology Examination. Although it is designed for six weeks, it can be reduced to a three-week course by collapsing each two-week period into one. Be sure to set aside enough time (at least two hours each day) to study. No matter which study schedule works best for you, the more time you spend studying, the more prepared and relaxed you will feel on the day of the exam.

It is important for you to discover a time and place for studying. Some students may set aside a certain number of hours every morning to study, while others may choose to study at night before going to sleep. Keep in mind that the most important factor is consistency. Work out a study routine and stick to it!

Week	Activity
1	Acquaint yourself with the Advanced Placement Biology Exam by reading the About the Book and About the Test sections. Study Chapter 1 of the AP Biology Course Review in this book.
2	Study Chapters 2 and 3 of the AP Biology Course Review. Pace yourself so that you can better comprehend what you are reading.
3	Organize and review your classroom notes.
4	Take AP Biology Practice Tests I & II in this book. After scoring your exams, review all incorrect answer explanations. Review subjects that were difficult for you by studying again the appropriate section of our topical review.
5	Take Practice Tests III and IV. After scoring your exams, review all incorrect answer explanations. For any types of questions or subjects that seem difficult to you, check your knowledge against the information in our review.
6	Take Practice Tests V and VI. After scoring them, review all incorrect answer explanations. Study any areas in which you consider yourself to be weak by using the AP Biology Course Review and any other reliable resources you have on hand. Review our practice tests once again to be sure you understand the problems that you originally answered incorrectly.

CHAPTER 1

THE MOLECULAR BASIS OF LIFE AND CELLS

A. BIOCHEMISTRY

Properties of Elements, Atoms, and Molecules

Element – An element is a substance which cannot be decomposed into simpler or less complex substances by ordinary chemical means.

Atoms – Each element is made up of one kind of atom. An atom is the smallest part of an element which can combine with other elements. Each atom consists of:

A) Atomic Nucleus – Small, dense center of an atom.

B) Proton – Positively charged particle of the nucleus.

C) Neutron – Electrically neutral particle of the nucleus.

D) Electron – Negatively charged particle which orbits the nucleus. In normal, neutral atoms, the number of electrons is equal to the number of protons.

Molecule – A group of atoms representing the smallest part of any compound capable of existing in a separate form and maintaining its own identity is a molecule. For instance, one atom of sodium (Na) combined with one atom of chloride (Cl) creates one molecule of salt (NaCl).

Atomic Weight – The total number of protons and neutrons in a nucleus is the atomic weight (mass number). This number approximates the total mass of the nucleus.

Atomic Number – The atomic number is equal to the number of protons in the nucleus of an element.

Isotope – Atoms of the same element that have a different number of neutrons are known as isotopes. All isotopes of the same element have essentially the same chemical properties but their physical properties may be affected.

Ions – Atoms or groups of atoms which have lost or gained electrons are called ions. One of the ions formed is always electropositive and the other electronegative.

Molecular Bonds and Forces

A molecule is composed of two or more atoms bonded together. These bonds include:

Covalent Bond – This involves the sharing of pairs of electrons between atoms. Covalent bonds may be single, double, or triple.

Polar Covalent Bond – A polar covalent bond is a bond in which the charge is distributed asymmetrically within the bond.

Non-Polar Covalent Bond – A non-polar covalent bond is a bond where the electrons are pulled exactly equally by two atoms.

Hydrogen Bond – A hydrogen bond is formed when a single hydrogen atom is shared between two electronegative atoms, usually nitrogen or oxygen.

Certain forces also affect molecules, and these include:

Van der Waals Forces – Van der Waals forces are weak linkages which occur between electrically neutral molecules or parts of molecules which are very close to each other.

Hydrophobic Interactions – Hydrophobic interactions occur between groups that are insoluble in water. These groups, which are non-polar, tend to clump together in the presence of water.

Properties and Functions of Water, Acids, and Bases (pH)

Acid – An acid is a compound which dissociates in water and yields hydrogen ions [H^+]. It is referred to as a proton donor. In aqueous form it conducts electricity, has a sour taste, turns blue litmus red, reacts with active metals to form hydrogen, and neutralizes bases.

Base – A base is a compound which dissociates in water and yields hydroxyl ions [OH^-]. Bases are proton acceptors. In aqueous form it conducts electricity, has a bitter taste, turns red litmus blue, feels soapy, and neutralizes acids.

The base that results when an acid donates its proton is called the conjugate base of the acid.

The acid that results when a base accepts a proton is called the conjugate acid of the base.

Water can act as either a weak acid or a weak base.

pH – A measure of H^+ ions in a solution. The pH scale ranges from $0-14$. A solution of pH O is very acidic (contains many H^+ ions) while a solution of pH 14 is extremely basic.

Carbon and Key Functional Groups

Carbon is the backbone of organic compounds and is found in enormous quantity in all living creatures. The atomic number of carbon is 6, so free atomic carbon has two electrons in its outer shell. Therefore, carbon almost always forms four bonds. Key functional groups consist of:

1) hydroxyl group   X-OH   Common in alcohols

2) amino group $X-NH_2$ Common in amino acids

3) carboxyl group   X-COOH   Common in amino acids and other organic molecules

4) methyl group $X-CH_3$ Common in organic molecules

5) aldehyde group   X-COH   Common in sugars

6) sulfhydryl group   X-S-H   Common in proteins

7) ketone group   X-CO-X   Common in sugars

8) phosphate group $X-H_2-PO_4$ Common energy carriers of cells

Note: The letter X is just a convenience used to represent an undesigned or unnamed molecule to which the functional group is attached. Some of the functional groups listed above will ionize in water, producing a charged condition.

Cell Constituents: Carbohydrates, Lipids, Proteins, Nucleic Acids

Carbohydrates – Carbohydrates are compounds composed of carbon, hydrogen, and oxygen, with the general molecular formula CH_2O. The principal carbohydrates include a variety of sugars.

A) Monosaccharides – A simple sugar or a carbohydrate which cannot be broken down into a simpler sugar. Its molecular formula is $C_6H_{12}O_6$, and the most common is glucose.

B) Disaccharide – A double sugar or a combination of two simple sugar molecules. Sucrose is a familiar disaccharide as are maltose and lactose.

OH—⬡—OH   +HO—⬡—OH ⟶ HO—⬡—O—⬡—OH
glucose   fructose   sucrose   +H₂O

Double sugar formation by dehydration synthesis.

C) Polysaccharide – A polysaccharide is a complex compound composed of a large number of glucose units. Examples of polysaccharides are starch, cellulose, and glycogen.

Lipids – Lipids are organic compounds that dissolve poorly, if at all, in water (hydrophobic). All lipids (fats and oils) are composed of carbon, hydrogen, and oxygen where the ratio of hydrogen atoms to oxygen atoms is greater than 2:1. A lipid molecule is composed of 1 glycerol and 2 fatty acids.

Phospholipid – A phospholipid is a variety of a substituted lipid which contains a phosphate group.

Proteins – All proteins are composed of carbon, hydrogen, oxygen, nitrogen, and sometimes phosphorus and sulfur. Approximately 50% of the dry weight of living matter is protein.

Amino Acids – The 20 amino acids are the building blocks of proteins.

$$H_2N - \overset{\overset{\displaystyle H}{|}}{\underset{\underset{\displaystyle R}{|}}{C}} - C\overset{\displaystyle O}{\underset{\displaystyle OH}{}}$$

An amino acid with R representing its distinctive side chain.

Polypeptides – Amino acids are assembled into polypeptides by means of peptide bonds. This is formed by a condensation reaction between the COOH groups and the NH_2 groups.

Primary Structure – The primary structure of protein molecules is the number of polypeptide chains and the number, type, and sequence of amino acids in each.

A – 6

Secondary Structure – The secondary structure of protein molecules is characterized by the same bond angles repeated in successive amino acids which gives the linear molecule a recurrent structural pattern.

Tertiary Structure – The three-dimensional folding pattern, which is super-imposed on the secondary structure, is called the tertiary structure.

Quaternary Structure – The quaternary structure is the manner in which two or more independently folded subunits fit together.

Nucleic Acids – Nucleic acids are long polymers involved in heredity and in the manufacture of different kinds of proteins. The two most important nucleic acids are deoxyribonucleic acid (DNA) and ribonucleic acid (RNA).

Nucleotides – These are the building blocks of nucleic acids. Nucleotides are complex molecules composed of a nitrogenous base, a 5-carbon sugar, and a phosphate group.

Structure of a nucleotide.

Deoxyribonucleic Acid (DNA) – Chromosomes and genes are composed mainly of DNA. It is composed of deoxyribose, nitrogenous bases, and phosphate groups.

cytosine   adenine   thymine   guanine

The four nitrogenous bases of DNA.

Ribonucleic Acid (RNA) – RNA is involved in protein synthesis. Unlike DNA, it is composed of the sugar ribose and the nitrogenous base uracil instead of thymine.

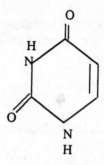

Uracil.

Properties of Chemical Reactions, Equilibrium, and Free Energy Changes

The four basic kinds of chemical reactions are: combination, decomposition, single replacement, and double replacement. ("Replacement" is sometimes called "metathesis.")

Combination can also be called synthesis. This refers to the formation of a compound from the union of its elements. For example:

$$Zn + S \rightarrow ZnS$$

Decomposition, or analysis, refers to the breakdown of a compound into its individual elements and/or compounds. For example:

$$C_{12}H_{22}O_{11} \rightarrow 12C + 11\,H_2O$$

The third type of reaction is called single replacement or single displacement. This type can best be shown by some examples where one substance is displacing another. For example:

$$Fe + CuSO_4 \rightarrow FeSO_4 + Cu$$

The last type of reaction is called double replacement or double displacement, because there is an actual exchange of "partners" to form new compounds. For example:

$$AgNO_3 + NaCl \rightarrow AgCl + NaNO_3$$

Equilibrium

When a system at equilibrium is disturbed by adding or removing one of the substances, all the concentrations will change until a new equilibrium point is reached with the same value of K_{eq}.

Increase in the concentrations of reactants shifts the equilibrium to the right, thus increasing the amount of products formed. Decreasing the concentrations of reactants shifts the equilibrium to the left and thus decreases the concentrations of products formed.

Free Energy

$$\Delta G = \Delta G^0 + 2.303RT \log Q$$

The symbol Q represents the mass action expression for the reaction. For gases, Q is written with partial pressures. ΔG is the free energy.

At equilibrium $Q = K_{eq}$ and the products and reactants have the same total free energy, such that $\Delta G = 0$.

$$\Delta G^0 = -2.303RT \log K_{eq} = -RT \ln K_{eq}$$

For the equation $2NO_2(g) \rightleftarrows N_2O_4(g)$,

$$\Delta G^0 = -2.303RT \log \left(\frac{P_{N_2O_4}}{\left(P_{NO_2}\right)^2} \right)_{eq}, \ K_c = \frac{[N_2O_4]}{[NO_2]^2}.$$

Properties of Enzymes, Coenzymes, and Cofactors

Enzymes – These are protein catalysts that lower the amount of activation energy needed for a reaction, allowing it to occur more rapidly. The enzyme binds with the substrate but resumes its original conformation after forming the enzyme-substrate complex.

Coenzymes – These are metal ions or non-proteinaceous organic molecules that bind briefly and loosely to some enzymes. The coenzyme is necessary for the catalytic reaction of such enzymes.

Cofactors – These are any organic or inorganic substance, especially an ion, that is required for the function of an enzyme.

B. CELLS

Classification and Characteristics of Cells: Prokaryotic vs. Eukaryotic, Plant vs. Animals

Prokaryote – Prokaryote refers to bacteria and blue-green algae. Prokaryotic cells have no nuclear membrane, and they lack membrane-bounded subcellular organelles such as mitochondria and chloroplasts. However, the membrane that bounds the cell is folded inward at various points and carries out many of the enzymatic functions of many internal membranes of eukaryotes.

Eukaryote – Eukaryote refers to all the protists, plants and animals. These are characterized by true nuclei; bounded by a nuclear membrane, and membrane-bounded subcellular organelles.

A Comparison of Eukaryotic and Prokaryotic Cells

Characteristic	Eukaryotic cells	Prokaryotic cells
Chromosomes	multiple, composed of nucleic acids and protein	single, composed only of nucleic acid
Nuclear membrane	present	absent
Mitochondria	present	absent
Golgi apparatus, endoplasmic reticulum, lysosomes, peroxisomes	present	absent
Photosynthetic apparatus	chlorophyll, when present is contained in chloroplasts	may contain chlorophyll
Microtubules	present	rarely present
Ribosomes	large	small
Flagella	have 9-2 tubular structure	lack 9-2 tubular structure
Cell wall	when present, does not contain muramic acid	contains muramic acid

Key Differences Between Plant and Animal Cells

Feature	Higher Plant Cell	Animal Cell
Membrane bound organelles	Many including mitochondria, large vacuoles, and chloroplasts	Many including mitochondria, lysosomes
Cell wall	Cellulose	None
Flagella or Cilia (when present)	Never present*	Microtubular (9 + 2 pattern)
Ability to engulf solid matter	Absent*	Present, extensive movable membranes
Centrioles	Absent*	Present

* Although absent in higher plants, these features are found in more primitive plants. Apparently they have been lost in the course of evolutionary change.

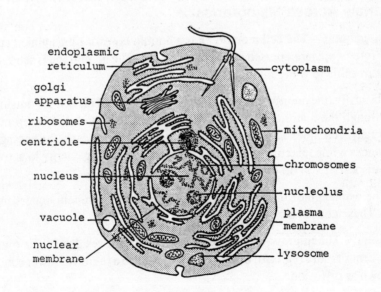

Typical animal cell.

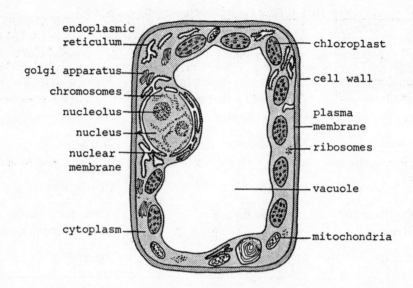

endoplasmic reticulum

golgi apparatus

chromosomes

nucleolus

nucleus

nuclear membrane

cytoplasm

chloroplast

cell wall

plasma membrane

ribosomes

vacuole

mitochondria

Typical plant cell.

Cell Membranes: Structure and Function, Movement of Materials Across Membranes

Cell Membrane – The cell membrane is a double layer of lipids which surrounds a cell. Proteins are interspersed in this lipid bilayer. The membrane is semipermeable; it is permeable to water but not to solutes.

Basically, the membrane keeps the inside of the cell in, and the outside out, and lets through the materials that the cell must exchange with the environment. The membrane's complex structure permits it to accept effortlessly the passage of some substances while rejecting others, and at times, expending energy to actively assist the transport of still others.

The exchange of materials between cell and environment occurs in different ways. These consist of:

Diffusion – The migration of molecules or ions as a result of their own random movements, from a region of higher concentration to a region of lower concentration is known as diffusion.

Osmosis – Osmosis is the movement of water through a semipermeable membrane. At constant temperature and pressure, the net movement of water is from the solution with lower concentration to the solution with higher concentration of osmotically active particles.

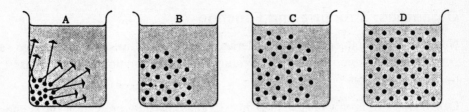

The sugar molecules, over a long period of time, will be distributed evenly in the water because of diffusion.

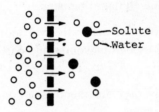

The process of osmosis.

Active Transport – The movement of ions and molecules against a concentration gradient is referred to as an active transport. The cell must expend energy to accomplish the transport. In passive transport, no energy is expended.

Endocytosis – Endocytosis is an active process in which the cell encloses a particle in a membrane-bounded vesicle, pinched off from the cell membrane. Endocytosis of solid particles is called phagocytosis.

Endocytosis in the amoeba.

Exocytosis – Exocytosis is the reverse of endocytosis. There is a discharge of vacuole-enclosed materials from a cell by the fusion of the cell membrane with the vacuole membrane.

Organelles: Structure and Function

Nucleus – A prominent, usually spherical or ellipsoidal membrane-bounded sac containing the chromosomes and providing physical separation between transcription and translation.

Cell Wall – This is only present in plant cells and is used for protection and support.

Centriole and Centrosome – These function in cell division. They are present only in animal cells.

Cilia and Flagella – These are hairlike extensions from the cytoplasm of a cell. They both show coordinated beating movements, which are the major means of locomotion and ingestion in unicellular organisms.

Nucleolus – The nucleolus is a generally oval body composed of protein and RNA. Nucleoli are produced by chromosomes and participate in the process of protein synthesis.

Peroxisomes – Peroxisomes are membrane-bounded organelles which contain powerful oxidative enzymes.

Vacuoles – Vacuoles are membrane-enclosed, fluid-filled spaces. They have their greatest development in plant cells where they store materials such as soluble organic nitrogen compounds, sugars, various organic acids, some proteins, and several pigments.

Chloroplasts – These are found only in the cells of plants and certain algae. Photosynthesis occurs in the chloroplasts.

Plastids – These structures are present only in the cytoplasm of plant cells. The most important plastid, chloroplast, contains chlorophyll, a green pigment.

Lysosomes – Lysosomes are membrane-enclosed bodies that function as storage vesicles for many digestive enzymes.

Endoplasmic Reticulum – The endoplasmic reticulum (ER) transports substances within the cell.

Ribosomes – These organelles are small particles composed chiefly of ribosomal-RNA and are the sites of protein synthesis.

Golgi Apparatus – The functions of the Golgi apparatus include storage, modification, and packaging of secretory products.

Organelles of Motility, Cytoskeleton

Cilia and flagella are fine, hairlike, movable organelles found on the surface of some cells, which provide locomotion. Cilia and flagella are structurally almost identical. They differ only in length, the number per cell, and their pattern of motion. They are composed of microtubules.

The cytoskeleton composes the internal structure of animal cells and is also composed of microtubules and actin microfilaments. The cytoskeleton controls the size, shape, and movement of the cell.

Cell Cycle: Interphase, Mitosis (Karyokinesis), Cytokinesis

CELL DIVISION

Mitosis – Mitosis is a form of cell division whereby each of two daughter nuclei receives the same chromosome complement as the parent nucleus. All kinds of asexual reproduction are carried out by mitosis; it is also responsible for growth, regeneration, and cell replacement in multicellular organisms.

A) Interphase – Interphase is no longer called the resting phase because a great deal of activity occurs during this phase. In the cytoplasm, oxidation and synthesis reactions take place. In the nucleus, DNA replicates itself and forms messenger RNA, transfer RNA, and ribosomal RNA.

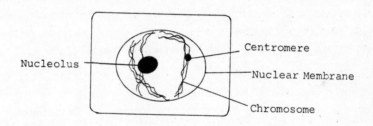

Interphase.

B) Prophase – Chromatids shorten and thicken during this stage of mitosis. The nucleoli disappear and the nuclear membrane breaks down and disappears as well. Spindle fibers begin to form. In an animal cell, there is also division of the centrosome and centrioles.

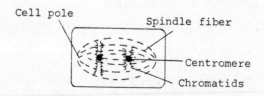

Late prophase in plant cell mitosis.

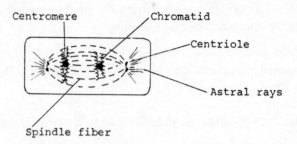

Prophase in animal cell mitosis.

C) Metaphase – During this phase, each chromosome moves to the equator, or middle of the spindle. The paired chromosomes attach to the spindle at the centromere.

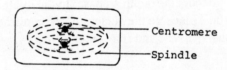

Metaphase in plant cell mitosis.

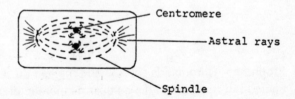

Metaphase in animal cell mitosis.

D) Anaphase – Anaphase is characterized by the separation of sister chromatids into a single-stranded chromosome. The chromosomes migrate to opposite poles of the cell.

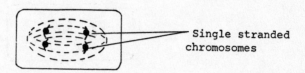

Anaphase in plant cell mitosis.

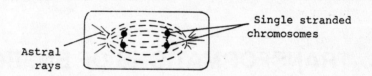

Anaphase in animal cell mitosis.

E) Telophase – During telophase, the chromosomes begin to uncoil and the nucleoli as well as the nuclear membrane reappear. In plant cells, a cell plate appears at the equator which divides the parent cell into two daughter cells. In animal cells, an invagination of the plasma membrane divides the parent cell.

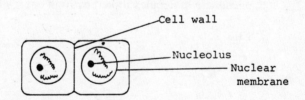

Late telophase in animal cell.

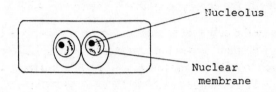

Nucleolus

Nuclear
membrane

Late telophase in plant cell.

Cytokinesis occurs at the end of mitosis and is the actual division of the cell into two daughter cells.

C. TRANSFORMATIONS OF ENERGY

Energy Currency of the Cell: ATP, Properties and Reactions

ATP (Adenosine triphosphate) is the smallest unit of energy present in a living cell. Each ATP molecule consists of three parts:

A) A 5-carbon ring sugar, the backbone of the molecule.

B) An adenosine base.

C) Three phosphate molecules linked by high energy bonds.

Each ATP molecule consists of three parts: an adenine base, a 5-carbon ring sugar named ribose, and three phosphates. Each phosphate is linked to the next via an oxygen atom in what is called a pyrophosphate bond. It is the high energy bond that is key in the energy available to the cell by ATP.

One of the cell's most essential reactions involves the recycling of ATP when its energy is spent. When the terminal pyrophosphate bond of ATP is broken, ATP is changed to ADP (adenosine diphosphate), or if its second pyrophosphate bond is broken, ATP is changed to AMP (adenosine monophosphate). The most common energy transferring reaction, in which ATP becomes ADP, can be summarized in its simplest terms as:

$$ATP + H_2O \rightleftarrows ADP + Pi + energy$$

or

$$A-P\sim P\sim P + H_2O \rightleftarrows A-P\sim P + Pi + energy$$

Transfer of Energy

When ATP is broken down, the terminal phosphate, along with its energy-rich bond, is often transferred to some substrate. If we let R stand for the substrate, this "phosphorylation" can be represented as:

$$A-P\sim P\sim P + R \rightleftarrows A-P\sim P + RP$$

Transferring the energy-rich phosphate group to the substrate is usually an intermediate step leading to other reactions. During synthesis (molecule building), for example, phosphorylation of a substrate may be the first step in linking two substrate molecules together. With its free energy increased, the phosphorylated substrate can more readily enter into reactions that are useful to the cell.

Types of Reactions: Coupled Reactions, Anabolism vs. Catabolism

Coupled reactions are very important in metabolism. They occur when the energy released from one reaction is used to drive another reaction to completion. These reactions are key in glycolysis, which is the breakdown of sugar.

Another type of reaction is anabolism, which is synthetic (or building) chemical activity that produces a more highly ordered chemical organization and a higher free energy state. The opposite of this is catabolism, which is chemical activity that decreases chemical organization and free energy therein.

Chemiosmosis

Chemiosmosis is a process which occurs in mitochondria, chloroplasts, and aerobic bacteria, where an electron transport system utilizes the energy of photosynthesis or oxidation to pump hydrogen ions across a membrane. This results in a proton concentration gradient that can be utilized to produce ATP.

Properties of Photosynthesis, Reactions of Photosynthesis, C_3 and C_4 Photosynthesis

Photosynthesis is the basic food-making process through which inorganic CO_2 and H_2O are transformed to organic compounds, specifically carbohydrates.

Chloroplasts absorb light energy and use CO_2 and H_2O to synthesize carbohydrates. Oxygen, which is formed as a by-product, is either eliminated into the air through the stomates, stored temporarily in the air spaces, or used in cellular respiration.

An overall chemical description of photosynthesis is the equation

$$6\ CO_2 + 6\ H_2O \xrightarrow[\text{chlorophyll}]{\text{light}} C_6H_{12}O_6 + 6\ O_2$$

Light Reaction (Photolysis) – A first step in photosynthesis is the decomposition of water molecules to separate hydrogen and oxygen components. This decomposition is associated with processes involving chlorophyll and light and is thus known as the light reaction.

Dark Reaction (CO_2 Fixation) – In this second phase, the hydrogen that results from photolysis reacts with CO_2 and carbohydrate forms. CO_2 fixation does not require light.

photolysis (light reaction

$$\text{light} \xrightarrow{\text{energy}} \text{chlorophyll}$$
$$2H_2O \xrightarrow{\text{energy}} 2H_2 + O_2$$

CO_2 fixation (dark reaction)

$$CO_2 + 2H_2 \dashrightarrow [CH_2O]$$
$$+ H_2O$$
$$\text{carbohydrates}$$

Photolysis and CO_2 fixation.

Plants are classified as either C_3 or C_4, depending on the pathways they use in photosynthesis. Photosynthesis in C_3 plants becomes inefficient when carbon dioxide gas is in low concentration. When this is the case, a process called photorespiration ensues, which wastes the products of the light reactions. The problem of photorespiration is avoided in the C_4 plants which have an alternate pathway involving a 4-carbon sugar. C_3 plants appear to be adapted to temperate climates while most C_4 plants are desert dwellers.

Respiration: Anaerobic and Aerobic, Glycolysis, Fermentation, the Krebs Cycle, Electron Transport

During cellular respiration, glucose must first be activated before it can break down and release energy. After activation, glucose enters into numerous reactions occurring in two stages. One stage is the anaerobic phase of cellular respiration and the other is the aerobic phase.

Glycolysis – Glycolysis refers to the breakdown of glucose which marks the start of the anaerobic reactions of cellular respiration. ATP is the energy source which activates glucose and initiates the process of glycolysis.

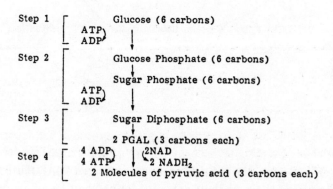

The major steps in glycolysis.

The steps in the figure above are summarized as follows:

Step 1 – Activation of glucose

Step 2 – Formation of sugar diphosphate

Step 3 – Formation and oxidation of PGAL, phosphoglyceraldehyde

Step 4 – Formation of pyruvic acid ($C_3H_4O_3$); net gain of two ATP molecules

If oxygen is not available after glycolysis, the cell undergoes a less efficient energy production pathway known as fermentation. Alcohol and carbon dioxide are often the products of this pathway.

If oxygen is available, then the cell will continue with a much more efficient aerobic pathway of energy production. The next step is the Krebs cycle (Citric Acid Cycle).

Krebs Cycle (Citric Acid Cycle) – The Krebs cycle is the final common pathway by which the carbon chains of amino acids, fatty acids, and carbohydrates are metabolized to yield CO_2. Pyruvic acid is converted to acetyl coenzyme A and, through a series of reactions, citric acid is formed.

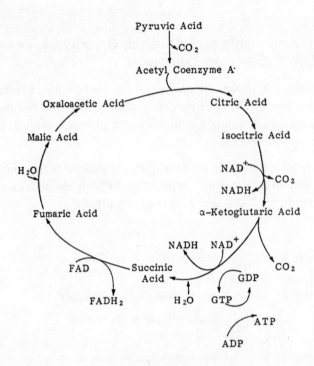

Pyruvic Acid

$\swarrow CO_2$

Acetyl Coenzyme A·

Oxaloacetic Acid   Citric Acid

Malic Acid   Isocitric Acid

H_2O NAD^+
                        $\searrow CO_2$
                    NADH

Fumaric Acid   α–Ketoglutaric Acid

          NADH NAD^+

FAD   Succinic CO_2
          Acid   GDP

$FADH_2$ H_2O GTP

                              ATP

                        ADP

Summary of the Krebs Cycle.

After the Krebs cycle, the reduced coenzymes formed ($NADH + FADH_2$) enter the electron transport chain. In these sets of reaction, the flow of electrons powers a proton pump in the mitochondria which creates the chemiosmotic difference.

This system consists of a series of enzymes and coenzymes which pick-up, hold, and then transfer hydrogen atoms among themselves until the hydrogen reaches its final acceptor, which is oxygen. Cytochromes are the enzymes and coenzymes involved in transferring hydrogen.

The cytochromes, together with other enzymes, split hydrogen atoms attached to compounds, such as $NADH_2$, into hydrogen ions and electrons. Each cytochrome then passes the hydrogen ions and electrons to another cytochrome in the series.

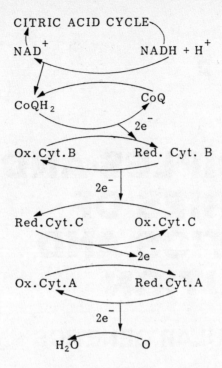

The respiratory chain.

The purpose of these reactions is to harvest energy that can be used to convert ADP to ATP. In this way the cell can build up energy reserves in the form of ATP.

CHAPTER 2

PRINCIPLES AND THEORIES OF GENETICS AND EVOLUTION

A. MOLECULAR GENETICS

DNA: The Genetic Material

Deoxyribonucleic acid, or DNA, is the genetic material of living organisms.

A series of experiments proved that DNA was the hereditary material. The first such evidence resulted from the transformation experiments of Fred Griffith in 1982, involving strains of pneumococcus.

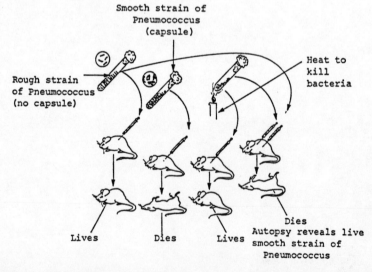

The experiments of Fred Griffith, which demonstrated the transfer of genetic information from dead, heat-killed bacteria to living bacteria of a different strain. Although neither the rough strain of pneumococcus nor heat-killed smooth strain pneumococci would kill a mouse, a combination of the two did. Autopsy of the dead mouse showed the presence of living, smooth strain pneumococci.

DNA is the transforming principle; therefore, it is the hereditary material. Strong evidence that DNA is the genetic material, came from the Hershey and Chase experiments with E. coli and the virus which attacks E. coli.

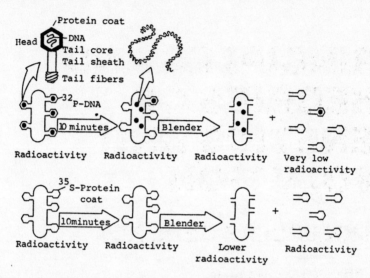

Hershey and Chase experiment demonstrating that only phage DNA enters the bacterial host cell after infection.

DNA: Structure, Properties, and Replication

The chemical composition of DNA:

* = site of attachment to deoxyribose

Structural formulas of purines (adenine and guanine), pyrimidines (thymine and cytosine), and a nucleotide.

Deoxyribonucleic acid is made up of a nitrogenous base, a five carbon sugar (deoxyribose) and phosphate groups. DNA may contain one of four nitrogenous bases which are the purines, adenine and guanine, and the pyrimidines, cytosine and thymine. Each nitrogenous base is attached to deoxyribose via a glycosidic linkage, and deoxyribose is attached to the phosphate group by an ester bond. This base sugar-phosphate combination is a nucleotide. There are four kinds of nucleotides in DNA, each containing one of the four nitrogenous bases. The nucleotides are joined by phosphate ester bonds into a chain. DNA is made up of two complementary chains of nucleotides.

The structure of DNA:

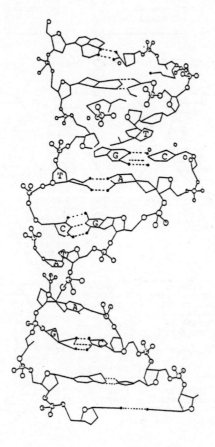

The DNA double helix. The purine (G=Guanine, A=Adenine) and pyrimidine (T=Thymine, C=Cytosine) base pairs are connected by hydrogen bonds.

By 1950, several properties of DNA were well established. Chargaff showed that the four nitrogenous bases of DNA did not occur in equal proportions; however, the total amount of purines equaled the total amount of pyrimidines (A + G = T + C). In addition, the amount of adenine equaled the amount of thymine (A = T), and likewise for guanine and cytosine (G = C). Pauling had suggested that, the structure of DNA might be some sort of an α-helix held together by hydrogen bonds. The final observation was made by Franklin and Wilkins. They inferred from x-ray diffraction studies that the nucleotide bases (which are planar molecules) were stacked one on top of the other like a pile of saucers.

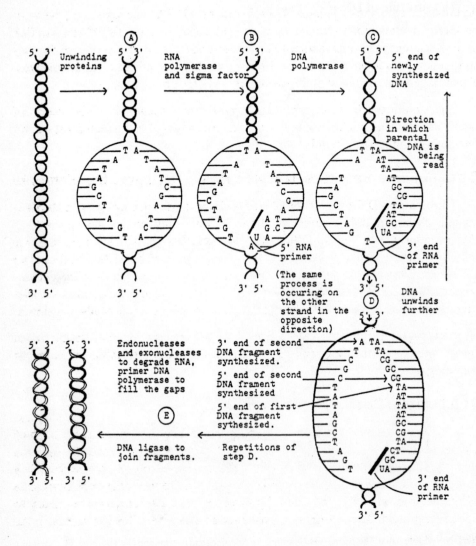

By studying the Watson-Crick model of DNA, Kornberg and his colleagues determined the mechanism of DNA replication. Since DNA is the genetic material of the cell, it must have the information to replicate itself built into it. The Watson-Crick model of DNA seems to offer a good explanation of how DNA is replicated. There are only four different nucleotides found in DNA. Chargaff's rule of specific base pairing states that these nucleotides (adenine, guanine, thymine, and cytosine) are ordered on both strands of the DNA helix so that adenine is always paired to thymine and cytosine to guanine. Thus, in order for DNA to be replicated, the helix need only be unwound to form a template. The nucleotide building blocks can line up in a sequence complementary to the order presented. In this way, two double-stranded helices identical to the original molecule are synthesized in the nucleus. This is a general overview of the mechanism of DNA replication.

DNA replicates by semi-conservative replication, such that the replication of one DNA molecule yields two hybrids, each composed of one parental strand and one newly synthesized strand.

Chromosome Structure, Nucleosomes, Transposable Elements

Chromatin is DNA that is complexed with proteins called histones. A chromosome is chromatin tightly coiled into large, highly visible bodies. This is the form they take during cell division. When the DNA is associated with histones, together they form tiny, beadlike globules called nucleosomes, each consisting of about 140 DNA base pairs wound twice around each cluster of histone. Between the nucleosomes is a string of DNA about 50 nucleotide pairs in length. Each chromosome, then, consists of one long DNA molecule wrapped around many beads of histones.

Transposable Elements – DNA sequences that appear to move from one part of the genome to another. A type of recombination can occur which is not based on homology.

The Genetic Code

The linear sequence of every protein a cell produces is encoded in the DNA of a specific gene on one of the chromosomes. But DNA does not make proteins directly — it can only direct the synthesis of RNA or copies of itself. Messenger RNA is the physical link between the gene and the protein. The messenger RNA molecule is synthesized on DNA, and faithfully incorporates the information necessary to specify a protein. The information contained in messenger RNA is written in the genetic code.

The fundamental characteristic of the genetic code is that it is a triplet code with three adjacent nucleotide bases, termed a codon, specifying each amino acid. The three nucleotides specify one of the 20 common amino acids.

The Genetic Code

First Position (5' end)	Second position	Third position (3' end) U	C	A	G
U	U	Phe	Phe	Leu	Leu
	C	Ser	Ser	Ser	Ser
	A	Tyr	Tyr	Terminator	Terminator
	G	Cys	Cys	Terminator	Trp
C	U	Leu	Leu	Leu	Leu
	C	Pro	Pro	Pro	Pro
	A	His	His	Glu NH_2	Glu NH_2
	G	Arg	Arg	Arg	Arg
A	U	Ileu	Ileu	Ileu	Met
	C	Thr	Thr	Thr	Thr
	A	Asp NH_2	Asp NH_2	Lys	Lys
	G	Ser	Ser	Arg	Arg
G	U	Val	Val	Val	Val
	C	Ala	Ala	Ala	Ala
	A	Asp	Asp	Asp	Asp
	G	Gly	Gly	Gly	Gly

RNA Structure, Properties, Transcription, mRNA Editing

RNA is made up of a nitrogenous base, a five carbon sugar (ribose) and a phosphate group. RNA may contain one of four nitrogenous bases: the purines, adenine and guanine, and the pyrimidines, uracil and cytosine.

There are three types of RNA. All are single-stranded and are transcribed from a DNA template by RNA polymerase in the nucleus.

A) Messenger RNA (mRNA) carries the genetic information coded for in the DNA to the ribosomes and is responsible for the translation of that information into a polypeptide chain.

B) Ribosomal RNA (rRNA) is an integral part of the ribosome, and its removal results in the destruction of the ribosome. rRNA interacts with the ribosomal protein and helps maintain the characteristic shape of the ribosome.

C) Transfer RNA (tRNA) is the smallest type of RNA. The function of tRNA is to insert the amino acid specified by the codon on mRNA into the polypeptide chain, and it is through the complementation of anticodon and codon that the appropriate amino acid is incorporated.

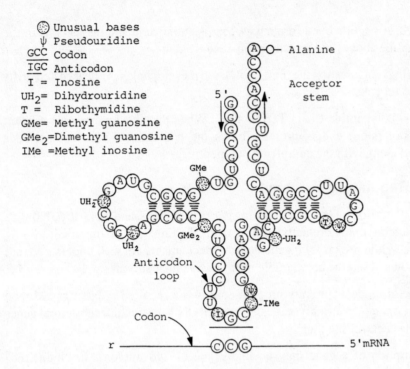

The complete nucleotide sequence of alanine tRNA showing the unusual bases and codon/anticodon position. Structure shown is two-dimensional.

In transcription, the chemical instructions encoded in DNA are copied into RNA. Only one strand of the double-stranded DNA is transcribed, and it is called the transcribed strand. The segment of the DNA molecule on which a single RNA molecule is transcribed is, in a sense, equivalent to a gene. RNA transcription is conservative — that is, the molecule being copied is conserved and not changed by the process of transcribing RNA.

Protein Synthesis, RNA Translation

Protein synthesis occurs as three steps:

Step 1 – Initiation: Initiation requires the smaller ribosomal subunit, an initiator tRNA, and the mRNA initiator codon (AVG), all of which form the initiation complex. When each component is in place, the larger ribosomal subunit joins the complex and a second amino acid can be inserted.

Step 2 – Elongation: The elongation of a polypeptide occurs through translocation - the formation of the peptide bond and the movement of a tRNA from the right to the left ribosomal pocket. As the polypeptide grows, a charged transfer RNA whose anticodon matches the mRNA codon in the right pocket becomes

A – 31

attached. A peptide bond forms between its amino acid and the last one in the polypeptide above, and translocation occurs again.

Following translocation the transfer RNA in the left pocket is released and drifts away to recycle.

Step 3 – Polypeptide Chain Termination: When the ribosome reaches a chain terminator (stop) codon, proteins block the pockets and the final tRNA is released along with the completed polypeptide.

Genetic Regulatory Systems

Of the 40,000 to 50,000 protein coding genes in humans, and 4,000 to 5,000 such genes in bacteria, many are active only part of the time, suggesting a controlling mechanism. While prokaryotic gene control mechanisms are well understood, not much is known about the comparable mechanisms in eukaryotes.

The operon model of prokaryotic gene control was reported by Jacob and Monod in 1961. The operon, in prokaryotes, is a region of DNA that includes structural genes and the genes controlling them.

Transcription of a gene may occur by one of two operons. In "inducible" transcription, the gene remains shut down until activated by an inducer substance. The repressible operon is controlled in the opposite manner. In this mechanism, the gene remains active until shut down by a repressor substance. Control regions on the gene generally consist of a promoter region (p), an operator region (o), and a regulator gene (i) which produces a repressor protein.

Mutations

a	b	c	d	e	f	Normal

a	b	c	e	f		Deletion (of segment d)

a	b	c	c	d	e	f	Duplication (of segment c)

a	e	d	c	b	f	Inversion (of segment b-e)

g	h	i	j	k	l	a	b	c	Translocation (of a-c to chromosome ghijkl).

Diagram illustrating the type of mutations that involve changes in the structure of the chromosome.

Mutation – Any inheritable change in a gene not due to segregation or to the normal recombination of genetic material. There are two major types of mutation:

A) Chromosomal Mutation – Caused by extensive chemical change in the structure of a chromosome.

B) Point Mutation (genic mutation) – Caused by a single change in molecular structure at a given locus.

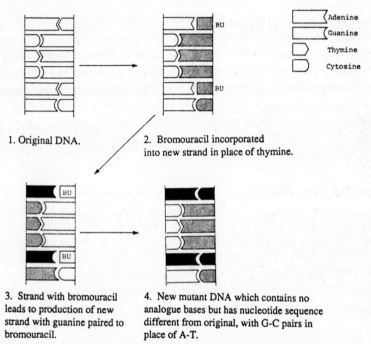

1. Original DNA.

2. Bromouracil incorporated into new strand in place of thymine.

3. Strand with bromouracil leads to production of new strand with guanine paired to bromouracil.

4. New mutant DNA which contains no analogue bases but has nucleotide sequence different from original, with G-C pairs in place of A-T.

Diagrammatic scheme of how an analogue of a purine or pyrimidine might interfere with the replication process and cause a mutation, an altered sequence of nucleotides in the DNA (indicated in black). The nucleotides of the new chain at each replication are indicated by the gray blocks. In this instance, two new G-C pairs are indicated.

Types of chromosomal mutations:

A) Deletion – A mutation in which a segment of the chromosome is missing.

B) Duplication – A mutation where a portion of a chromosome breaks off and is fused onto the homologous chromosome.

C) Translocation – A mutation where segments of two non-homologous chromosomes are exchanged.

D) Inversion – A mutation where a segment is removed and reinserted in the same location, but in the opposite direction.

Point mutations usually involve the substitution of one nucleotide for another, and the deletion of nucleotides from the sequence and their addition to the sequence. Point mutations can result from exposure to x-ray, gamma rays, ultraviolet rays and other types of radiation, from errors in base pairing during replication, and from interaction with chemical mutagens.

Recombinant DNA, DNA Cloning, Hybridization, Sequencing DNA

Recombinant DNA is a term for the laboratory manipulation of DNA in which DNA molecules or fragments from various sources are severed and combined enzymatically and reinserted into living organisms. **Gene cloning** is a recent technique whereby pieces of DNA from any source are spliced into plasmid DNA, cultured in growing bacteria, purified, and recovered in quantity. **Gene sequencing** is a method for determining the specific sequence of nucleotides in a gene. **Hybridization** is a process whereby specific DNA or RNA sequences can be located by a complementary DNA or RNA probe which has been labelled with radioactivity so that it can be detected.

Viruses: DNA and RNA

Viruses are minute, biologically active particles made up of a nucleic acid core, or covering of protein, and sometimes an enzyme or two. Some viruses have the standard double-stranded DNA, others double-stranded RNA or single-stranded RNA. Viruses enter the host, disrupt its DNA, undergo their own DNA or RNA replication, and transcribe their genes into mRNA for producing viral proteins. All materials – synthesizing enzymes, ribosomes, and energy – are provided by the host.

B. HEREDITY

Meiosis

Meiosis – Meiosis consists of two successive cell divisions with only one duplication of chromosomes. This results in daughter cells with a haploid number of chromosomes or one-half of the chromosome number in the original cell. This process occurs during the formation of gametes and in spore formation in plants.

A) Spermatogenesis – This process results in sperm cell formation with four immature sperm cells with a haploid number of chromosomes.

B) Oogenesis – This process results in egg cell formation with only one immature egg cell with a haploid number of chromosomes, which becomes mature and larger as yolk forms within the cell.

First Meiotic Division –

A) Interphase I – Chromosome duplication begins to occur during this phase.

B) Prophase I – During this phase, the chromosomes shorten and thicken and synapsis occurs with pairing of homologous chromosomes. Crossing-over between non-sister chromatids will also occur. The centrioles will migrate to opposite poles and the nucleolus and nuclear membrane begin to dissolve.

C) Metaphase I – The tetrads, composed of two doubled homologous chromosomes, migrate to the equatorial plane during metaphase I.

D) Anaphase I – During this stage, the paired homologous chromosomes separate and move to opposite poles of the cell. Thus, the number of chromosome types in each resulting cell is reduced to the haploid number.

E) Telophase I – Cytoplasmic division occurs during telophase I. The formation of two new nuclei with half the chromosomes of the original cell occurs.

F) Prophase II – The centrioles that had migrated to each pole of the parental cell, now incorporated in each haploid daughter cell, divide, and a new spindle forms in each cell. The chromosomes move to the equator.

G) Metaphase II – The chromosomes are lined up at the equator of the new spindle, which is at a right angle to the old spindle.

H) Anaphase II – The centromeres divide and the daughter chromatids, now chromosomes, separate and move to opposite poles.

I) Telophase II – Cytoplasmic division occurs. The chromosomes gradually return to the dispersed form and a nuclear membrane forms.

Mendelian Genetics and Laws, Probability

By studying one single trait at a time in garden peas, Gregor Mendel, in 1857, was able to discover the basic laws of genetics.

A) Definitions

1) A gene is the part of a chromosome that codes for a certain hereditary trait.

2) A chromosome is a filamentous or rod-shaped body in the cell nucleus that contains the genes.

3) A genotype is the genetic makeup of an organism, or the set of genes that it possesses.

4) A phenotype is the outward, visible expression of the hereditary makeup of an organism.

5) Homologous chromosomes are chromosomes bearing genes for the same characters.

6) A homozygote is an organism possessing an identical pair of alleles on homologous chromosomes for a given character or for all given characters.

7) A heterozygote is an organism possessing different alleles on homologous chromosomes for a given character or for all given characters.

8) Crossing over means that paired chromosomes may break and their fragments reunite in new combinations.

9) Translocations are the shifting of gene positions in chromosomes that may result in a change in the serial arrangement of genes. In general, it is the transfer of a chromosome fragment to a non-homologous chromosome.

10) Linkage is the tendency of two or more genes on the same chromosome to cause the traits they control to be inherited together.

An Abstract of the Data Obtained by Mendel from His Breeding Experiments with Garden Peas

Parental Characters	First Generation	Second Generation	Ratios
Yellow seeds × green seeds	all yellow	6022 yellow:2001 green	3.01:1
Round seeds × wrinkled seeds	all round	5474 round:1850 wrinkled	2.96:1
Green pods × yellow pods	all green	428 green:152 yellow	2.82:1
Long stems × short stems	all long	787 long:277 short	2.84:1
Axial flowers × terminal flowers	all axial	651 axial:207 terminal	3.14:1
Inflated pods × constricted pods	all inflated	882 inflated:299 constricted	2.95:1
Red flowers × white flowers	all red	705 red:224 white	3.15:1

The 3:1 ratio that resulted from this data enabled Mendel to recognize that the offspring of each plant had two factors for any given characteristic instead of a single factor.

11) Alleles are types of alternative genes that occupy a given locus on a chromosome; they are matching genes that can control contrasting characters.

12) Genetic mutation is a change in an allele or segment of a chromosome that may give rise to an altered genotype, which often leads to the expression of an altered phenotype.

13) A codon is a sequence of three adjacent nucleotides that codes for a single amino acid.

B) Laws of Genetics

    1) Law of Dominance – Of two contrasting characteristics, the dominant one may completely mask the appearance of the recessive one.

    2) Law of Segregation and Recombination – Each trait is transmitted as an unchanging unit, independent of other traits, thereby giving the recessive traits a chance to recombine and show their presence in some of the offspring.

    3) Law of Independent Assortment – Each character for a trait operates as a unit and the distribution of one pair of factors is independent of another pair of factors linked on different chromosomes.

Patterns of Inheritance, Chromosomes, Genes, and Alleles

In 1900, Walter Sutton compared the behavior of chromosomes with the behavior of the hereditary characters that Mendel had proposed and formulated the chromosome principle of inheritance.

The Chromosome Principle of Inheritance:

A) Chromosomes and Mendelian factors exist in pairs.

B) The segregation of Mendelian factors corresponds to the separation of homologous chromosomes during the reduction division stage of meiosis.

C) The recombination of Mendelian factors corresponds to the restoration of the diploid number of chromosomes at fertilization.

D) The factors that Mendel described as passing from parent to offspring correspond to the passing of chromosomes into gametes which then unite and develop into offspring.

E) The Mendelian idea that two sets of characters present in a parent assort independently, corresponds to the random separation of the two sets of chromosomes as they enter a different gamete during meiosis.

Sutton's chromosome principle of inheritance states that the hereditary characters, or factors, that control heredity are located in the chromosomes. By 1910, the factors of heredity were called genes.

Gene Interactions, Linkage, and Mapping

Linkage, the occurrence of genes in linkage groups, and crossing over were discovered when expected Mendelian ratios failed to appear. In a testcross (cross of a known double heterozygote with a double homozygote to determine linkage relationships) with garden peas, Bateson and Punnet found odd ratios and seemingly impossible combinations of traits in the progeny. In the testcross AaBb × aabb, the predictable Mendelian ratio in the progeny would be 1:1:1:1: or 25% AaBb, 25% Aabb, 25% aaBb, and 25% aabb. If the alleles are on different chromosomes, no other combinations are possible. If, however, the genes in the above testcross were linked on the same chromosome, then the progeny would be 1:1 or 50% AaBb and 50% aabb. If they are fully linked, no other combinations are possible.

Recombination frequencies are used to construct genetic maps. Map distances are measured in recombination percentages between loci (a specific place on a chromosome where a gene is located). Recombination genetic maps identify the chromosome on which a locus is found, and relative locations of genes and distances between them.

Genetic Defects in Humans

When individuals have a chromosome number greater or fewer than the normal 46 chromosomes, certain defects may develop.

A) Down's Syndrome – The presence of 47 chromosomes instead of 46 makes the individual mentally retarded. The extra chromosome results from nondisjunction during the formation of the egg cell.

B) Turner's Syndrome – The absence of one chromosome makes this individual a short, sterile female having underdeveloped ovaries and breasts. Nondisjunction in meiosis results in an offspring lacking a sex chromosome (45 chromosomes — 44 + X).

C) Klinefelter's Syndrome – A male is born possessing 47 chromosomes (44 + XXY) making him tall and sterile with underdeveloped testes. This is also the result of nondisjunction of the sex chromosomes.

C. EVOLUTION

The Origin of Life

Soon after the Earth's formation, when conditions were quite different from those existing today, a period of spontaneous chemical synthesis began in the warm ancient seas. During this era, amino acids, sugars, and nucleotide bases — the structural subunits of some of life's macromolecules — formed spontaneously from the hydrogen-rich molecules of ammonia, methane, and water. Such spontaneous synthesis was only possible because there was little oxygen in the atmosphere. The energy for synthesis came in the form of lightening, ultra-violet light, and higher energy radiations. These molecules polymerized due to their high concentrations, and eventually the first primitive signs of life arose.

Evidence for Evolution

The evidence for evolution can be explained by these eight facts:

A) Comparative Anatomy – Similarities of organs in related organisms show common ancestry.

B) Vestigial Structures – Structures of no apparent use to the organism, may be explained by descent from forms that used these structures.

C) Comparative Embryology – The embryo goes through developmental stages in common with other types of species.

D) Comparative Physiology – Many different organisms have similar enzymes. Mammals have similar hormones.

E) Taxonomy – All organisms can be classified into kingdom, phylum, class, order, family, genus, and species. This commonness in classification seems to indicate relationships between organisms.

F) Biogeography – Natural barriers, such as oceans, deserts, and mountains, which restrict the spread of species to other favorable environments. Isolation frequently produces many variations of species.

G) Genetics – Gene mutations, chromosome segment rearrangements, and chromosome segment doubling produce variations and new species.

H) Paleontology – Present individual species can be traced back to origins through skeletal fossils.

Natural Selection

The Darwin-Wallace theory of natural selection states that a significant part of evolution is dictated by natural forces, which select for survival those organisms that can respond best to certain conditions. Since more organisms are born than can be accommodated by the environment, a limited number is chosen to live and reproduce. Variation is characteristic of all animals and plants, and it is this variety which provides the means for this choice. Those individuals who are chosen for survival will be the ones with the most and best adaptive traits. These include the ability to compete successfully for food, water, shelter, and other essential elements; the ability to reproduce and perpetuate the species; and the ability to resist adverse natural forces, which are the agents of selection.

The Hardy-Weinberg Law

The Hardy-Weinberg Law states that in a population at equilibrium, both gene and genotype frequencies remain constant from generation to generation.

Factors Affecting Allele Frequencies: Genetic Drift, Bottleneck Effect, Founder Effect

Changes in allele frequency are caused by migration, mutation, and genetic drift. Genetic drift refers to the absence of natural selection. Random changes in gene frequencies occur, including the random loss of alleles. Random changes in the gene pool can be produced by catastrophic events where there are few survivors. Following such events, allele frequencies in the population may become quite different through chance alone. Such events are sometimes called population bottlenecks. A bottleneck effect can be created where a few people colonize a new territory. This is known as a founder effect.

Mechanisms of Speciation: Isolating Mechanisms, Allopatry, Sympatry, Adaptive Radiation

The following three mechanisms are fundamental in speciation:

A) Allopatric speciation is the formation of new species through the geographic isolation of groups from the parent population (as occurs through colonization or geological disruption).

B) Sympatric speciation occurs within a population and without geographical isolation. It is rare in animals, but not in plants.

C) Adaptive radiation is the formation of new species arising from a common ancestor resulting from their adaptation to different environments.

Evolutionary Patterns: Gradualism, Punctuated Equilibrium

Gradualism is the belief that evolution occurs very gradually, or in such small increments as to be nearly unobservable. In contrast, **punctuated equilibrium** describes evolutionary change occurring in relatively sudden spurts, following long periods of minor change or no change.

CHAPTER 3

ORGANISMAL AND POPULATION BIOLOGY

A. THE INTERRELATIONSHIP OF LIVING THINGS

Taxonomy and Systematics

Taxonomy is the science of classification, and systematics is the science of evolutionary relationships.

The Five Kingdom Classification Scheme

The largest taxon is the kingdom. In the scheme of classification, there are five kingdoms: the Monera, Protista, Fungi, Plantae, and Animalia. They are subdivided into descending levels of taxa: phylum (division for plantae), class, order, family, genus, species, and subspecies (sometimes varieties and races).

B. DIVERSITY AND CHARACTERISTICS OF THE KINGDOM MONERA

Monera

Bacteria and blue-green algae are included in the kingdom Monera. These cells, having no nuclear membrane and only a single, circular chromosome, are termed prokaryotes. Both blue-green algae and bacteria lack membrane bounded subcellular organelles, such as mitochondria and chloroplasts.

Protista

Most protists are unicellular. Some are composed of colonies. All protists are eukaryotes. That is, they are characterized by nuclei bounded by a nuclear membrane. Examples of protists are flagellates, the protozoans, and slime molds.

Algae — photosynthetic protists — range from single-celled to complex multicellular forms. Most are aquatic, living in all of the earth's waters, and in a few instances on land. Included are the vast, floating, microscopic marine phytoplankton and the marine seaweeds and kelps.

Fungi

Fungi are multicellular, nonmotile heterotrophs, lacking tissue organization except in their reproductive structures. Some are coenocytic (made up of a multinucleated mass of cytoplasm without subdivision into cells), while others are cellular. Most have chitinous cell walls, while a few have walls of cellulose.

C. THE KINGDOM PLANTAE

Diversity, Classification, Phylogeny

Plants are characterized as being multicellular and photosynthetic, most with tissue, organ, and system organization (roots, stems, leaves, flowers). They develop from protected embryos, contain chlorophylls a and b, produce starches, and have walls of cellulose.

Plants are classified into five categories:

A) Nonvascular Plants – Characterized as having little or no organized tissue for conducting water and food (xylem and phloem), e.g. mosses.

B) Seedless Vascular Plants – Characterized as having organized vascular tissue, limited root development, many primitive traits, chlorophyll a and b, e.g. ground pine, horsetails, and ferns.

C) Seed-Producing Vascular Plants – Characterized as having seeds, organized vascular tissues, and extensive root, stem and leaf development.

D) Gymnosperms – Characterized as having seeds without surrounding fruit, tracheids only for water conduction, terrestrial, e.g. maidenhair tree, Cycads, Gnetum, pines, redwoods, firs, junipers, larch, cypress, and hemlock.

E) Angiosperms – Characterized as having flowers, seeds with surrounding fruit, e.g. magnolia, cabbage, tobacco, cotton, iris, orchids, and grains.

Adaptations to Land

To accommodate to land, plants (except for the nonvascular plants) grew much larger due to the presence of vascular tissue, including the woody, hardened xylem, and tough accompanying supporting tissues. Apparently large size provided a novel way of adapting to the terrestrial environment. Also, deep root systems allowed the plants to obtain water and allowed them to grow larger.

The Life Cycle (Life History): Alternation of Generations in Moss, Fern, Pine, Angiosperms (Flowering Plants)

Alternation of generations in plants is characterized by the existence of two phases in the life of a single individual, including a diploid spore-producing phase and a haploid, gamete-producing phase.

Plant	Alternation of Generation
moss	separate generations, gametophyte usually dominant, swimming sperm, egg, and zygote protected.
fern	separate generations, dominant sporophyte, minute gametophyte, swimming sperm.
pine	nonmotile sperm, wind-pollinated.
angiosperms	nonmotile sperm, both wind- and insect-pollinated.

Anatomy, Morphology, and Physiology of Vascular Plants

A mature vascular plant possesses several distinct cell types which group together in tissues. The major plant tissues include epidermal, parenchyma, sclerenchyma, chlorenchyma, vascular, and meristematic.

Summary of Plant Tissues

Tissue	Location	Functions
Epidermal	Root	Protection-Increases absorption area
	Stem	Protection-Reduces H_2O loss
	Leaf	Protection-Reduces H_2O loss Regulates gas exchange
Parenchyma	Root, stem, leaf	Storage of food and H_2O
Sclerenchyma	Stem and leaf	Support
Chlorenchyma	Leaf and young stems	Photosynthesis
Vascular a. Xylem	Root, stem, leaf	Upward transport of fluid
b. Phloem	Leaf, root stem	Downward transport of fluid
Meristematic	Root and stem	Growth; formation of xylem, phloem, and other tissues

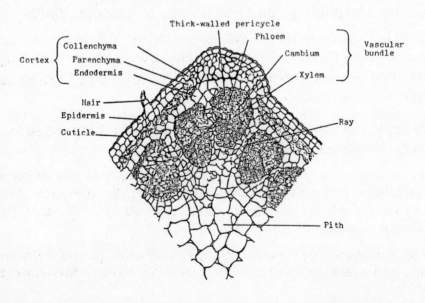

A sector of a cross section of a stem from a herbaceous dicot, alfalfa.

Reproduction and Growth in Seed Plants: Seed Formation, Seed Structure, Germination, Growth

Mitosis and differentiation in the primary endosperm and zygote will produce the embryo, which along with the food supply and seed coat, make up the seed.

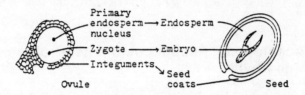

Development of an ovule into a seed.

For many seeds, germination (the emergence from dormancy) simply requires the uptake of water, suitable temperatures, and the availability of oxygen, but others require special conditions, such as exposure to freezing temperatures, fire, abrasion, or exposure to animal digestive enzymes. The growth of the plant is dependent on three key hormones.

Plant Hormones: Types, Functions, Effects on Plant Growth

Auxins – These plant growth regulators stimulate the elongation of specific plant cells and inhibit the growth of other plant cells.

Gibberellins – In some plants, gibberellins are involved in the stimulation of flower formation. They also increase the stem length of some plant species and the size of fruits. Gibberellins also stimulate the germination of seeds.

Cytokinins – Cytokinins increase the rate of cell division and stimulate the growth of cells in a tissue culture. They also influence the shedding of leaves and fruits, seed germination, and the pattern of branch growth.

Environmental Influences on Plants and Plant Responses to Stimuli: Tropisms, Photoperiodism

Auxin is involved in tropisms, or growth responses. They involve positive phototropism (bending or growing toward light), negative phototropism (bending or growing away from light), geotropism (influenced by gravity), and thigmotropism (mechanical or touch).

Photoperiodism is any response to changing lengths of night or day. Flowering in plants is often photoperiodic, but so far no specific hormonal mechanism has been found.

D. THE KINGDOM ANIMALIA

Diversity, Classification, Phylogeny

The phylogenetic tree representing animal evolution reveals the separate origins of metazoans and parazoans. An early major split produced the protostomes and deuterostomes, which include the higher invertebrates and the vertebrates, respectively.

Survey of Acoelomate, Pseudocoelomate, Protostome, and Deuterostome Phyla

The acoelomates are animals that have no coelom (body cavity.) They include the phylum Platyhelminthes (flatworms). In acoelomate animals, the space between the body wall and the digestive tract is not a cavity, as in higher animals, but is filled with muscle fibers and a loose tissue of mesenchymal origin called parenchyma, both derived from the mesoderm.

The pseudocoelomates consist of the following phyla: Nematoda, Rotifera, Gastrotricha, Nematomorpha, and Acanthocephala. These animals have a body cavity which is not entirely lined with peritoneum.

A major division in animal evolution produced the protostomes and deuterostomes. Each has bilateral symmetry, a one way gut, and a true coelom (eucoelomate). During primitive gut formation in protostome embryos, the blastopore forms the mouth of the animal, while in deuterostomes this is the anal area.

Structure and Function of Tissues, Organs, and Systems Especially in Vertebrates

ANIMAL TISSUES

The cells that make up multicellular organisms become differentiated in many ways. One or more types of differentiated cells are organized into tissues. The basic tissues of a complex animal are the epithelial, connective, nerve, muscle, and blood tissues.

Summary of Animal Tissues

Tissue	Location	Functions
Epithelial	Covering of body Lining internal organs	Protection Secretion
Muscle		
Skeletal	Attached to skeleton bones	Voluntary movement
Smooth	Walls of internal organs	Involuntary movement
Cardiac	Walls of heart	Pumping blood
Connective		
Binding	Covering organs, in tendons and ligaments	Holding tissues and organs together
Bone	Skeleton	Support, protection, movement
Adipose	Beneath skin and around internal organs	Fat storage, insulation, cushion
Cartilage	Ends of bone, part of nose and ears	Reduction of friction, support
Nerve	Brain	Interpretation of impulses, mental activity
	Spinal cord, nerves, ganglions	Carrying impulses to and from all organs
Blood	Blood vessels, heart	Carrying materials to and from cells, carrying oxygen, fighting germs, clotting

THE HUMAN DIGESTIVE SYSTEM

The digestive system of man consists of the alimentary canal and several glands. This alimentary canal consists of the oral cavity (mouth), pharynx, esophagus, stomach, small intestine, large intestine, and the rectum.

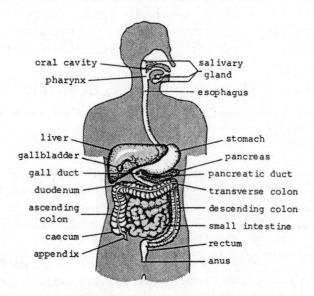

Human digestive system. (The organs are slightly displaced, and the small intestine is greatly shortened.)

A) **Oral Cavity (mouth)** – The mouth cavity is supported by jaws and is bound on the sides by the teeth, gums, and cheeks. The tongue binds the bottom and the palate binds the top. Food is pushed between the teeth by the action of the tongue so it can be chewed and swallowed. Saliva is the digestive juice secreted that begins the chemical phase of digestion.

B) **Pharynx** – Food passes from the mouth cavity into the pharynx where the digestive and respiratory passages cross. Once food passes the upper part of the pharynx, swallowing becomes involuntary.

C) **Esophagus** – Whenever food reaches the lower part of the pharynx, it enters the esophagus and peristalsis pushes the food further down the esophagus into the stomach.

D) **Stomach** – The stomach has two muscular valves at both ends: the cardiac sphincter which controls the passage of food from the esophagus into the stomach, and the pyloric sphincter which is responsible for the control of the passage of partially digested food from the stomach to the small intestine. Gastric juice is also secreted by the gastric glands lining the stomach walls. Gastric juice begins the digestion of proteins.

E) Pancreas – The pancreas is the gland formed by the duodenum and the under surface of the stomach. It is responsible for producing pancreatic fluid which aids in digestion. Sodium bicarbonate, an amylase, a lipase, trypsin, chymotrypsin, carboxypeptidase, and nucleases are all found in the pancreatic fluid.

F) Small Intestine – The small intestine is a narrow tube between 20 and 25 feet long divided into three sections: the duodenum, the jejunum, and the ileum. The final digestion and absorption of disaccharides, peptides, fatty acids, and monoglycerides is the work of villi, small fingerlike projections, which line the small intestine.

G) Liver – Even though the liver is not an organ of digestion, it does secrete bile which aids in digestion by neutralizing the acid chyme from the stomach and emulsifying fats. The liver is also responsible for the chemical destruction of excess amino acids, the storage of glycogen, and the breakdown of old red blood cells.

H) Large Intestine – The large intestine receives the liquid material that remains after digestion and absorption in the small intestine have been completed. However, the primary function of the large intestine is the reabsorption of water.

INGESTION AND DIGESTION IN OTHER ORGANISMS

A) Hydra – The hydra possesses tentacles which have stinging cells (nematocysts) which shoot out a poison to paralyze the prey. If successful in capturing an animal, the tentacles push it into the hydra's mouth. From there, the food enters the gastric cavity. The hydra uses both intracellular and extracellular digestion.

B) Earthworm – As the earthworm moves through soil, the suction action of the pharynx draws material into the mouth cavity. Then from the mouth, food goes into the pharynx, the esophagus, and then the crop which is a temporary storage area. This food then passes into a muscular gizzard where it is ground and churned. The food mass finally passes into the intestine; any undigested material is eliminated through the anus.

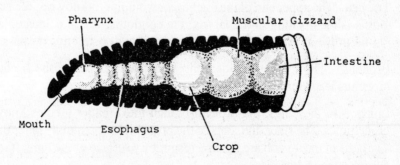

The digestive system of the earthworm.

C) Grasshopper – The grasshopper is capable of consuming large amounts of plant leaves. This plant material must first pass through the esophagus into the crop, a temporary storage organ. It then travels to the muscular gizzard where food is ground. Digestion takes place in the stomach. Enzymes secreted by six gastric glands are responsible for digestion. Absorption takes place mainly in the stomach. Undigested material passes into the intestine, collects in the rectum, and is eliminated through the anus.

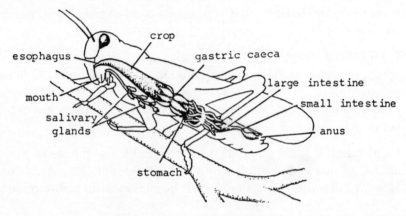

The digestive system of the grasshopper.

RESPIRATION IN HUMANS

The respiratory system in humans begins as a passageway in the nose. Inhaled air then passes through the pharynx, the trachea, the bronchi, and the lungs.

A) Nose – The nose is better adapted to inhale air than the mouth. The nostrils, the two openings in the nose, lead into the nasal passages which are lined by the mucous membrane. Just beneath the mucous membrane are capillaries which warm the air before it reaches the lungs.

B) Pharynx – Air passes via the nasal cavities to the pharynx where the paths of the digestive and respiratory systems cross.

C) Trachea – The upper part of the trachea, or windpipe, is known as the larynx. The glottis is the opening in the larynx; the epiglottis, which is located above the glottis, prevents food from entering the glottis and obstructing the passage of air.

D) Bronchi – The trachea divides into two branches called the bronchi. Each bronchus leads into a lung.

E) Lungs – In the lungs, the bronchi branch into smaller tubules known as the bronchioles. The finer divisions of the bronchioles eventually enter the alveoli. The cells of the alveoli are the true respiratory surface of the lung. It is here that gas exchange takes place.

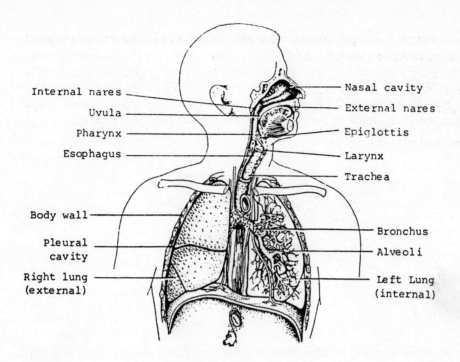

The human respiratory system.

RESPIRATION IN OTHER ORGANISMS

Protozoa

A) Amoeba – Simple diffusion of gases between the cell and water is sufficient to take care of the respiratory needs of the amoeba.

B) Paramecium – The paramecium takes in dissolved oxygen and releases dissolved carbon dioxide directly through the plasma membrane.

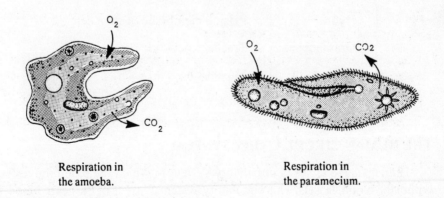

Respiration in the amoeba.

Respiration in the paramecium.

Hydra – Dissolved oxygen and carbon dioxide diffuse in and out of two cell layers through the plasma membrane.

Grasshopper – The grasshopper carries on respiration by means of spiracles and tracheae. Blood plays no role in transporting oxygen and carbon dioxide. Muscles of the abdomen pump air into and out of the spiracles and the tracheae.

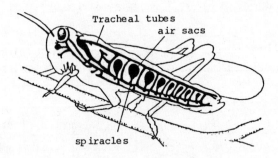

Respiration in the grasshopper.

Earthworm – The skin of the earthworm is its respiratory surface. Oxygen from the air diffuses into the capillaries of the skin and joins with hemoglobin dissolved in the blood plasma. This oxyhemoglobin is released to the tissue cells. Carbon dioxide from the tissue cells diffuses into the blood. When the blood reaches the capillaries in the skin again, the carbon dioxide diffuses through the skin into the air.

Comparison of Various Respiratory Surfaces Among Organisms

Organism	Respiratory Surface Present
Protozoan	Plasma membrane
Hydra	Plasma membrane of each cell
Grasshopper	Tracheae network
Earthworm	Moist skin
Human	Air sacs in lungs

THE HUMAN CIRCULATORY SYSTEM

Humans have a closed circulatory system in which the blood moves entirely within the blood vessels. This circulatory system consists of the heart, blood, veins, arteries, capillaries, lymph, and lymph vessels.

Heart

The heart is a pump-like muscle covered by a protective membrane known as the pericardium and divided into four chambers. These chambers are the left and right atria, and the left and right ventricles.

A) Atria – The atria are the upper chambers of the heart that receive blood from the superior and inferior vena cava. This blood is then pumped to the lower chambers or ventricles.

B) Ventricles – The ventricles have thick walls as compared to the thin walls of the atria. They must pump blood out of the heart to the lungs and other distant parts of the body.

The heart also contains many important valves. The tricuspid valve is located between the right atrium and the right ventricle. It prevents the backflow of blood into the atrium after the contraction of the right ventricle. The bicuspid valve, or mitral valve, is situated between the left atrium and the left ventricle. It prevents the backflow of blood into the left atrium after the left ventricle contracts.

TRANSPORT MECHANISMS IN OTHER ORGANISMS

A) Protozoans – Most protozoans are continually bathed by food and oxygen because they live in water or another type of fluid. With the process of cyclosis or diffusion, digested materials and oxygen are distributed within the cell, and water and carbon dioxide are removed. Proteins are transported by the endoplasmic reticulum.

B) Hydra – Like the protozoans, materials in the hydra are distributed to the necessary organelles by diffusion, cyclosis, and by the endoplasmic reticulum.

C) Earthworm – The circulatory system of the earthworm is known as a "closed" system because the blood is confined to the blood vessels at all times. A pump that forces blood to the capillaries consists of five pairs of aortic loops. Contraction of these loops forces blood into the ventral blood vessel. This ventral blood vessel transports blood toward the rear of the worm. The dorsal blood vessel forces blood back to the aortic loops at the anterior end of the worm.

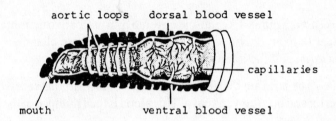

"Closed" circulatory system of the earthworm.

THE HUMAN ENDOCRINE SYSTEM

The major glands of the human endocrine system include the thyroid gland, parathyroid glands, pituitary gland, pancreas, adrenal glands, pineal gland, thymus gland, and the sex glands.

A) Thyroid Gland – The thyroid gland is a two-lobed structure located in the neck. It is responsible for the secretion of the hormone thyroxin. Thyroxin increases the rate of cellular oxidation and influences growth and development of the body.

B) Parathyroid Glands – The parathyroid glands are located in back of the thyroid gland. They secrete the hormone parathormone which is responsible for regulating the amount of calcium and phosphate salts in the blood.

C) Pituitary Gland – The pituitary gland is located at the base of the brain. It consists of three lobes: the anterior lobe; the intermediate lobe, which is only a vestige in adulthood; and the posterior lobe.

   1) Hormones of the Anterior Lobe

      a) growth hormone – stimulates growth of bones.

      b) thyroid-stimulating hormone (TSH) – stimulates the thyroid gland to produce thyroxin.

      c) prolactin – regulates development of the mammary glands of a pregnant female and stimulates secretion of milk in a woman after childbirth.

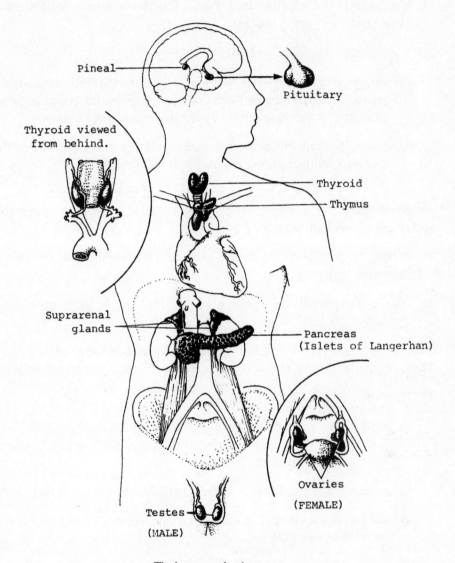

The human endocrine system.

d) adrenocorticotropic hormone (ACTH) – stimulates the secretion of hormones by the cortex of the adrenal glands.

e) follicle-stimulating hormone (FSH) – this hormone acts upon the gonads, or sex organs.

f) luteinizing hormone (LH) – in the male, LH causes the cells in the testes to secrete androgens. In females, LH causes the follicle in an ovary to change into the corpus luteum.

2) Hormones of the Intermediate Lobe – This lobe secretes a hormone that has no known effect in humans.

3) Hormones of the Posterior Lobe

   a) vasopressin (ADH) – this hormone causes the muscular walls of the arterioles to contract, thus increasing blood pressure. It regulates the amount of water reabsorbed by the nephrons in the kidney.

   b) oxytocin – this hormone stimulates the muscle of the walls of the uterus to contract during childbirth. It induces labor.

D) Pancreas – The pancreas is both an endocrine and an exocrine gland. As an endocrine gland, the islets of Langerhans, scattered through the pancreas, secrete insulin and glucagon.

1) Insulin – Acts to lower the level of glucose in the bloodstream. Glucose is converted to glycogen.

2) Glucagon – Increases the level of glucose in the blood by helping to change liver glycogen into glucose.

E) Adrenal Glands – The two adrenal glands are located on top of each kidney. They are composed of two regions: the adrenal cortex and the adrenal medulla.

1) Hormones of the Adrenal Cortex

   a) cortisones – regulate the change of amino acids and fatty acids into glucose. They also help to suppress reactions that lead to the inflammation of injured parts.

   b) cortins – regulate the use of sodium and calcium salts by the body cells.

   c) sex hormones – they are similar in chemical composition to hormones secreted by sex glands.

2) Hormones of the Adrenal Medulla

   a) epinephrine – this hormone is responsible for the release of glucose from the liver, the relaxation of the smooth muscles of the bronchioles, dilation of the pupils of the eye, a reduction in the clotting time of blood, and an increase in the heartbeat rate, blood pressure, and respiration rate.

b) norepinephrine – this hormone is responsible for the constriction of blood vessels.

F) Pineal Gland – The pineal gland is attached to the brain above the cerebellum. It is responsible for the production of melatonin whose role in humans is uncertain.

G) Thymus Gland – The thymus gland is located under the breastbone. Although there is no convincing evidence for its role in the human adult, it does secrete thymus hormone in infants which stimulates the formation of an antibody system.

H) Sex Glands – These glands include the testes of the male and the ovaries of the female.

1) Testes – Luteinizing hormone stimulates specific cells of the testes to secrete androgens. Testosterone, which controls the development of male secondary sex characteristics, is the principal androgen.

2) Ovaries – Estrogen is secreted from the cells which line the ovarian follicle. This hormone is responsible for the development of female secondary sex characteristics.

Human Endocrine Glands and Their Functions

Gland	Hormone	Function
Pituitary Anterior lobe	Growth hormone	Stimulates growth of skeleton
	FSH	Stimulates follicle formation in ovaries and sperm formation in testes
	LH	Stimulates formation of corpus luteum in ovaries and secretion of testosterone in testes
	TSH	Stimulates secretion of thyroxin from thyroid gland
	ACTH	Stimulates secretion of cortisone and cortin from adrenal cortex
	Prolactin	Stimulates secretion of milk in mammary glands
Posterior lobe	Vasopressin (ADH)	Controls narrowing of arteries and rate of water absorption in kidney tubules
	Oxytocin	Stimulates contraction of smooth muscle of uterus
Thyroid	Thyroxin	Controls rate of metabolism and physical and mental development
	Calcitonin	Controls calcium metabolism
Parathyroids	Parathormone	Regulates calcium and phosphate level of blood

Human Endocrine Glands and Their Functions (Cont'd)

Gland	Hormone	Function
Islets of Langerhans Beta cells	Insulin	Promotes storage and oxidation of glucose
Alpha cells	Glucagon	Releases glucose into bloodstream
Thymus	Thymus hormone	Stimulates formation of antibody system
Adrenal Cortex	Cortisones	Promote glucose formation from amino acids and fatty acids
	Cortins	Control water and salt balance
	Sex hormones	Influence sexual development
Medulla	Epinephrine (adrenalin) or norepinephrine (noradrenalin)	Releases glucose into bloodstream, increases rate of heartbeat, increases rate of respiration, reduces clotting time, relaxes smooth muscle in air passages
Gonads Ovaries, follicle cells	Estrogen	Controls female secondary sex characteristics
Corpus luteum cells	Progesterone	Helps maintain attachment of embryo to mother
Testes	Testosterone	Controls male secondary sex characteristics

THE NERVOUS SYSTEM

The nervous system is a system of conduction that transmits information from receptors to appropriate structures for action.

Neurons – The unit of structure that conducts electrochemical impulses over a certain distance. In many neurons, the nerve impulses are generated in the dendrites. These impulses are then conducted along the axon, which is a long fiber. A myelin sheath covers the axon.

A) Sensory Neurons – Sensory neurons conduct impulses from receptors to the central nervous system.

B) Interneurons – Interneurons are always found within the spinal cord and the brain. They form the intermediate link in the nervous system pathway.

C) Motor Neurons – Motor neurons conduct impulses from the central nervous system to the effectors which are muscles and glands. They will bring about the responses to the stimulus.

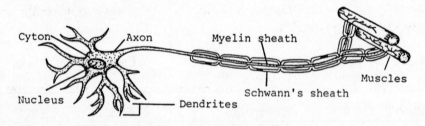

The structure of a neuron.

Nerve Impulse – The signal that is transmitted from one neuron to another. When a neuron is not stimulated, the outside of the neuron is positively charged and the inside is negatively charged. However, when there is neuronal stimulation, the inside of the neuron is temporarily positively charged and the outside is temporarily negatively charged. This marks the beginning of the generation and flow of the nerve impulse.

Synapse – The junction between the axon of one neuron and the dendrite of the next neuron in line. An impulse is transmitted across the synaptic gap by a specific chemical transmitter which is acetylcholine. When a nerve impulse reaches the end brush of the first axon, the end brush secretes acetylcholine into the synaptic gap.

It is here that the acetylcholine changes the permeability of the dendrite's membrane of the second neuron. As soon as the acetylcholine is no longer needed, it is decomposed by the enzyme acetylcholinesterase.

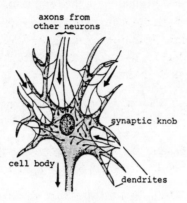

Nerve impulse across a synapse.

Reflex Arc – The unit of function of the nervous system. It is formed by a sequence of sensory neurons, interneurons, and motor neurons which conduct the nerve impulses for the given reflex.

THE HUMAN NERVOUS SYSTEM

A) The Central Nervous System – The brain and the spinal cord comprise the central nervous system. The brain is divided into three regions: the forebrain, the midbrain, and the hindbrain. Each of these regions has a specific function attributed to the particular lobe.

1) Brain

a) Forebrain – The cerebrum is the most prominent part of the forebrain and is divided into two hemispheres. It also has four major areas known as the sensory area, motor area, speech area, and association area. The thalamus, hypothalamus, pineal gland, and part of the pituitary gland are also part of the human forebrain.

b) Midbrain – The midbrain is one of the smallest regions of the human brain. Its function is to relay nerve impulses between the two other brain regions: the forebrain and the hindbrain. It also aids in the maintenance of balance.

c) Hindbrain – The medulla oblongata and the cerebellum are the two main regions of the hindbrain.

1) medulla oblongata – controls reflex centers for respiration and heartbeat, coughing, swallowing, and sneezing.

2) cerebellum – coordinates locomotor activity in the body initiated by impulses originating in the forebrain.

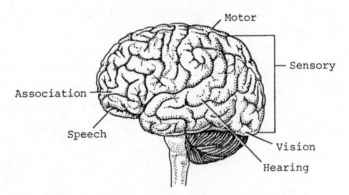

The major areas of the cerebrum.

2) Spinal Cord – The spinal cord runs from the medulla down through the backbone. Throughout its length, it is enclosed by three meninges and by the spinal column vertebrae. Running vertically in the spinal cord center is a narrow canal filled with cerebrospinal fluid.

The spinal cord contains control centers for reflex acts below the neck, and it provides the major pathway for impulses between the peripheral nervous system and the brain. It is also a connecting center between sensory and motor neurons.

B) The Peripheral Nervous System – The peripheral nervous system is composed of nerve fibers which connect the brain and the spinal cord (central nervous system) to the sense organs, glands, and muscles. It can be subdivided into the somatic nervous system and the autonomic nervous system.

1) Somatic Nervous System – The somatic nervous system consists of nerves which transmit impulses from receptors to the central nervous system and from the central nervous system to the skeletal muscles of the body.

2) Autonomic Nervous System – The autonomic nervous system is composed of sensory and motor neurons which run between the central nervous system and various internal organs, such as the heart, glands, and intestines. It regulates internal responses which keep the internal environment constant. It is subdivided into two smaller systems.

a) sympathetic system – a branch of the autonomic nervous system with motor neurons arising from the spinal cord. This system accelerates heartbeat rate, constricts arteries, slows peristalsis, relaxes the bladder, dilates breathing passages, dilates the pupil, and increases secretion.

b) parasympathetic system – a branch of the autonomic nervous system which consists of fibers arising from the brain. The effectors on the organs innervated by the parasympathetic system are opposite to the effects of the sympathetic system.

Actions of the Autonomic Nervous System

Organ Innervated	Sympathetic Action	Parasympathetic Action
Heart	Accelerates heartbeat	Slows heartbeat
Arteries	Constricts arteries	Dilates arteries
Lungs	Dilates bronchial passages	Constricts bronchial passage
Digestive Tract	Slows peristalsis rate	Increases peristalsis rate
Eye	Dilates pupil	Constricts pupil
Urinary Bladder	Relaxes bladder	Constricts bladder

THE NERVOUS SYSTEM OF OTHER ORGANISMS

Protozoans – Protozoans have no nervous system; however, their protoplasm does receive and respond to certain stimuli.

Hydra – The hydra possesses a simple nervous system, which has no central control, known as a nerve net. A stimulus applied to a specific part of the body will generate an impulse which will travel to all body parts.

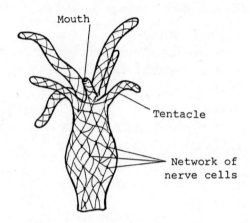

Nerve net of the hydra.

Earthworm – The earthworm possesses a central nervous system which includes a brain, a nerve cord (which is a chain of ganglions), sense organs, nerve fibers, muscles, and glands.

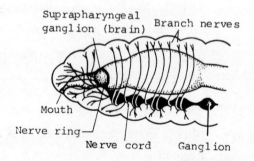

Nervous system of the earthworm.

Grasshopper – The grasshopper's nervous system consists of ganglia bundled together to form the peripheral nervous system. The ganglia of the grasshopper are better developed than in the earthworm.

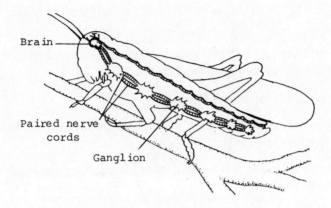

Brain

Paired nerve
cords

Ganglion

Nervous system of the grasshopper.

SKELETAL SYSTEM IN HUMANS

The cartilage, ligaments, and a skeleton composed of 206 bones make up the skeletal system of man.

A) Cartilage – Cartilage is a soft material present at the ends of bones, especially at joints.

B) Ligaments – Ligaments are strong bands of connective tissue which bind one bone to another bone.

C) Skeleton – The bones of the skeleton serve five important functions:

   1) Allowing movements of parts of the body.

   2) Supporting various organs of the body.

   3) Supplying the body with red blood cells and some white blood cells.

   4) Protecting internal organs.

   5) Storing calcium and phosphate salts.

The skeleton is composed of the vertebral column, the skull, the limbs, the breastbone, and the ribs.

THE MUSCULAR SYSTEM IN HUMANS

A) Kinds of Muscles

   1) Smooth Muscle – Smooth muscle is found in the walls of the hollow organs of the body.

2)   Cardiac Muscle – Cardiac muscle is the muscle that comprises the walls of the heart.

3)   Skeletal Muscle – Skeletal muscles are muscles attached to the skeleton. They are also known as striated muscles.

B) The Structure of Muscles and Bones

Bones move only when there is a pull on the muscles attached to the bone. A single skeletal muscle consists of:

1)   Tendon – A tendon is a band of strong, connective tissue which attaches muscle to bone.

2)   Origin – The origin is one end of the muscle which is attached to a bone that does not move when the muscle contracts.

3)   Insertion – The insertion is the other end of the muscle which is attached to a bone that moves when the muscle contracts.

4)   Belly – The belly is the thickened part of the muscle which contracts and pulls.

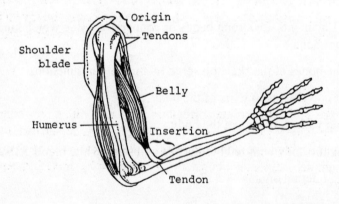

The mechanism of movement of the upper arm.

C) Skeletal Muscle Activation

The nervous system controls skeletal muscle contraction. End brushes of motor neurons come in contact with muscle fibers at the motor end plate, a synapse-like junction. Muscle contraction occurs when acetylcholine is discharged on the muscle fiber surface after the impulse reaches the motor end plate.

D) Structure of a Muscle Fiber

Skeletal muscle is composed of long fibers whose cytoplasm possesses alternating light and dark bands. These bands are part of fibrils which lie parallel to one another. The dark bands are termed A-bands and the light bands are the I-bands. The H-band bisects the A-bands while the Z-line bisects the I-band.

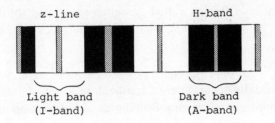

Single muscle fibril.

E) Chemical Composition of Muscle Contraction

Thick filaments that make up the A-band are composed of the protein myosin. The thin filaments extend in either direction from the Z-line and are composed of the protein actin. The actin and myosin proteins form filaments that alternate and overlap each other. During muscle contraction, the actin filaments are pulled inward between myosin filaments, causing the muscle to shorten. Energy for this process is released from ATP molecules after external impulses stimulate the muscle fiber.

Homeostasis

Homeostasis is the automatic maintenance of a steady state within the bodies of all organisms. It is the tendency of organisms to constantly maintain the conditions of their internal environment by responding to both internal and external changes. The kidney, for example, maintains a constant environment by excreting certain substances and conserving others.

Immune Response

The immune system recognizes and eliminates antigens, foreign substances, or organisms that enter the body. Ameboid, phagocytic cells, the macrophages, and the neutrophils engulf particles and invaders, massing at infection sites and causing pus. B-cell lymphocytes and T-cell lymphocytes, the most common immune system cells, participate in the immune response, both humoral (involving antibodies or immunoglobulins) and cell mediated (cellular). When activated, T-cells contain free ribosomes, while the B-cells have bound ribosomes (rough endoplasmic reticulum).

The immune system is widespread. Central lymphoid tissue includes the bone marrow and thymus, while peripheral lymphoid tissue includes the lymph nodes, spleen, adenoids, tonsils, and Peyer's patches.

Reproduction and Development: Gametogenesis, Fertilization, Embryogeny Development

In animals, meiosis is also called gametogenesis, and the four products of meiosis are called gametocytes, cells that must undergo cellular differentiation to become gametes.

In external fertilization, eggs and sperm are shed into water, where fertilization and development usually occur. Internal fertilization is almost universal on land and is also common in the aquatic realm. Both kinds of fertilization involve the fusion of gametes.

The three developmental stages are:

A) Oviparity – Involves independent, external development of an embryo, usually in an egg case. It almost always follows external fertilization, but may also follow internal.

B) Ovoviviparity – Follows internal fertilization, and refers to the retention of the embryo within the uterus, where little if any exchange goes on between mother and offspring.

C) Viviparity – Follows internal fertilization and is the retention of the embryo in the uterus, where the mother's circulatory system provides nourishment, gas exchange, and waste removal.

STAGES OF EMBRYONIC DEVELOPMENT

The earliest stage in embryonic development is the one-cell, diploid zygote which results from the fertilization of an ovum by a sperm.

A) Cleavage – A series of mitotic divisions of the zygote which result in the formation of daughter cells called blastomeres.

B) Morula – A solid ball of 16 blastomeres.

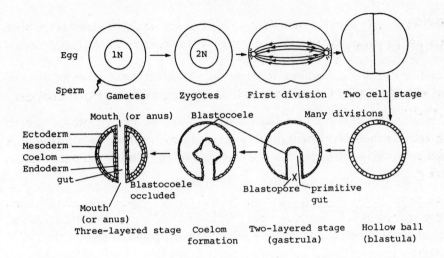

Early embryonic development in animals.

C) Blastula – A hollow ball; a fluid-filled cavity at the center of the sphere is the blastocoel.

D) Gastrula – The cells of the blastula have differentiated into two, and then three embryonic germ layers, forming a gastrula. Early forms of all major structures are laid down in the gastrula period. After this period, the developing organism is called a fetus.

Derivatives of the Primary Germ Layers

Primary Germ Layer	Derivatives
Endoderm	Inner lining of alimentary canal and respiratory tract; inner lining of liver; pancreas; salivary, thyroid, parathyroid, thymus glands; urinary bladder; urethra lining
Mesoderm	Skeletal system; muscular system; reproductive system; excretory system; circulatory system; dermis of skin; connective tissue
Ectoderm	Epidermis; sweat glands; hair; nails; skin; nervous system; parts of eye, ear, and skin receptors; pituitary and adrenal glands; enamel of teeth

Behavior

LEARNED BEHAVIOR

Conditioning, habits, and imprinting are all specific types of behavior that are learned and acquired as the result of individual experiences.

A) Conditioning – Conditioned behavior is a response caused by a stimulus different from that which originally triggered the response. Experiments conducted by Pavlov on dogs demonstrate conditioning of behavior.

> ringing of bell (stimulus 1) → barking (response 1)
>
> food + ringing of bell (stimulus 2) → saliva flow (response 2)
>
> ringing of bell (stimulus 1) → saliva flow (response 2)

Conditional behavior of Pavlov's dog.

B) Habits – Habit behavior is learned behavior that becomes automatic and involuntary as a consequence of repetition. When an action is constantly repeated, the amount of thinking is reduced because impulses pass through the nerve pathways more quickly. The behavior soon becomes automatic.

C) Imprinting – Imprinting involves the establishment of a fixed pathway in the nervous system by the stimulus of the very first object that is seen, heard, or smelled by the particular organism. The research of Konrad Lorenz with newly-hatched geese demonstrated this type of learning.

INNATE BEHAVIOR

Taxis, reflexes, and instincts are all specific types of behavior that are inborn and involuntary.

A) Taxis – Taxis is the response to a stimulus by automatically moving either toward or away from the stimulus.

1) Phototaxis – Photosynthetic microorganisms move toward light of moderate intensity.

2) Chemotaxis – Organisms move in response to some chemical.

E. coli bacteria congregate near specific chemical.

B) Reflex – A reflex is an automatic response to a stimulus in which only a part of the body is involved; it is the simplest inborn response.

The knee jerk is a stretch reflex that is a response to a tap on the tendon below the knee cap which stretches the attached muscle. This tapping activates stretch receptors. Stretching a spindle fiber triggers nerve impulses.

C) Instinct – An instinct is a complex behavior pattern which is unlearned and automatic and is often beneficial in adapting the individual to its environment.

Nest-making of birds and web-spinning by spiders are examples of instinctive behavior.

1) Instinct of self-preservation – This is characterized by "fight or flight" behavior of animals.

2) Instinct of species-preservation – This is characterized by the instinctive behavior of the animal not to escape or fight, but to find a safer area for habitation.

3) Releasers – The releasers are signals which possess the ability to trigger instinctive acts.

VOLUNTARY BEHAVIOR

Voluntary behavior includes activities under direct control of the will, such as learning and memory.

A) Learning – Intelligence measures the ability to learn and properly establish new patterns of behavior. Humans demonstrate the highest degree of intelligence among all animals. This is due, in part, to the highly developed cerebrum which contains a great quantity of nerve pathways and neurons.

B) Memory – All learning is dependent upon one's memory. Memory is essential for all previous learning to be retained and used.

E. PRINCIPLES OF ECOLOGY

Population Dynamics and Growth Patterns, Biotic Potential, Limiting Factors

Ecology can be defined as the study of the interactions between groups of organisms and their environment. The term autecology refers to studies of individual organisms or populations, of single species and their interactions with the environment. Synecology refers to studies of various groups of organisms that associate to form a functional unit of the environment.

A population has characteristics which are a function of the whole group and not of the individual members; among these are population density, birth rate, death rate, age distribution, and biotic potential.

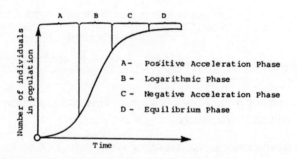

A typical S-shaped growth curve of a population in which the total number of individuals is plotted against time.

A) Population Density – The number or mass of individuals per area or volume of habitable space.

B) Maximum Birth Rate – This is the largest number of organisms that could be produced per unit time under ideal conditions, when there are no limiting factors.

C) Minimum Mortality – This is the number of deaths which would occur under ideal conditions; death due to old age.

D) Biotic Potential (reproductive potential) – The ability of a population to increase in numbers when the age ratio is stable and all environmental conditions are optimal.

Ecosystems and Communities: Energetics and Energy Flow, Productivity, Species Interactions, Succession

The energy cycle starts with sunlight being utilized by green plants on earth. The kinetic energy of sunlight is transformed into potential energy stored in chemical bonds in green plants. The potential energy is released in cell respiration and is used in various ways.

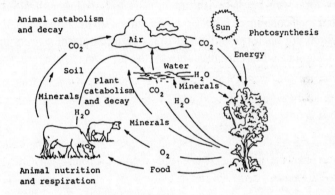

The energy cycle. This diagram shows the relationships between plants and animals and the nonliving materials of the earth. The energy of the sunlight is the only thing that is not returned to its source.

Some of the food synthesized by green plants are broken down by the plants for energy, releasing carbon dioxide and water. Bacteria and fungi break down the bodies of dead plants, using the liberated energy for their own metabolism. Carbon dioxide and water are then released and recycled.

Species interact in the following ways:

A) Contest competition is the active physical confrontation between two organisms which allows one to win the resources.

B) Scramble competition is the exploitation of a common vital resource by both species.

C) Mutualism is a type of relationship where both species benefit from one another. An example is nitrogen-fixing bacteria that live in nodules in the roots of legumes.

D) Commensalism is a relationship between two species in which one specie benefits while the other receives neither benefit nor harm; for example, epiphytes grow on the branches of forest trees.

E) Parasitism is a relationship where the host organism is harmed. Parasites can be classified as external or internal.

F) Succession is a fairly orderly process of changes of communities in a region. It involves replacement of the dominant species within a given area by other species.

Every food web begins with the autotrophic organisms (mainly green plants) being eaten by a consumer. The food web ends with decomposers, the organisms of decay, which are bacteria and fungi that degrade complex organic materials into simple substances which are reusable by the producers (green plants).

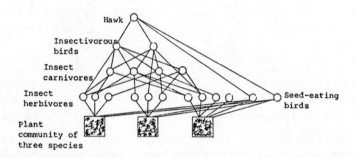

Hypothetical food web. It is assumed that there are 300 species of plants, 10 species of insect herbivores, two bird herbivores, two bird insectivores, and one hawk. In a real community, there would not only be more species at each trophic level, but also many animals that feed at more than one level, or that change levels as they grow older. Some general conclusions emerge from even an oversimplified model like this however. There is an initial diversity introduced by the number of plants. This diversity is multiplied at the plant-eating level. At each subsequent level the diversity is reduced as the food chains converge.

Herbivores consume green plants, and may be acted upon directly by the decomposers or fed upon by secondary consumers, the carnivores. The successive levels in the food webs of a community are referred to as trophic levels.

A pyramid with the producers at the base, and the primary consumers at the apex, can show how energy is being supplied by the producers. It can also show a decrease of energy from the base of the apex accompanied by a decrease in numbers of organisms.

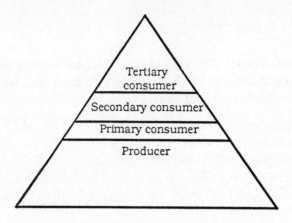

Pyramid of energy and numbers.

Biosphere and Biomes

Types of Habitats:

There are four major habitats: marine, estuarine, fresh water, and terrestrial. A biome is a large community characterized by the kinds of plants and animals present.

Types of Biomes:

A) The Tundra Biome – A tundra is a band of treeless, wet, arctic grassland stretching between the Arctic Ocean and polar ice caps and the forests to the south. The main characteristics of the tundra are low temperatures and a short growing season.

B) The Forest Biomes

1) The northern coniferous forest stretches across North America and Eurasia just south of the tundra. The forest is characterized by spruce, fir and pine trees and by such animals as the wolf, the lynx, and the snowshoe hare.

2) The moist coniferous forest biome stretches along the west coast of North America from Alaska south to central California. It is characterized by great humidity, high temperatures, high rainfall, and small seasonal ranges.

3) The temperature deciduous forest biome was found originally in eastern North America, Europe, parts of Japan and Australia, and the southern part of South America. It is characterized by moderate temperatures with distinct summers and winters and abundant, evenly distributed rainfall. Most of this forest region has now been replaced by cultivated fields and cities.

4) The tropical rain forests stretch around low lying areas near the equator. Dense vegetation, annual rainfall of 200 cm. or more, and a tremendous variety of animals characterize this area.

C) The Grassland Biome – The grassland biome usually occupies the interiors of continents, the prairies of the western United States, and those of Argentina, Australia, southern Russia and Siberia. Grasslands are characterized by rainfalls of about 25 to 75 cm. per year and they provide natural pastures for grazing animals.

D) The Chaparral Biome – The chaparral biome is found in California, Mexico, the Mediterranean, and Australia's south coast. It is characterized by mild temperatures, relatively abundant rain in winter, very dry summers and trees with hard, thick evergreen leaves.

E) The Desert Biome – The desert is characterized by rainfall of less than 25 cm. per year, and sparse vegetation that consists of greasewood, sagebrush, and cactus. Such animals as the kangaroo rat and the pocket mouse are able to live there.

F) The Marine Biome – Although the saltiness of the open ocean is relatively uniform, the concentration of phosphates, nitrates, and other nutrients vary widely in different parts of the sea and at different times of the year. All animals and plants are represented except amphibians, centipedes, millipedes, and insects. Life may have originated in the intertidal zone of the marine biome, which is the zone between the high and low tide.

The marine biome is made up of four zones:

1) The intertidal zone supports a variety of organisms because of the high and low tide.

2) The littoral zone is beyond the intertidal zone. It includes many species of aquatic organisms especially producers.

3) The open sea zone – The upper layer of this zone supports a tremendous amount of producers and therefore many consumers. The lower layer, though, supports only a few scavengers and their predators.

4) The ocean floor contains bacteria of decay and worms.

G) Freshwater Zones – Freshwater zones are divided into standing water-lakes, ponds and swamps, and running water, rivers, creeks, and springs. Freshwater zones are characterized by an assortment of animals and plants. Aquatic life is most prolific in the littoral zones of lakes. Freshwater zones change much more rapidly than other biomes.

Biogeochemical Cycles

Biogeochemical cycles involve the movement of mineral ions and molecules in and out of ecosystems. Most ions enter the living realm at the producer level.

The Nitrogen Cycle – Outside of life, nitrogen occurs in exchange pools and reservoirs. The largest reservoir is N_2 in the atmosphere, but is only available to nitrogen–fixers. Nitrate ions are made available in soil.

Phosphorus and Calcium Cycles – Cycles of calcium and phosphorus occur between living organisms and water. The two ions are taken up in soluble phosphate and calcium ions. Phosphates are used in producing ATP, nucleic acids, phospholipids, and tooth and shell materials, while calcium is essential to bone and shell development and in membrane activity.

The Carbon Cycle – Carbon is an essential part of nearly all the molecules of life. The principal exchange pool on land consists of carbon dioxide gas while the source in the water is dissolved carbon dioxide gas and carbonate ion. A large reservoir occurs in the form of limestone and fossil fuels.

THE ADVANCED PLACEMENT EXAMINATION IN

BIOLOGY

TEST I

THE ADVANCED PLACEMENT EXAMINATION IN

BIOLOGY

ANSWER SHEET

1. Ⓐ Ⓑ Ⓒ Ⓓ Ⓔ
2. Ⓐ Ⓑ Ⓒ Ⓓ Ⓔ
3. Ⓐ Ⓑ Ⓒ Ⓓ Ⓔ
4. Ⓐ Ⓑ Ⓒ Ⓓ Ⓔ
5. Ⓐ Ⓑ Ⓒ Ⓓ Ⓔ
6. Ⓐ Ⓑ Ⓒ Ⓓ Ⓔ
7. Ⓐ Ⓑ Ⓒ Ⓓ Ⓔ
8. Ⓐ Ⓑ Ⓒ Ⓓ Ⓔ
9. Ⓐ Ⓑ Ⓒ Ⓓ Ⓔ
10. Ⓐ Ⓑ Ⓒ Ⓓ Ⓔ
11. Ⓐ Ⓑ Ⓒ Ⓓ Ⓔ
12. Ⓐ Ⓑ Ⓒ Ⓓ Ⓔ
13. Ⓐ Ⓑ Ⓒ Ⓓ Ⓔ
14. Ⓐ Ⓑ Ⓒ Ⓓ Ⓔ
15. Ⓐ Ⓑ Ⓒ Ⓓ Ⓔ
16. Ⓐ Ⓑ Ⓒ Ⓓ Ⓔ
17. Ⓐ Ⓑ Ⓒ Ⓓ Ⓔ
18. Ⓐ Ⓑ Ⓒ Ⓓ Ⓔ
19. Ⓐ Ⓑ Ⓒ Ⓓ Ⓔ
20. Ⓐ Ⓑ Ⓒ Ⓓ Ⓔ

21. Ⓐ Ⓑ Ⓒ Ⓓ Ⓔ
22. Ⓐ Ⓑ Ⓒ Ⓓ Ⓔ
23. Ⓐ Ⓑ Ⓒ Ⓓ Ⓔ
24. Ⓐ Ⓑ Ⓒ Ⓓ Ⓔ
25. Ⓐ Ⓑ Ⓒ Ⓓ Ⓔ
26. Ⓐ Ⓑ Ⓒ Ⓓ Ⓔ
27. Ⓐ Ⓑ Ⓒ Ⓓ Ⓔ
28. Ⓐ Ⓑ Ⓒ Ⓓ Ⓔ
29. Ⓐ Ⓑ Ⓒ Ⓓ Ⓔ
30. Ⓐ Ⓑ Ⓒ Ⓓ Ⓔ
31. Ⓐ Ⓑ Ⓒ Ⓓ Ⓔ
32. Ⓐ Ⓑ Ⓒ Ⓓ Ⓔ
33. Ⓐ Ⓑ Ⓒ Ⓓ Ⓔ
34. Ⓐ Ⓑ Ⓒ Ⓓ Ⓔ
35. Ⓐ Ⓑ Ⓒ Ⓓ Ⓔ
36. Ⓐ Ⓑ Ⓒ Ⓓ Ⓔ
37. Ⓐ Ⓑ Ⓒ Ⓓ Ⓔ
38. Ⓐ Ⓑ Ⓒ Ⓓ Ⓔ
39. Ⓐ Ⓑ Ⓒ Ⓓ Ⓔ
40. Ⓐ Ⓑ Ⓒ Ⓓ Ⓔ

41. Ⓐ Ⓑ Ⓒ Ⓓ Ⓔ
42. Ⓐ Ⓑ Ⓒ Ⓓ Ⓔ
43. Ⓐ Ⓑ Ⓒ Ⓓ Ⓔ
44. Ⓐ Ⓑ Ⓒ Ⓓ Ⓔ
45. Ⓐ Ⓑ Ⓒ Ⓓ Ⓔ
46. Ⓐ Ⓑ Ⓒ Ⓓ Ⓔ
47. Ⓐ Ⓑ Ⓒ Ⓓ Ⓔ
48. Ⓐ Ⓑ Ⓒ Ⓓ Ⓔ
49. Ⓐ Ⓑ Ⓒ Ⓓ Ⓔ
50. Ⓐ Ⓑ Ⓒ Ⓓ Ⓔ
51. Ⓐ Ⓑ Ⓒ Ⓓ Ⓔ
52. Ⓐ Ⓑ Ⓒ Ⓓ Ⓔ
53. Ⓐ Ⓑ Ⓒ Ⓓ Ⓔ
54. Ⓐ Ⓑ Ⓒ Ⓓ Ⓔ
55. Ⓐ Ⓑ Ⓒ Ⓓ Ⓔ
56. Ⓐ Ⓑ Ⓒ Ⓓ Ⓔ
57. Ⓐ Ⓑ Ⓒ Ⓓ Ⓔ
58. Ⓐ Ⓑ Ⓒ Ⓓ Ⓔ
59. Ⓐ Ⓑ Ⓒ Ⓓ Ⓔ
60. Ⓐ Ⓑ Ⓒ Ⓓ Ⓔ

61. Ⓐ Ⓑ Ⓒ Ⓓ Ⓔ
62. Ⓐ Ⓑ Ⓒ Ⓓ Ⓔ
63. Ⓐ Ⓑ Ⓒ Ⓓ Ⓔ
64. Ⓐ Ⓑ Ⓒ Ⓓ Ⓔ
65. Ⓐ Ⓑ Ⓒ Ⓓ Ⓔ
66. Ⓐ Ⓑ Ⓒ Ⓓ Ⓔ
67. Ⓐ Ⓑ Ⓒ Ⓓ Ⓔ
68. Ⓐ Ⓑ Ⓒ Ⓓ Ⓔ
69. Ⓐ Ⓑ Ⓒ Ⓓ Ⓔ
70. Ⓐ Ⓑ Ⓒ Ⓓ Ⓔ
71. Ⓐ Ⓑ Ⓒ Ⓓ Ⓔ
72. Ⓐ Ⓑ Ⓒ Ⓓ Ⓔ
73. Ⓐ Ⓑ Ⓒ Ⓓ Ⓔ
74. Ⓐ Ⓑ Ⓒ Ⓓ Ⓔ
75. Ⓐ Ⓑ Ⓒ Ⓓ Ⓔ
76. Ⓐ Ⓑ Ⓒ Ⓓ Ⓔ
77. Ⓐ Ⓑ Ⓒ Ⓓ Ⓔ
78. Ⓐ Ⓑ Ⓒ Ⓓ Ⓔ
79. Ⓐ Ⓑ Ⓒ Ⓓ Ⓔ
80. Ⓐ Ⓑ Ⓒ Ⓓ Ⓔ

81. Ⓐ Ⓑ Ⓒ Ⓓ Ⓔ
82. Ⓐ Ⓑ Ⓒ Ⓓ Ⓔ
83. Ⓐ Ⓑ Ⓒ Ⓓ Ⓔ
84. Ⓐ Ⓑ Ⓒ Ⓓ Ⓔ
85. Ⓐ Ⓑ Ⓒ Ⓓ Ⓔ
86. Ⓐ Ⓑ Ⓒ Ⓓ Ⓔ
87. Ⓐ Ⓑ Ⓒ Ⓓ Ⓔ
88. Ⓐ Ⓑ Ⓒ Ⓓ Ⓔ
89. Ⓐ Ⓑ Ⓒ Ⓓ Ⓔ
90. Ⓐ Ⓑ Ⓒ Ⓓ Ⓔ
91. Ⓐ Ⓑ Ⓒ Ⓓ Ⓔ
92. Ⓐ Ⓑ Ⓒ Ⓓ Ⓔ
93. Ⓐ Ⓑ Ⓒ Ⓓ Ⓔ
94. Ⓐ Ⓑ Ⓒ Ⓓ Ⓔ
95. Ⓐ Ⓑ Ⓒ Ⓓ Ⓔ
96. Ⓐ Ⓑ Ⓒ Ⓓ Ⓔ
97. Ⓐ Ⓑ Ⓒ Ⓓ Ⓔ
98. Ⓐ Ⓑ Ⓒ Ⓓ Ⓔ
99. Ⓐ Ⓑ Ⓒ Ⓓ Ⓔ
100. Ⓐ Ⓑ Ⓒ Ⓓ Ⓔ

101. Ⓐ Ⓑ Ⓒ Ⓓ Ⓔ
102. Ⓐ Ⓑ Ⓒ Ⓓ Ⓔ
103. Ⓐ Ⓑ Ⓒ Ⓓ Ⓔ
104. Ⓐ Ⓑ Ⓒ Ⓓ Ⓔ
105. Ⓐ Ⓑ Ⓒ Ⓓ Ⓔ
106. Ⓐ Ⓑ Ⓒ Ⓓ Ⓔ
107. Ⓐ Ⓑ Ⓒ Ⓓ Ⓔ
108. Ⓐ Ⓑ Ⓒ Ⓓ Ⓔ
109. Ⓐ Ⓑ Ⓒ Ⓓ Ⓔ
110. Ⓐ Ⓑ Ⓒ Ⓓ Ⓔ
111. Ⓐ Ⓑ Ⓒ Ⓓ Ⓔ
112. Ⓐ Ⓑ Ⓒ Ⓓ Ⓔ
113. Ⓐ Ⓑ Ⓒ Ⓓ Ⓔ
114. Ⓐ Ⓑ Ⓒ Ⓓ Ⓔ
115. Ⓐ Ⓑ Ⓒ Ⓓ Ⓔ
116. Ⓐ Ⓑ Ⓒ Ⓓ Ⓔ
117. Ⓐ Ⓑ Ⓒ Ⓓ Ⓔ
118. Ⓐ Ⓑ Ⓒ Ⓓ Ⓔ
119. Ⓐ Ⓑ Ⓒ Ⓓ Ⓔ
120. Ⓐ Ⓑ Ⓒ Ⓓ Ⓔ

ADVANCED PLACEMENT
BIOLOGY EXAM I

SECTION I

120 Questions
90 minutes

DIRECTIONS: For each question, there are five possible choices. Select the best choice for each question. Blacken the correct space on the answer sheet.

1. The most efficient oxygenation of blood is found in a:

   (A) bass   (D) robin

   (B) frog   (E) turtle

   (C) snake

2. The main abiotic source of carbon in the environment for the carbon cycle comes from:

   (A) carbon dioxide in the air

   (B) carbon dioxide in water

   (C) carbon monoxide in the air

   (D) carbon monoxide in water

   (E) carbonates in the air and water

3.   The optimum pH and body site for amylase activity is:

(A) 2, stomach   (D) 8, stomach

(B) 5, small intestine (E) 10, small intestine

(C) 7, oral cavity

4.   In the process of photosynthesis, free oxygen is liberated from:

(A) carbohydrates   (D) chlorophyll

(B) carbon dioxide   (E) water

(C) carbon monoxide

5.   Hydrolysis of lipid molecules yields:

(A) amino acids and water

(B) amino acids and glucose

(C) fatty acids and glycerol

(D) glucose and glycerol

(E) glycerol and water

6.   Prokaryotic cells lack:

(A) a cell membrane   (D) a nuclear membrane

(B) cytoplasm   (E) ribosomes

(C) a DNA molecule

7. The useful energy flowing through a food chain is available mostly to the:

(A) herbivores

(D) secondary consumers

(B) primary consumers

(E) tertiary consumers

(C) producers

8. Enzymes affect biochemical reactions by:

(A) blocking their end product formation

(B) destroying all substances produced in the reactions

(C) raising the temperature of the reaction's environment

(D) reversing their direction

(E) accelerating the reaction rates

9. Simple squamous tissue is a type of which of the following kinds of tissue?

(A) connective

(D) nerve

(B) epithelial

(E) vascular

(C) muscle

10. The ten-inch human-body tube accepting swallowed food is the:

(A) esophagus

(D) pharynx

(B) larynx

(E) trachea

(C) nasal cavity

11. The majority of ATP molecules derived from nutrient metabolism are generated by (the):

(A) anaerobic fermentation and glycolysis

(B) fermentation and electron transport chain

(C) glycolysis and substrate phosphorylation

(D) Krebs cycle and electron transport chain

(E) substrate phosphorylation

12. Mitosis functions in many organism life cycle events except:

(A) body cell replacement   (D) growth

(B) development   (E) wound healing

(C) gametogenesis

13. The Mendelian law that describes the behavior of two or more gene pairs is the law of:

(A) codominance

(B) dominance

(C) independent assortment

(D) segregation of genes

(E) recombination

14. The scientific name *Escherichia coli* refers to this bacterium's:

(A) class and family

(B) family and order

(C) genus and species

(D) kingdom and phylum

(E) order and phylum

15. Two parents are heterozygous and display respective blood types A and B. If they mate, the probability of producing an offspring with blood type O is:

(A) 0% (D) 75%

(B) 25% (E) 100%

(C) 50%

16. An organism with genotype AaBb can produce a variety of different sex cell genotypes equaling:

(A) 1 (D) 8

(B) 2 (E) 16

(C) 4

17. Fill in the missing blanks for the photosynthetic reaction:

Water + ___(1)___   Carbohydrate + ___(2)___

(A) (1) carbon dioxide, (2) oxygen

(B) (1) chlorophyll, (2) oxygen

(C) (1) light, (2) carbon dioxide

(D) (1) oxygen, (2) water

(E) (1) sugar, (2) light

18. Two pink flowers (Rr), of the species Japanese four-o'clock plant, mate. Assuming incomplete dominance, the chance of obtaining a red-colored offspring is:

(A) 0% (D) 75%

(B) 25% (E) 100%

(C) 50%

19.   The human condition of colorblindness is:

   (A)   caused by a recessive allele

   (B)   equally common in both sexes

   (C)   expressed by a heterozygous genotype in females

   (D)   inherited by males from their fathers

   (E)   produced by a homozygous genotype in males

20.   All are common forms of energy used in metabolism except:

   (A)   chemical   (D)   light

   (B)   heat   (E)   nuclear

   (C)   kinetic

21.   A DNA strand in a double helix has a base sequence of
      ATACGT. The base sequence of its DNA complement is:

   (A)   ACGUAU   (D)   TGCATA

   (B)   ATACGT   (E)   UAUGCA

   (C)   TATGCA

22.   RNA is made by the process of:

   (A)   duplication   (D)   transcription

   (B)   fermentation   (E)   translation

   (C)   replication

23. Select the light with the shortest wavelength absorbed during photosynthesis:

(A) blue

(D) red

(B) green

(E) yellow

(C) orange

24. Genes control body chemistry by ultimately specifying the structure of:

(A) carbohydrates

(D) proteins

(B) lipids

(E) water

(C) phospholipids

25. The gene that turns structural genes off and on in an operon is the:

(A) cistron

(D) regulator

(B) operator

(E) repressor

(C) promotor

26. The variable portion of a DNA nucleotide is at its:

(A) base

(D) ribose

(B) deoxyribose

(E) sugar

(C) phosphate group

27. Select the cell type containing the highest concentration of mitochondria.

(A) erythrocyte

(D) neuron

(B) leukocyte

(E) skin

(C) muscle

28. Viral replication, in which the host cell bursts following each cycle, is termed:

(A) conjugation

(D) transduction

(B) lysogenic

(E) transformation

(C) lytic

29. The smallest, most specific category of classification is the:

(A) class

(D) phylum

(B) family

(E) species

(C) genus

30. The largest, most general category of classification is the:

(A) class

(D) phylum

(B) genus

(E) species

(C) kingdom

31. The function of phloem is to:

(A) cover and protect

(B) convert nutrients from the soil

(C) strengthen and support

(D) store reserve materials

(E) transport organic solutes

32. Which one of the following five components of a reflex arc carries out the organism's response?

(A) effector

(D) sensory neuron

(B) motor neuron

(E) spinal cord

(C) receptor

33.	In the binomial *Quercus alba*, the first term represents the organism's:

(A) class

(D) phylum

(B) genus

(E) species

(C) order

34.	The largest number of known species is represented by the phylum:

(A) Arthropoda

(D) Platyhelminthes

(B) Annelida

(E) Porifera

(C) Echinodermata

35.	A human birth defect produced by a dominant allele of a gene is:

(A) albinism

(D) high cholesterol

(B) diabetes mellitus

(E) low melanin levels

(C) hemophilia

36.	The root system of plants can function for all of the following except:

(A) absorption

(D) transpiration

(B) anchorage

(E) transport

(C) storage

37.	What type of leaf structures and environmental conditions promote gas exchange in plants?

(A) Cortex, heat

(D) Stomata, heat

(B) Cortex, cold

(E) Stomata, normal temperatures

(C) Mesophyll, high humidity

38. The falling off of a leaf from the branch of a deciduous tree is termed:

(A) absorption

(B) abscission

(C) evaporation

(D) perennial

(E) transpiration

39. Phloem conducts:

(A) ions

(B) glucose

(C) glycogen

(D) minerals

(E) sucrose

40. The skin performs all of the following human body functions except:

(A) identification of an individual

(B) protection

(C) sensation

(D) storage

(E) temperature regulation

41. The biceps brachii produce movements by pulling on:

(A) bones

(B) joints

(C) muscles

(D) nerves

(E) skin

42. Which of the following can be said to be a semiconservative process?

(A) conjugation

(D) translation

(B) DNA replication

(E) translocation

(C) RNA transcription

43. Neurons that conduct signals away from the central nervous system are classified as:

(A) afferent

(D) motor

(B) associative

(E) sensory

(C) internuncial

44. The innermost layer of the eye is the:

(A) choroid coat

(D) retina

(B) cornea

(E) sclera

(C) pupil

45. Which of the following is not a polymer?

(A) DNA

(D) RNA

(B) glycogen

(E) starch

(C) glucose

46. Select the nonpathogenic bacterium:

(A) *Clostridium*

(D) *Staphylococcus*

(B) *Escherichia*

(E) *Treponema*

(C) *Salmonella*

47. The largest number of chambers is found in the heart of a(n):

(A) amphibian (D) reptile

(B) bird (E) shark

(C) fish

48. Which law explains the inhalation and exhalation of air in terms of pressure changes?

(A) Archimedes' law (D) Dalton's law

(B) Aristotle's law (E) Mendel's law

(C) Boyle's law

49. Evolution is a process exhibited by a(n):

(A) cell (D) organ system

(B) tissue (E) population

(C) organ

50. Graded variations in a species trait over a geographic distribution is a(n):

(A) cline (D) mutation

(B) genus (E) polymorphism

(C) inbreeding

51. *Paramecium caudatum* is best classified into the kingdom:

(A) Animalia (D) Plantae

(B) Fungi (E) Protista

(C) Monera

52. Which of the following is a single bacterial cell?

   (A) *Diplobacillus*

   (B) *Gonococcus*

   (C) *Staphylococcus*

   (D) *Streptococcus*

   (E) *Streptomyces*

53. Bacteria that can effectively carry out metabolism in the presence or absence of oxygen are described as:

   (A) aerobic

   (B) anaerobic

   (C) facultative anaerobes

   (D) fermentative microbes

   (E) glycolytic

54. Each is an important assumption for maintenance of a Hardy-Weinberg equilibrium in a population except:

   (A) asexual reproduction

   (B) random mating among members

   (C) large population size

   (D) lack of emigration or immigration

   (E) absence of new mutations

55. All the roles and associations of a species in its community comprise its:

   (A) cycle

   (B) ecosystem

   (C) habitat

   (D) niche

   (E) population

56. The free-swimming coelenterate larva is the:

(A) coral

(D) planula

(B) hydra

(E) polyp

(C) medusa

57. Plant-eaters can digest plant cell walls due to their utilization of which enzyme?

(A) amylase

(D) pepsin

(B) cellulase

(E) trypsin

(C) chymotrypsin

58. A coelom is a(n):

(A) body cavity bounded by mesoderm, in which the viscera are suspended.

(B) digestive tract, which is endodermal in origin

(C) outer skin that is ectodermal in origin

(D) specialized region of the higher forebrain

(E) one of the four mammalian heart chambers

59. Which of the following types of organisms occupies the trophic level of least biomass?

(A) herbivores

(D) secondary consumers

(B) plants

(E) tertiary consumers

(C) primary consumers

60. The invertebrate phylum phylogenetically closest to the chordates is:

(A) Annelida

(D) Echinodermata

(B) Arthropoda

(E) Mollusca

(C) Cnidaria

61. An insect metamorphic life cycle occurs in which of the following sequences?

(A) adult-pupa-larva-egg

(B) egg-larva-pupa-adult

(C) larva-adult-egg-pupa

(D) pupa-egg-larva-adult

(E) pupa-adult-larva-egg

62. Highest pressure of circulating blood is found in a(n):

(A) arteriole

(D) vein

(B) artery

(E) venule

(C) capillary

63. Which of the following is part of a human's axial skeleton?

(A) clavicle

(D) rib

(B) fibula

(E) scapula

(C) humerus

64. Glial cells:

    (A) conduct signals   (D) support neurons

    (B) contribute to movement (E) transport oxygen

    (C) cover the skin

65. Select the disease caused by a protozoa:

    (A) chicken pox   (D) measles

    (B) common cold   (E) smallpox

    (C) malaria

66. Which of the following has a vitamin as a building block?

    (A) apoenzyme   (D) mineral

    (B) coenzyme   (E) protein

    (C) holoenzyme

67. The filtering of inhaled debris that travels through the upper respiratory tract occurs through the action of:

    (A) cilia   (D) phagocytes

    (B) goblet cells   (E) villi

    (C) Leidig cells

68. Substances in the blood are transported across the nephron tubules by mechanisms in the process of:

    (A) bulk flow   (D) reabsorption

    (B) filtration   (E) secretion

    (C) osmosis

69. A person receives the results of a hematocrit during a series of blood tests. A hematocrit is the:

(A) abundance of white blood cells in blood

(B) concentration of sugar in the blood

(C) level of circulating antibodies

(D) percentage of blood cellular material by volume

(E) typing of the blood by the ABO scheme

70. An insect is captured and studied in a laboratory. This insect has a pair of short, rigid wings, and a pair of thin veined wings. It also has chewing mouthparts. The insect will most likely be classified as a member of which of the following orders?

(A) Diptera

(B) Hemiptera

(C) Homoptera

(D) Lepidoptera

(E) Orthoptera

DIRECTIONS: The following groups of questions have five lettered choices followed by a list of diagrams, numbered phrases, sentences, or words. For each numbered diagram, phrase, sentence, or word choose the heading which most directly applies. Blacken the correct space on the answer sheet. Each heading may be used once, more than once, or not at all.

Questions 71 - 74 refer to the biomolecule diagrams below.

A

```
        H
        |
        C = O
        |
   H -  C - OH
        |
  HO -  C - H
        |
   H -  C - OH
        |
   H -  C - OH
        |
   H -  C - OH
        |
        H
```

B

```
        H
        |
   H -  C - OH
        |
   H -  C - OH
        |
   H -  C - OH
        |
        H
```

C

```
           R   O
           |   ||
   H.      |   ||
     N  -  C - C - OH
   H⁄       |
           H
```

D

```
  -O
   ‖
-O-P-O- CH₂        N base
   ‖          O
   O
         H H
       H     H
       OH   OH
```

E

71.   carbohydrates

72.   lipids

73.   nucleic acid

74.   proteins

Questions 75 - 78 refer to the diagram below.

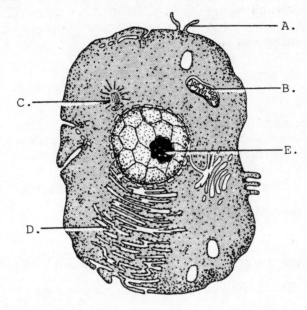

75.   organize cell division

76.   movement of the cell

77.   internal transport

78.   extraction of energy from nutrients

Questions 79-82 refer to stages of meiosis I.

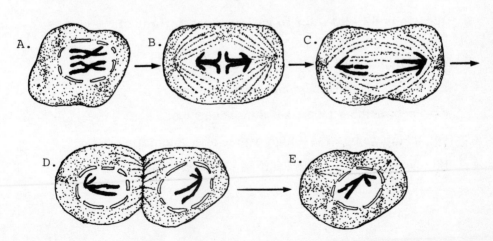

79.   anaphase

80.   metaphase

81.   prophase

82.   telophase

Questions 83 - 86 refer to the listed descriptions of skeletal muscle contraction types.

(A)   a short, individual contraction and relaxation

(B)   a sustained maximal response

(C)   an accumulation of an abundance of lactic acid

(D)   the merging of separate responses into a powerful output

(E)   tonic activity, as occurring in muscle tone

83.   fatigue

84.   simple twitch

85.   summation

86.   tetanus

Questions 87 - 90 refer to various descriptions of:

(A)   members have jointed appendages

(B)   flatworms; members lack segmentation

(C)   each member possesses a muscular foot

(D)   members possess a high degree of segmentation

(E)   closest phylogenetic relatives of the chordates

87.   Annelida

88.   Arthropoda

89.   Echinodermata

90.   Mollusca

Questions 91-93 refer to the drawing below.

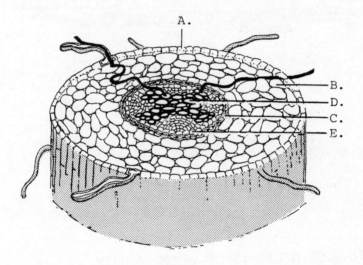

91.   The root cortex

92.   The root epidermis

93.   Root xylem

Questions 94-97 refer to the drawing below.

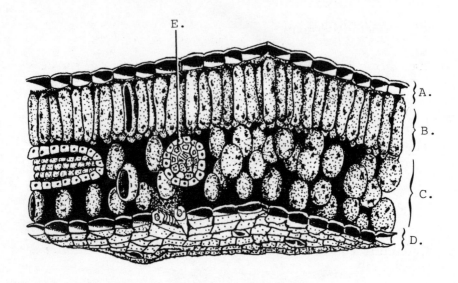

94. Gas exchange occurs here.

95. A waxy cuticle is thicker here.

96. A higher chloroplast density occurs here.

97. The highest humidity would be here.

Questions 98-101 refer to the diagram below.

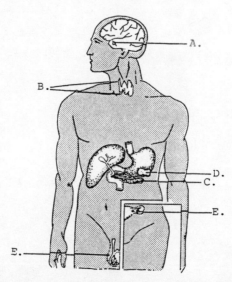

98. A sudden change in an organism's amount of extracellular fluid will be corrected by this organ.

99. A person eats three candy bars. Within minutes, this endocrine gland effects blood glucose homeostasis.

100. Substances that cause vasoconstriction change the diameters of blood vessels in order to assist in increasing blood pressure. Such substances are produced by this gland.

101. Secrete aldosterone

Questions 102 - 105 refer to the organelles of a cell and their function(s).

(A) site of mRNA translation

(B) contains a circular arrangement of 18 microtubules that surround 2 microtubules

(C) contains a circular arrangement of 27 microtubules

(D) site of rRNA synthesis

(E) contains a circular arrangement of nine microtubules surrounding two microtubules

102. centriole

103. cilium

104. nucleolus

105. ribosome

Questions 106 - 107 refer to the drawing below.

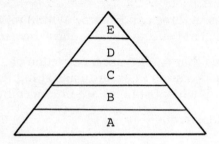

106.   Producer biomass

107.   Secondary consumer

DIRECTIONS: The following questions refer to experimental or laboratory situations or data. Read the description of each situation. Then choose the best answer to each question. Blacken the correct space on the answer sheet.

Question 108 - 110 refer to the diagram below.

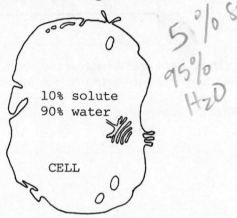

108.   The intracellular environment is best described as:

(A) hypertonic   (D) osmotic

(B) hypotonic   (E) permeable

(C) isotonic

109.   The extracellular environment will:

(A) gain water   (D) lose solute

(B) gain solute   (E) remain unchanged

(C) lose water

110.   Over time the cell will:

(A) become more hypertonic intracellularly

(B) enlarge and experience lysis

(C) experience crenation

(D) lose motility

(E) lose solute

Questions 111 - 112 refer to the genetic grid below.

	A B	A b	a B	a b
A b	1	2	3	4
a b	5	6	7	8

111.   In the genetic cross, what is the percentage of genetic recom-
       binations that are heterozygous for both loci?

   (A) 0   (D) 75

   (B) 25   (E) 100

   (C) 50

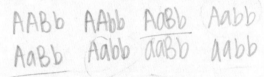

112.   A genotype that is not produced among offspring from this
       cross:

   (A) AABb   (D) Aabb

   (B) AAbb   (E) aaBB

   (C) AaBb

Questions 113-115 refer to the diagram below.

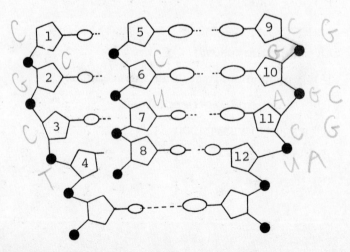

28

113. DNA nucleotide base #2 is cytosine. The RNA base of #6 is:

(A) Adenine   (D) Thymine

(B) Cytosine   (E) Uracil

(C) Guanine

114. RNA base #7 is uracil. The DNA base at #3 is:

(A) Adenine   (D) Thymine

(B) Cytosine   (E) Uracil

(C) Guanine

U A

115. The DNA-base sequence from 1 to 4 is CGCT. The RNA base sequence from 5 to 8 is:

(A) ACGG   (D) GCGA

(B) CGCT   (E) TCGC

(C) CGCU

Questions 116 - 118 refer to the drawing below:

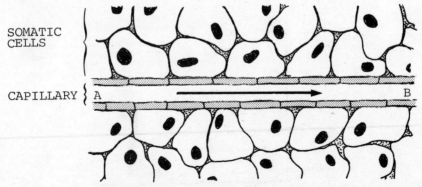

Arrow indicates the direction of blood flow.

29

116. Region A:

(A) accepts carbon dioxide from cells

(B) receives blood from an arteriole

(C) receives nutrients from cells

(D) has a lower blood pressure than the blood pressure at B

(E) transports blood to an artery

117. Region B:

(A) accepts oxygen from cells

(B) gives carbon dioxide to cells

(C) has a higher blood pressure than the blood pressure at A

(D) transports blood to a venule

(E) unloads nutrients to cells

118. The best word to describe this blood vessel's function is:

(A) circulation

(B) exchange

(C) flow

(D) pressurization

(E) transport

Questions 119 - 120 refer to the graph below.

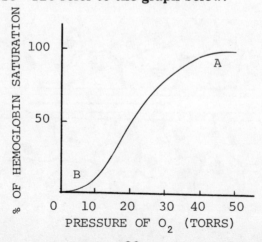

30

119.   Point A on the graph reflects blood hemoglobin behavior at the:

    (A) heart   (D) tissue cell

    (B) kidney   (E) vein

    (C) lung

120.   Point B on the graph reflects blood hemoglobin behavior at the:

    (A) artery   (D) lung

    (B) kidney   (E) tissue cell

    (C) liver

ADVANCED PLACEMENT
BIOLOGY EXAM I

SECTION II

DIRECTIONS: Answer each of the following four questions in essay format. Each answer should be clear, organized, and well-balanced. Diagrams may be used in addition to the discussion, but a diagram alone will not suffice. Suggested writing time per essay is 22 minutes.

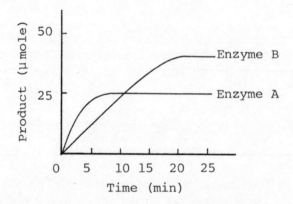

Graph I = condition of enzyme A, catalysis = 37°C, pH = 3, excess substrate A. Condition of enzyme B, catalysis = 37°C, pH = 8, excess substrate B.

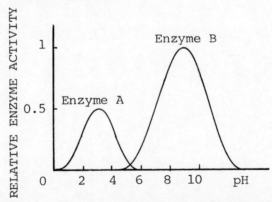

Graph II = condition of enzyme A, catalysis = 37°C, excess substrate A. Condition of enzyme B, catalysis = 37°C, excess substrate B.

1. Enzymes are important reaction catalyzers.

    A) From the data given in Graph I, discuss the effectiveness of each enzyme.
    B) Give an example of an enzyme in the human body that corresponds with enzyme A and another that corresponds with enzyme B.
    C) Describe the effects an excess of carbon dioxide in the body would have on enzyme effectiveness.

2. Describe the different types of mutations and the process of translation. Include a discussion of how point mutations affect proper protein synthesis at the level of translation.

3. Defend the accuracy of this ecologically-based statement: "Energy flows through an ecosystem but materials cycle." Offer examples of food chains and biogeochemical cycles to support its accuracy.

4. Describe how each of the following animal and plant cells perform its unique function by specialized structural traits.

    Animal = erythrocyte, neuron, muscle fiber.
    Plant = epidermal cell, tracheids, parenchyma cells.

ADVANCED PLACEMENT
BIOLOGY EXAM I

ANSWER KEY

1.	D	31.	E	61.	B	91.	B
2.	A	32.	A	62.	B	92.	A
3.	C	33.	B	63.	D	93.	D
4.	E	34.	A	64.	D	94.	D
5.	C	35.	C	65.	C	95.	A
6.	D	36.	D	66.	B	96.	B
7.	C	37.	E	67.	A	97.	E
8.	E	38.	B	68.	E	98.	A
9.	B	39.	E	69.	D	99.	C
10.	A	40.	D	70.	E	100.	D
11.	D	41.	A	71.	A	101.	D
12.	C	42.	B	72.	B	102.	C
13.	C	43.	D	73.	D	103.	B
14.	C	44.	D	74.	C	104.	D
15.	B	45.	C	75.	C	105.	A
16.	C	46.	B	76.	A	106.	A
17.	A	47.	B	77.	D	107.	C
18.	B	48.	C	78.	B	108.	A
19.	A	49.	E	79.	C	109.	C
20.	E	50.	A	80.	B	110.	B
21.	C	51.	E	81.	A	111.	B
22.	D	52.	B	82.	D	112.	E
23.	A	53.	C	83.	C	113.	B
24.	D	54.	A	84.	A	114.	D
25.	B	55.	D	85.	D	115.	C
26.	A	56.	D	86.	B	116.	B
27.	C	57.	B	87.	D	117.	D
28.	C	58.	A	88.	A	118.	B
29.	E	59.	E	89.	E	119.	C
30.	C	60.	D	90.	C	120.	E

ADVANCED PLACEMENT
BIOLOGY EXAM I

DETAILED EXPLANATIONS
OF ANSWERS

SECTION I

1.   (D)
Four-chambered hearts, found in birds and mammals, separate the left side (oxygenated) and right side (deoxygenated) blood. Thus, blood exiting the left ventricle remains high in oxygen. Fish, amphibians, and reptiles lack such separation.

2.   (A)
In the carbon cycle, CO_2 circulates between the living and nonliving sectors of an ecosystem. CO_2 composes .04 of one percent of the atmosphere. Producers (plants) in food chains fix it into the protoplasm of plants via photosynthesis. The element moves through the other trophic levels by nutrition and eating. Respiration by consumers and decomposers returns it to the abiotic sector, air, from the biotic sector, living organisms.

3.   (C)
The interior of the stomach has a pH of 2, which is necessary for the action of pepsin. The surface of the skin has a pH of 5.0 to 5.5. The pH range of the small intestine's lumen includes pH = 8. None of the listed organs has a pH of 10. The pH of the oral cavity is usually about 7, which is necessary for the action of salivary amylase, which begins the digestion of starch.

4.   (E)
During the light reactions of photosynthesis, the process of photoly-

sis occurs. By light activation, water, a reactant, is chemically broken into hydrogen and oxygen gasses. The hydrogen is incorporated into the carbon and oxygen atoms of CO_2, the other photosynthesis reactant, to build a sugar. The liberated oxygen from photolysis is the source of the other product, O_2, along with the synthesized sugar.

5.   (C)

Hydrolysis is a type of chemical digestion. Amino acids are the digested building blocks of proteins. Glucose is a subunit of carbohydrates. Water molecules are required to split chemical bonds in hydrolysis but are not produced in the process.

6.   (D)

Prokaryotic cells (bacteria and blue-green algae) are more primitive, less complex cells than eukaryotic cells of other species. They do, however, possess all the listed structures except D. Their genetic material is not encased in a well-defined nuclear membrane.

7.   (C)

Producers intercept rays of light energy from the sun to conduct photosynthesis. With each succession through the link of a food chain, less energy becomes available for use to run the metabolic processes of the organisms in a particular trophic level. Thus there is less energy available to the herbivores, the primary consumers, than there is to the plants, the producers. Succeedingly less remains for the secondary and tertiary consumers.

8.   (E)

Enzymes are organic catalysts, speeding up chemical reactions of living systems by accelerating the attainment of reaction equilibria without shifting their positions. Enzymes are specific with the reaction they catalyze and the substrates they bind to. They do not participate in the reaction itself.

9.   (B)

Epithelial tissue covers the free surfaces of the body. For example, simple (one cell layer) squamous (flat, platelike) epithelial tissue can be found on the surface of the skin, and acts as a protective barrier. Muscle tissue consists of muscle fibers, and contains no simple squamous tissue. Nerve tissue is made almost entirely of neurons and neuroglial cells. Connective tissue, such as bone, blood, and tendons, contain cells that are separated by and suspended in some sort of matrix. Vascular tissue is not a valid tissue category.

10.   (A)

The other choices are respiratory tract structures.

11.   (D)

Only a small fraction of ATP molecules is produced from anaerobic process of fermentation or glycolysis. Once pyruvic acid is formed, its entry into the aerobic Krebs cycle unleashes most of the original glucose molecule's energy. Krebs cycle reactions yield high energy electrons (oxidation) that are then shuttled down a series of transport acceptors located in the inner mitochondrial membrane until they finally combine with oxygen and H^+ to form water. During electron transport, a proton gradient is generated across the inner mitochondrial membrane. The collapse of this proton gradient provides energy for the production of ATP molecules from ADP molecules and inorganic phosphates.

12.   (C)

Mitosis produces body cells whereas meiosis is the cell division process yielding gametes or sex cells: gametogenesis.

13.   (C)

The law of independent assortment states that if two genes are on different chromosomes, then, during meiosis, they may end up in the same gamete, or they may not.

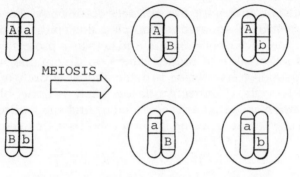

possible gamete genotypes

As one can see, allele A does not <u>have</u> to end up in the same gamete that allele B ends up in. Linked genes, genes whose alleles are located on the same chromosome, cannot undergo independent assortment:

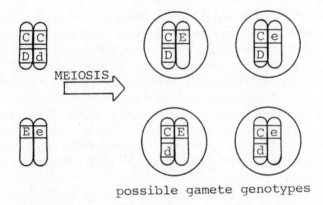

possible gamete genotypes

In this case, alleles C and D and alleles C and d are linked, and cannot be sorted independently. Each pair <u>must</u> end up in the same gamete. The only instance in which linked genes can be sorted independently is one in which "crossing over" occurs.

Mendel's law of dominance and of segregation do not necessarily involve two or more gene pairs. The other choices do not represent Mendelian laws.

14.   (C)

This is an example of a specific organism's binomial, taxonomical classification. The first name identifies the organism's genus. It is capitalized. The species name begins with a lower-cased letter. Both names, usually derived from Latin or Greek, are underlined or italicized. The genus and species are the two most exact taxonomic categories. Other choices are broader, taxonomic categories from the following hierarchy:

Kingdom (broadest, most general)
   Phylum or Division (in plants)
      Class
         Order
            Family
               Genus
                  Species (most exact, specific)

15.   (B)

The blood type A parent is $I^A i$ and the blood type B parent is $I^B i$. Use of probability shows any one of the four blood types occurring among offspring with equal probability, thus 25% for O, A, B, or AB. Also, using a Punnett square,

	I^A	i
I^B	$I^A I^B$	$I^B i$
i	$I^A i$	ii

it is found that the 4 genotypes, $I^A I^B$ (AB), $I^A i$ (A), $I^B i$ (B), ii (O) occur in equal ratio.

16.   (C)

Mendel's law of Independent Assortment leads to AB, Ab, aB, and ab combinations in produced sex cells. The answer has a mathematical base, accounting for all possible combinations when one gene is selected from each of the two pairs.

17.   (A)

During photosynthesis, the gas carbon dioxide reacts with the hydrogen from water. Photolysis, or the chemical splitting of water by light, releases oxygen as one of the products. CO_2 reacts with the available hydrogen in coupled dark reactions to make the other product, a sugar such as glucose - $C_6H_{12}O_6$.

18.   (B)

The outcome of the mating of two pink flowers of genotype Rr, assuming incomplete dominance, is best displayed by a Punnett square:

	R	r
R	RR	Rr
r	Rr	rr

All offspring of genotype RR will be red; all offspring of genotype Rr will be pink; all those of genotype rr will be white. Therefore, there is a 25% chance that any offspring will be red.

19.   (A)

Many of the better-known sex-linked human conditions, such as hemophilia and colorblindness, are caused by recessive alleles. Sex-linked (X-linked) genes are located on the X-chromosome. Thus males, whose sex chromosomes are X and Y, have only one such gene. Assuming that there are only two alleles for this X-linked gene, males are genotypically either C—(normal) or c—(i.e. - colorblind). The Y-chromosome does not offer a second gene in this case. Males can thus not be homozygous. For females, whose sex chromosomes are X and X, three genotypes are possible: CC, Cc and cc. A woman of genotype Cc is a carrier of the disease but does not express the recessive effect of colorblindness. She can, however, pass on her recessive allele to her offspring. In order to produce a colorblind female (cc), a female carrier would have to mate with a colorblind male (c—). Each parent offers a C allele on the X chromosome for a c genotype in the offspring. This is unlikely and an infrequent event.

An example of a common cross is:

Cc x C—
(female)   (male)

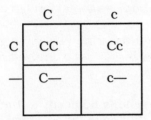

One-half of the males produced are color blind. One-half of the females produced are carriers.

20.   (E)
Chemical energy is stored in the bonds of biomolecules: sugars, lipids, etc. Heat is a by-product of any chemical conversion of metabolism. Light energy drives photosynthesis, and kinetic energy is the energy of motion. Animals move, generating this kinetic energy from conversion of chemical bond energy.

21.   (C)
Two base-pairing rules must be memorized for DNA strand complementarity: A-T and G-C. Thus, the given DNA strand of six bases dictates only one possible complement.

22.   (D)
This is a rote-memory question. Duplication or replication refer to DNA copying. Translation is the RNA direction of protein synthesis.

23.   (A)
The choices are ranked in order of increasing wavelength: blue-green-yellow-orange-red. Green is reflected; red light has a wavelength range of 650 nm - 700 nm in the visible spectrum. Since light of 680 nm and 700 nm wavelength is utilized by photosystem II and photosystem I respectively, the light with the shortest wavelength absorbed during photosynthesis is blue light.

24.   (D)

  DNA serves as a template for RNA synthesis and RNA serves as a template for protein synthesis. Proteins participate in a wide variety of body chemistry. For instance, as enzymes they catalyze nearly all chemical reactions in biological systems., They serve as transport molecules such as oxygen-carrying molecules, hemoglobins and myoglobins. They protect our bodies against foreign pathogens in the form of antibodies.

25.   (B)

  In an operon, the operator gene is adjacent to the first of several consecutive structural genes that code for enzymes that are needed for a particular metabolic pathway. These structural genes are often arranged in the same order that the enzymes which they code for are used in the pathway.   The promotor is located next to the operator gene, opposite the side of the linked structural genes and is the location at which the RNA polymerase, which generated the mRNA that is necessary for enzyme synthesis, binds. The regulator gene is at another location on the chromosome. This location can be near or far from the operon that it regulates.

26.   (A)

  The base: adenine (A), cytosine (C), guanine (G) or thymine (T) varies from nucleotide to nucleotide building block in a DNA strand. Any DNA nucleotide is occupied by only one of these bases for four possible nucleotide structures. The other choices are constant in the nucleotide. Ribose is a component of an RNA nucleotide.

27.   (C)

  Muscle cells are the engines of an animal, developing contractile pulling forces to produce work. Mitochondria, cell powerhouses to extract energy from nutrients, are most in demand here.

28.   (C)

  Viruses are obligate, intracellular parasites. This means that they must enter host cells and use materials that are found within the cell to reproduce. Viruses lack ribosomes and ATP-generating systems. Thus, viruses enter host cell and use host cell's ribosomes and other organelles for their reproductive needs. Eventually, enough viral protein and viral DNA are produced, the new viruses are assembled, and the cell ruptures, releasing the new viruses. This frees new viral particles at the conclusion of this lytic life cycle. In an alternative lysogenic life cycle, viruses instead incorporate their DNA into the host's chromosome and remain latent. The virus may later start a lytic cycle. Transduction refers to the process of a virus taking along some of a host cell's DNA and injecting this DNA, along with its own viral DNA, into a new host cell. The other choices do not refer to viral life cycles.

29.   (E)
The hierarchy of classification levels is, from most general down to most restrictive:

Kingdom
　　Phylum
　　　　Class
　　　　　　Order
　　　　　　　　Family
　　　　　　　　　　Genus.
　　　　　　　　　　　　Species

30.   (C)
Refer to the scheme in #29.

31.   (E)
Phloem is one of two types of plant vascular tissue. It transports organic solutes, especially sugars, both upward and downward throughout the plant body. Xylem, the other plant vascular tissue type, transports water and dissolved minerals upward through the plant from their absorption site in the roots. Epidermal tissue covers the plant. Large parenchymal cells in roots and leaves store certain substances as a reserve. Supportive cells include collenchyma and sclerenchyma, which have thick cell walls.

32.   (A)
Effector is a term for an organ of response, usually a muscle or gland. A reflex arc is a neural circuit producing the automatic, unconscious behavior of a reflex. Knee jerks and shivering are examples of reflexes. A receptor is a small sensory organ or a bare or modified nerve end, and is the first portion of the reflex arc. The receptor detects the

stimulus that is being reacted to. A sensory neuron extends from the receptive site to the spinal cord. A motor neuron transmits a signal from the spinal cord to the effector in response to the stimulus.

33.   (B)

In the binomial system of organism nomenclature, the first taxonomic name is the genus name and is capitalized. It is followed by the species name, which begins with a lowercase letter. This is the scientific name of the white oak tree.

34.   (A)

There are more arthropod species than species of any other phylum. Arthropods include such well-known groups as arachnids, crustaceans, and insects.

Annelids are segmented worms; echinoderms include sea urchins and sea anemones; poriferans are sponges; phylum Platyhelminthes represents the flatworms.

35.   (C)

Two well-known examples of recessive sex-linked traits in human beings are red-green color blindness and hemophilia. These recessive sex-linked traits occur in a higher frequency in men than in women.

Albinism is an autosomal recessive disease. An individual heterozygous for albinism appears normal because one normal gene can be sufficient for making enough of the functional enzyme that make melanin pigments. Albinism is associated with low melanin levels.

Diabetes mellitus is characterized by an elevated level of glucose in blood and urine and arises from a deficiency of insulin. The causes for the disease are not clear but there is evidence that this defect has molecular basis such as abnormally formed insulin.

High cholesterol level is a genetic disease resulting from a mutation at a single autosomal locus coded for the receptor for LDL (low-density lipoprotein). Whether a trait is dominant or recessive does not apply to this disease because the heterozygotes suffer from a milder problem than the homozygotes. The heterozygotes possess functional LDL receptors though they are present at a deficient level.

36.   (D)
Transpiration is the loss of water through stomata (microscopic openings) in leaves to the atmosphere by evaporation. All other choices are root functions.

37.   (E)
Stomata control gas exchange by either opening or closing, thereby regulating the amount of air that enters the leaf. At normal temperatures, stomata are open. At high temperatures, the guard cells that surround each stoma will expand, thereby sealing each stoma.

38.   (B)
Leaves develop an abscission layer at their branch attachment, severing their articulation to the plant's main stem. All other choices are nondescriptive or irrelevant.

39.   (E)
Sucrose is the major carbohydrate molecule transported in plants and most of the movement of carbohydrates is through the phloem. Xylem transports ions (minerals). The other carbohydrates listed are not common in plants and glucose is not a transported molecule.

40.   (D)
A study of skin structure reveals skin's ability to perform all but one of the listed capabilities, i.e., fingerprints for identification, blood vessels to vent body heat, receptors to sense stimuli, and layers to protect.

41.   (A)
The biceps brachii are the muscles on the ventral portion of the upper arm that pull and bend the forearm. Bones are the rigid bars that yield to skeletal muscles' pulling force. Movable joints allow a source of mobility between articulating bones. Skeletal muscles are stimulated by nerves. Lacking this stimulation they will not respond.

42.   (B)

DNA replication is referred to as a semiconservative process because, after replication is completed, each of the two daughter molecules of DNA contains one strand from the parent DNA molecule. The other strand of each daughter molecule is assembled from nucleotides present in the nucleus.

43.   (D)

Sensory or afferent neurons send signals toward the central nervous system (CNS). Associative, or internuncial, neurons are within the CNS. Motor or efferent neurons, with axons outside and directed away from the CNS, send signals out to peripheral points.

44.   (D)

The retina contains the receptor cells that receive and register incoming light rays. The choroid is a middle layer of darkly pigmented and highly vascularized tissue. This structure provides blood to the eye and absorbs light to prevent internal reflection that may blur the image. The outer sclera (white of the eye) includes the transparent cornea. The pupil is an opening in the donut-shaped, colored iris interior to the cornea. The size of the pupil is regulated by the contraction and relaxation of the iris, which controls the amount of light admitted into the eye.

45.   (C)

A polymer is a long complex molecule formed by the bonding of simpler, repetitive subunits. DNA and RNA are polymers of nucleotides. Glycogen and starch are polysaccharides. Polysaccharides are polymers of simple sugars, including glucose. Glucose is a subunit, not a polymer.

46.   (B)

Clostridium can cause botulism and tetanus. *Staphylococcus* can infect a wound and cause blood poisoning. *Salmonella* can cause food poisoning. *E. coli* lives normally in the human large intestine, or colon. *Treponema* can cause syphilis.

47.   (B)

A four-chambered heart with a complete separation of sides is a characteristic of mammals and birds, the warm-blooded vertebrates. Most of the cold-blooded vertebrates do not have a completely separated four-chambered heart. For instance, fish (e.g. shark) has a two-chambered heart composed of an atrium and a ventricle. No mixing of oxygenated and deoxygenated blood occurs in fish because of a single pathway of blood circulation: gills➡ systemic circulation ➡ heart ➡ gills. Amphibians and reptiles, with the exception of crocodilians, have a 3-chambered heart composed of a left and a right atrium and an incompletely divided ventricle. Blood in amphibians and reptiles circulates through a pulmonary and a systemic pathway in each cycle with very little mixing of oxygenated and deoxygenated blood.

48.   (C)

Boyle's law states that air pressure is inversely proportional to volume. As the chest cavity increases due to the flattening of the diaphragm and rib elevation, internal pressure drops below that of the atmosphere, causing an inrush of air.

49.   (E)

Evolution is a theory that groups of organisms change morphologically and physiologically over the course of many generations. The result of evolution is that the descendants are different from their ancestors. To be more specific, evolution involves a change in allele frequencies in a population's gene pool over successive generations. Therefore, a population can evolve but an individual cannot. The other choices represent different levels of organization of an organism.

50.   (A)

This is a strict definition of cline. For example, north-south clines in average body size are found in many birds and mammals. These species are larger in the colder climate and smaller in the warmer climate. Genus is a taxonomy unit and mutations are a source of genetic variation. Inbreeding refers to the matings of closely related individuals, which increases the percentage of homozygosity within a population. Two or more morphologically distinct forms in a population constitute polymorphism.

51.   (E)
  Paramecium caudatum, according to the currently recognized five-kingdom system, is a protist. Animalia includes all multicellular animals. Fungi consists of unicellular and multicellular organisms that do not possess chlorophyll and that require the presence of organic matter in order to survive. Monera is made up of bacteria and blue-green algae. Plantae consists of unicellular and multicellular plants.

52.   (B)
  "Di" means two, "strepto" means chain, and "staphylo" means bunch. *Gonococcus* is a single, spherically shaped bacterial cell that causes gonorrhea.

53.   (C)
  Facultatively anaerobic organisms, under normal aerobic condi-tions, will use oxygen in their metabolism, as humans do; in the absence of sufficient oxygen, however, such organisms can metabo-lize molecules other than oxygen. Aerobic is a term applied to situations involving oxygen, while anaerobic is a term applied to those involving no oxygen. Fermentation is the synthesis of alcohol via the glycolytic pathway. Glycolysis is the series of anaerobic metabolic reactions that converts glucose to pyruvic acid.

54.   (A)
  All other choices represent important conditions necessary to main-tain a consistency of a population's gene frequencies. Sexual repro-duction with random mating is an important requirement for this equilibrium.

55.   (D)
  This is the definition of "niche." It involves factors of prey, predators, habitat, and behavior of the organism. The more the niches of two species overlap, the more intense the interspecific competition be-tween the two.

56.   (D)
  Coral and hydra are types of adult coelenterates. Medusa and polyp

are alternating adult body forms of many coelenterate species.

57.   (B)
Use of cellulases allows herbivores to digest cellulose, the major component of cell walls. Amylase (ptyalin) breaks starch down into monosaccharides and disaccharides. Chymotrypsin, trypsin, and pepsin all are proteases.

58.   (A)
This is an exact definition of the coelom.

59.   (E)
Energy flows from plants (producers) to herbivores (primary consumers) to additional levels of consumers in the food chain. Each succeeding link has less remaining available energy. Generally, only about 10% of the energy from one trophic level is passed to the next level. Loss of energy can be attributed to the respiration of organisms and the consequent dissipation of heat, due to the inability of most animals to digest the cellulose of plants.

60.   (D)
Based on embryonic development, two evolutionary lines have been distinguished. One evolutionary line, containing the phyla Annelida, Arthropoda, and Mollusca, is called Protostomia; the blastopores of protostomes develop into mouths. The other evolutionary line, containing the phyla Chordata and Echinodermata, is called Deuterostomia; the blastopores of members of these phyla become anuses. Coelenterates are more primitive than either the Protostomia or the Deuterostomia.

61.   (B)
The metamorphic life cycle of insects can be exemplified by the life cycle of a butterfly. From a butterfly's egg will hatch a caterpillar (the larval form). The caterpillar, after a period of growth, will spin itself a cocoon and become a pupa. Eventually, the adult butterfly will emerge from the cocoon.

62.   (B)

Blood flows through the circulatory system due to a pressure gradient. The blood will flow from a region of higher pressure to one of lower pressure. Therefore, blood pressure must be greatest at the beginning of blood's circuit, namely, the aorta, or artery.

63.   (D)

The rib is one of the twelve pairs of bones which form a rib cage to protect the lungs and heart. Along with the skull and vertebral column, the rib cage forms the axial skeleton. The bones of the paired appendages, the pectoral and pelvic girdles belong to the appendicular skeleton.

64.   (D)

Glial cells bind neurons together. They offer nerve cells support, protection, and nutritional supply.

65.   (C)

Malaria is caused by protozoans of the genus *Plasmodium*, of the class Sporozoa. The other choices represent diseases caused by viruses.

66.   (B)

All enzymes are composed primarily of protein. The more complex enzymes have non-protein portions called cofactors; the protein portion of the enzyme is called an apoenzyme. If the cofactor is an easily separated organic molecule, it is called a coenzyme. Many coenzymes are related to vitamins. An enzyme deprived of its vitamin is thus incomplete, leading to the nonexecution of a key step in metabolism. Holoenzyme refers to the RNA polymerase, with its core enzyme and sigma subunit associated together.

67.   (A)

Cilia line the upper respiratory tract, waving against air inflow to filter out unneeded debris. Villi are fingerlike extensions of the membranes of cells lining the small intestine. They increase surface area to facilitate absorption of digested nutrients. Goblet cells line the same region and secrete mucus. Leidig cells are in the male testis.

68.   (E)
Filtration first moves blood plasma substances from the glomerulus (capillary) into the cuplike Bowman's capsule at the nephron's origin. After monitoring these solute concentrations (e.g. - glucose, sodium, etc.), reabsorption returns them to the blood from the nephron tubule at high percentage rates. Secretion is a third step, moving materials from the blood (peritubular capillaries) to the *distal convoluted tubule* for exit and elimination.

69.   (D)
Hematocrit is the percentage of blood cells in blood by volume. For males this value is normally 47 ± 5; for females, it is 42 ± 5. Thirty-two percent is abnormally low, indicating anemia - a diminished capacity of the blood carry oxygen.

70.   (E)
The traits that are mentioned in question 70 are those of Orthopterans. Members of this order include grasshoppers and cockroaches. Dipterans, such as houseflies, have one pair of wings and sucking mouthparts. Hemipterans, the true bugs, have one pair of wings that are thicker proximally and membranous distally, and a pair of wings that are totally membranous. Homopterans have either no wings, or two pairs of arched wings. Lepidopterans, such as butterflies, have two pairs of scale-covered wings and sucking mouthparts.

71. (A)   72. (B)   73 (D)   74. (C)

Formula A shows a hexose sugar molecule. By adding the number of different types of atoms from the structural formula, the molecular formula is $C_6H_{12}O_6$ with six carbon atoms in a chain. This particular hexose sugar is glucose, a monosaccharide. Formula B shows but three carbon atoms in a chain. The base ratio is also not 1-2-1 for carbon-hydrogen-oxygen as it is in simple carbohydrates. B is the alcohol, glycerol, that bonds to fatty acids in a lipid or fat molecule. Formula C is an amino acid, the repetitive building block for protein macromolecules. The amino group, NH_2, and carboxylic acid group, COOH, are identifying functional groups of an amino acid. D is the structural formula for a nucleotide, the building block subunit for nucleic acids such as DNA and RNA. Its identifying components are a 5-carbon sugar, a varying nitrogen base, and a phosphate group. Formula E is a carbon skeleton for a steroid molecule.

75. (C)   76. (A)   77. (D)   78. (B)
Centrioles are paired, cylinder- shaped organelles at right angles to one another. Located near the cell nucleus, they coordinate cell division. Cilia are numerous, hairlike projections of the cell membrane that beat in synchrony to propel the cell in movement. The endoplasmic reticulum is a winding, tubular system that establishes a channel for internal transport. The mitochondrion is an oblong, football-shaped organelle with a double membrane. Its inner one forms pockets, or cristae and has enzymes to run cell respiration. The darkstaining spherical nucleolus in the nucleus does not appear as an answer.

79. (C)   80. (B)   81. (A)   82. (D)
Note, from the illustration, the synapsis (pairing and attraction) of homologous chromosomes. This phenomenon, along with their crossing over denotes prophase of meiosis one. The chromosomes align along the central plane of the spindle in metaphase. The homologs separate in anaphase with cell cytokinesis and complete separation into two daughter cells occurring in telophase. Illustration E is a meiosis II stage.

83. (C)   84. (A)   85. (D)   86. (B)
As a skeletal muscle receives separate, well-spaced stimuli of sufficient intensity, it will twitch to each stimulus. Each twitch is marked by contraction and relaxation of a muscle. As the rate of stimulation increases, the muscle does not have sufficient time to totally relax between contractions. The muscle then contracts from an already partially contracted position, producing greater force than that produced by a normal twitch. Rapid signals from the nervous system causes this to occur quite often; each time, the muscle is more contracted prior to the next signal and subsequent contraction. In this way, the contractions summate to produce a greater muscular force than the force of a single simple twitch. At some point, the neural impulses arrive at so rapid a rate, that the muscle has absolutely no time to relax between impulses. It is at this point that the muscle has reached a state of tetanus. Overwhelming a muscle with demands beyond its ability to receive new nutrients and oxygen yields accumu-

lation of lactic acid, the source of muscle fatigue. Muscles, when not producing movement, still remain somewhat taut or maintain muscle tone (tonus).

87. (D)   88. (A)   89. (E)   90. (C)
Annelids, or segmented worms, display a high degree of somatic segmentation. Arthropods ("arthro" = joint + "pods" = foot), such as insects, have many jointed appendages. There is a great deal of evidence to support the statement that echinoderms are more closely related to chordates than any other group (both undergo cleavage that is indeterminate and radial during their ontogenies; both are deuterostomes). All mollusks possess a muscular foot.

91.   (B)
The cortex is the deep layer of primarily parenchymal cells beneath the outer epidermis. The cells store starch.

92.   (A)
Epidermis is the single-layered outer covering tissue. Root epidermis has no waxy cuticle, since its function is water absorption.

93.   (D)
Tubular cells of xylem are in the stele, or vascular conducting core. The starlike points formed by thick-walled xylem cells alternate with patches of phloem. Xylem plays a role in water and mineral transportation.

94.   (D)
The lower epidermal layer has stomata for gas exchange.

95.   (A)
Cuticle is the waxy layer that covers both the upper and the lower epidermis of a leaf and it is generally thicker on the upper epidermis. This protects the internal tissue of a leaf from excessive water loss, from fungal infection, and from mechanical injury.

96. (B)

Internal parenchyma cells contain chloroplasts for photosynthesis. Parenchyma cells lie in the entire region between the upper and the lower epidermis, forming a soft tissue called mesophyll. The upper palisade mesophyll is composed of vertically arranged cylindrical parenchyma cells. The lower spongy mesophyll consists of irregularly shaped parenchyma cells. Intercellular spaces in these two mesophyll layers are important in communication with the stomata for gas exchange. A high chloroplast density occurs within the upper palisade mesophyll. This gives the plant an advantage, in that the upper surface of a leaf facing the sun's rays receives most sunlight and photosynthesis can then be carried out at its optimal rate.

97. (E)

The intercellular space in the spongy mesophyll usually has a 100% humidity. A thin film of water is formed on the surface of mesophyll cells. Gases dissolve in this water film before they enter into the cells..

98. (A)

The pituitary gland is composed of an anterior and a posterior lobe. The stalk of the posterior lobe is connected to the hypothalamus. Antidiuretic hormone (ADH) is produced in the hypothalamus and stored in the posterior pituitary. Upon nervous stimulation from the hypothalamus, the posterior pituitary releases ADH which acts on kidney tubule to reabsorb water.

99. (C)

The pancreas secretes insulin to lower blood sugar and maintain equilibrium.

100. (D)

The adrenal glands produce adrenaline. This hormone is a well-known constrictor of blood vessels. The principle demonstrated here is that of negative feedback: a stimulus met by a response that reverses the trend of the stimulus. A dropping blood pressure must be opposed and corrected. Constricting blood vessels forces the same amount of blood to travel through a region of decreased volume; this causes a rise in blood pressure.

101.   (D)
The hormone aldosterone is secreted by the adrenal cortex to promote sodium reabsorption in the kidney.

102. (C)   103. (B)   104. (D) 105. (A)
The answers to questions 102 - 105 are self-explanatory.

106.   (A)
Producers are the first organisms in a food chain. Through photosynthesis, producers obtain a great amount of energy to build the highest biomass among the different trophic levels in an ecosystem.

107.   (C)
After the producer tier at A, the primary consumers (herbivores) are next, at B. Secondary consumers, carnivores, compose the next food chain link and biomass tier.

108.   (A)
Since the extracellular environment has a 95% concentration of water, the extracellular environment has a greater concentration of water than the intracellular environment does. Therefore, water will flow into the cell due to osmosis. Terms of "tonicity" refer to the solute concentration in water. The inside cell setting has a higher, hyper, solute concentration. Its solute concentration is not less, hypo, or equal, iso, to the extracellular environment.

109.   (C)
As explained in solution 108, water will flow into the cell. Therefore, the extracellular environment will lose water.

110.   (B)
Osmosis is the movement of water from an area of a higher level of concentration to one of lower concentration through a semipermeable membrane. Water concentration is higher outside, 95%, and lower inside. The semipermeable membrane allows it to flow in.

The cell will eventually rupture, or lyse.

111. (B)   112. (E)

The complete Punnett square for the cross AaBb x Aabb appears as follows:

	AB	Ab	aB	ab
Ab	AABb 1	AAbb 2	AaBb 3	Aabb 4
ab	AaBb 5	Aabb 6	aaBb 7	aabb 8

Only cells 3 and 5 (1/4 of total) show genetic recombinations heterozygous for both gene pairs. Of all the genotypes listed in problem 112, only aaBB does not appear in the Punnett square, and therefore is not a possible genotype for the offspring.

113.   (B)

If base #2 is cytosine, then base #10 must be guanine, which is the complement of cytosine. Base #6, therefore, must be complement of base #10 (guanine), or cytosine.

114.   (D)

If RNA base #7 is uracil, than DNA base #11 must be adenine. (Uracil, not thymine, is the RNA complement of adenine). DNA base #3 must therefore be thymine.

115.   (C)

A left-hand DNA sequence CGCT is complementary to a right-hand base sequence of GCGA. This is transcribed as CGCU.

116.   (B)
Blood pressure is greatest in blood vessels closest to the heart and decreases as blood flow away from the heart.   Aorta, arteries, anterioles, capillaries, venules, veins and vena cava are the blood vessels arranged in order of decreasing blood pressure.   Such a gradient of blood pressure is essential for a one-direction flow and is a result of friction between the flowing blood and the wall of blood vessels.

117.   (D)
This is the capillary end that connects with a venule, is lower in pressure, and receives carbon dioxide and waste products from cells for transport to venules and veins.

118.   (B)
This term best indicates the two-way traffic of molecules between the blood and the cells around the capillary.

119.   (C)
Blood, and its hemoglobin, flows to the lung to load up oxygen. Hence, hemoglobin is highly saturated with oxygen.

120.   (E)
Blood hemoglobin is relatively unsaturated at tissue cells, as it liberates the oxygen to these cells.

SECTION II

ESSAY I

Enzymes are globular proteins with distinct surface geometries that result from the folding of the amino acid sequence of a polypeptide chain. They accelerate the rate of a reaction toward equilibrium without changing the position of equilibrium. This catalyzation makes reactions occur within a reasonable time frame as well as within physiological constraints such as temperature. Enzymes lower the activation energy of a reaction, which allows the reaction to proceed.

Contained within the conformation of the enzyme is an active site, which is specific to a certain substrate, or substance, upon which the enzyme acts. The substrate binds to the active site through weak non-covalent bonds such as hydrogen bonds, van der Waals forces, and hydrophobic interactions that arise between the active groups of the amino acid and the substrate. This binding of the substrate to the active site distorts the geometry of the enzyme and changes the concentrations of the reactants, which make the reaction progress.

Graph I shows the rate of two enzyme-catalyzed reactions. The steeper slope for enzyme A indicates a faster rate of catalyzation than B. Enzyme A has a lower level of production than enzyme B and therefore gives less product. Since the effectiveness of a catalyst is determined by how quickly a reaction reaches its equilibrium, and the maximum production corresponds to a property of the reaction that is unaffected by enzymes, enzyme A is more efficient than B.

Enzyme A exhibits its maximum activity at a pH of 3; enzyme B's greatest activity was at 8, according to Graph II. An enzyme sharing the same characteristics given of enzyme A is pepsin, which is found in the stomach. It breaks down proteins by cleaving the peptide bond between amino acids. Gastric juice containing hydrochloric acid is secreted during digestion, making the environment of the stomach acidic, and activating pepsinogen to become pepsin.

Enzyme B is characteristic of trypsin, an enzyme found in the small intestine. It is secreted by the pancreas and is also a proteolytic enzyme. Its inactive form is trypsinogen. When bile is secreted, full of bicarbonate ions, in the small intestine, it makes the chyme from the stomach alkaline. This pH change stimulates enterokinases to cleave trypsinogen to become trypsin.

As carbon dioxide levels increase in the bloodstream, the pH of the blood lowers from its approximately neutral pH. This acidic change affects the structure and activity of enzymes. As evidenced in Graph II, different enzymes have varying activities and efficiencies at similar pH. Therefore, making the blood more acidic could result in a denaturation of enzymes, and thus, a loss of specific 3-D conformation. The consequence is a loss of function, since it can no longer bind the specific substrate. Acidic pHs could also activate enzymes that function more efficiently at lower pHs. Cumulative effects can be felt on the level of metabolism, whose function hinges on the positive and negative activity of enzymes to alter the availability of substrates.

ESSAY II

A mutation is a change in the base sequence of a gene which leads to the formation of a new allele. Mutations can be classified into point mutations, chromosomal mutations, and genomic mutations.

Point mutations affect small regions of a chromosome. Substitution is a point mutation in which nucleotides are replaced by different ones. Substitution can occur spontaneously through the mispairing of bases during DNA replication. Deletion causes a gene to have several bases less than normal while addition gives the opposite result.

Chromosomal mutations affect larger regions of a chromosome and are usually initiated by a breakage in the DNA backbone. Translocation is the interchange of chromosome segments between two nonhomologous chromosomes. It is different from crossing-over, a normal genetic process that gives rise to variation. Crossing-over involves two homologous chromosomes. Deletion results in a karyotype that has lost a segment or segments of chromosome. A segment of chromosome without a centromere does not attach to any chromosome, it does not move with the spindle fiber during cell division, and is not incorporated into either daughter cell. A karyotype has a chromosome longer than normal because of a duplicate segment at its end. When a broken segment reattaches to its original position in a reversed order, a change in genetic order results without loss or gain of total gene count. This is called inversion. Translocation and inversion cause new groupings of genes. The favorable groupings are conserved by natural selection while the unfavorable ones are selected out.

Genomic mutations involve changes in the number of chromosomes present in the karyotype. This abnormality results from nondisjunction, a process in which homologous chromosomes fail to separate and move to opposite poles during cell division. This results in one daughter cell receiving an extra chromosome while the other daughter cell receives one less. Trisomy is a condition where three chromosomes of one type are present in the nucleus. For example, Down's Syndrome is also called Trisomy 21 because of the presence of three chromosome #21 in the nucleus of the affected individual.

Messenger RNA (mRNA) is synthesized in the nucleus by the process of transcription from a DNA template. mRNA is transported to the cytoplasm for translation. Ribosomes convert the nucleotide sequence of mRNA into the amino acid sequence of the polypeptide chain. Ribosomes consist of two subunits. The smaller subunit is responsible for binding the mRNA, and the larger subunit contains the enzymes that catalyze the formation of peptide bonds. An mRNA molecule associated with several ribosomes is called a polysome, which allows protein to be synthesized before mRNA becomes degraded. Protein synthesis is invariably started at AUG, the start codon. mRNA is translated from the 5' end to the 3' end and polypeptides are synthesized from the left amino end to the right carboxyl end. Specific amino acids are brought to the mRNA ribosome complex by tRNA which has an anticodon complementary to the triplet codon of the mRNA. Three nucleotides of the mRNA sequence are read at a time. The translocation of a ribosome along the mRNA, three nucleotides to the right, is an energy requiring process. During protein synthesis, adjacent amino acids are linked together by peptide bonds formed through dehydration. Once a stop codon is encountered, protein synthesis stops and the complex of mRNA, ribosome, and nascent polypeptide chain dissociates. There are three stop codons among the 64 triplet codes and these stop codons have no tRNA anticodon complementary to them. Some nascent polypeptides have to be modified before they become useful.

A single base substitution changes a codon to another of the 64 possible genetic codes. Substitution at the third base of a codon may not cause any effect because some amino acids are coded by multiple codons which differ from each other at the third base position, e.g., both UUA and UUG code for leucine. However, the results of an amino acid substitution can be serious. The degree of seriousness increases when the exchanged amino acid belongs to a different charge group, such as a nonpolar amino acid being substituted by a polar one, etc. Also, amino acid substitution at the functional site of a protein, such as the active site of an enzyme, has a more serious effect than at other locations. Sometimes, base substitution changes a coding codon to a stop codon, and this leads to premature termination of protein synthesis. The effect is deleterious if this extra stop codon is close to the start codon.

Deletion or addition of a three base multiple results in the deletion or insertion of amino acids. The significance depends on the location of the deletion or insertion. If it happens on a region such that the functional site of a protein is malformed, this mutation can be deleterious. Deletion or addition of bases not at a multiple of three causes a reading-frame shift, where the order in which the mRNA is read is shifted according to the number of amino acids. Starting from the site of mutation, the mRNA is translated into a different polypeptide chain. Proteins formed this way cannot carry out their normal functions. This type of mutation is usually deleterious as well.

ESSAY III

The original source of energy to any ecosystem of the planet earth is the sun. Its radiant energy is trapped by chlorophyll-containing organisms. Through photosynthesis, such organisms convert the energy of light into the stored energy of chemical bonds of sugars, the product of photosynthesis. From this initial process, different populations of organisms in a community are integrated by nutrition, their feeding levels in a food chain.

Photosynthetic organisms serve in the role of producers and as initiators of the food chain. Subsequent links in the chain denote different levels of consumers; primary consumer, secondary consumer, etc. and decomposers. Primary consumers in a given food chain are herbivores, plant eaters. Subsequent links are carnivores, eating animal flesh, and decomposers correspond to the end of a food chain and consist of bacteria and fungi, the organisms of decay. As a concrete example: oak tree (producer) - insect (primary consumer, herbivore) - snake (secondary consumer, carnivore) - owl (tertiary consumer, carnivore, etc.) An example from a pond could be: alga - minnow - sunfish - bass - grizzly bear.

Each chemical energy transfer, link by link in the food chain, is accompanied by a major conversion to heat. Useful chemical energy remains stored in chemical bonds. It becomes a component of organism protoplasm, available by feeding through predation at the next food chain link. Because of the large heat conversion at each step, up to 90% of the number of links in a food chain is limited usually to four or five. Heat is dissipated and cannot be changed back to a usable chemical form. Therefore, constant useful energy input is required from sunlight and photosynthesis converting light energy into chemical energy for utilization by other organisms. Turn off the sunlight to the earth and food chains will eventually run down.

Materials, on the other hand, are reusable. A copper atom can remain a copper atom if it is not changed chemically. It can continually be reshuttled between the biotic and abiotic sectors of the environment. Current shortage of materials, as with certain metals, stems from a lack in efficiency of recycling to keep such materials available. Without such recycling, a metal is kept tied up in one sector. It is thus unavailable while taken out of circulation.

One simple example of such a biogeochemical cycle is the phosphorous cycle. Phosphorous is an essential element to protoplasm (bio), but has its abiotic reservoir in the earth's geological structure (geo). Acceleration of mining and erosion practices has released large amounts of rock phosphorous into waterways as phosphate ions. The dissolved phosphate is used by plants and reaches animals through the food chain. Phosphatizing bacteria work on plant and animal carcasses and wastes, returning the phosphorous to the waterways. This return and subsequent reuse is characteristic of circulation or a cycle.

ESSAY IV

Erythrocytes synthesize, store, and transport hemoglobins. Hemoglobin binds O_2 and CO_2. Its affinities for these gases depend on the pH of the medium modulated by CO_2 concentration. Mature erythrocytes are enucleated in higher vertebrates and are biconcave disk-shaped. These special structures give the erythrocyte a larger surface area to accommodate hemoglobin and the diffusion of gases across its membrane. Hemoglobin embedded in the stroma of an erythrocyte is advantageous in that no free floating hemoglobin is present to disturb the osmotic relationship between blood and tissue fluid.

Neurons are capable of conducting and transmitting electric impulses rapidly. They consist of a cell body containing the nucleus and other organelles, extensively branched dendrites, and a long, single axon which may branch at its terminal end. Dendrites receive and direct impulses to the cell body. Axons transmit impulses away from the cell body. The capability of forming synapsis with other neurons and target organs allows neurons to form long conducting pathways to various parts of the body. Axons in the central nervous system are wrapped in glial cells while those outside the central nervous system are enveloped in Schwann cells. These two kinds of cells form myelin sheaths that speed up the conduction of impulses in the axon.

Muscles in our body can be differentiated into skeletal muscles, smooth muscles, and cardiac muscles. Skeletal muscle fiber is cylindrical, coenocytic (with many nuclei) with a striated appearance. Bundles of skeletal muscle fibers attach to bone and are responsible for rapid action under the control of the voluntary nervous system. Smooth muscle fibers are nonstriated, thin, and elongated cells, and they form a sheet of muscle tissue which serves as the walls of the viscera and blood vessels. Cardiac muscle fibers are striated, their activity similar to smooth muscle. Muscle fiber is composed of a sheath enclosing numerous myofibrils. The contractile materials of the myofibril are thin actin and thick myosin filaments. A special structure called a crossbridge is formed when the globular heads of myosin filament are in contact with the actin molecule of the thin filament. This crossbridge is responsible for the sliding together of the thick and thin filaments which produce muscle contraction.

Epidermal cells are relatively flat with a thicker outer cell wall. These irregularly shaped cells interlock to form the surface tissue of stem, roots, and leaf with no intercellular spaces for the prevention of water loss. Some epidermal cells are specialized to perform different functions. For example, guard cells are sausage-shaped epidermal cells that regulate the size of stomata. Epidermal cells of root tissue are devoid of cuticle and have hairlike processes to facilitate water absorption.

Tracheids and vessel cells are the two main elements of the xylem of plant's vascular tissue. Tracheids are elongated, tapering cells with pits on their cell walls. These pits are particularly numerous at the tapering ends and the vertically linked pattern of the tracheids form an upward transport system for water and dissolved substances. Tracheids have liquefied secondary cell walls which serve as a supportive structure for the plant.

Parenchyma cells are relatively unspecialized vegetative cells found in roots, stems, and leaves. They have a thin primary cell wall and usually lack a secondary cell wall. They are capable of cell division. Parenchyma cells in leaves contain a high density of chloroplasts and are photosynthetic. Those located in stem and root serve to store nutrients and water. Turgid parenchyma cells help to give shape and support to the plant. Their large vacuoles take in water and push against the cell wall to maintain turgidity.

THE ADVANCED PLACEMENT EXAMINATION IN

BIOLOGY

TEST II

ADVANCED PLACEMENT BIOLOGY EXAM II

SECTION I

120 Questions
90 Minutes

DIRECTIONS: For each of the following questions or incomplete sentences, there are five choices. choose the answer which is most correct. Darken the corresponding space on your answer sheet.

1.   In the electron transport chain, the final electron acceptor is

    (A) oxygen   (D) the mitochondrion

    (B) water   (E) hydrogen ion

    (C) cytochrome oxidase

2.   All of the following statements about the light reactions of photosynthesis are true, EXCEPT

    (A) Carbon dioxide fixation occurs - i.e. CO_2 is reduced to organic compounds

    (B) Light is absorbed by photosystems

    (C) Water is split into oxygen, hydrogen ions and electrons

    (D) Light reactions consist of cyclic and noncyclic photo-phosphorylation

    (E) ATP is synthesized

69

3. The levels of organization of study of an organism, starting with the most microscopic level, is shown by which of the following sequences?

    (A) Cells...organs...chemicals...tissues...systems...organism

    (B) Chemicals...cells...tissues...organs...systems...organism

    (C) Systems...organs...tissues...cells...chemicals...organism

    (D) Organism...systems...tissues...organs...chemicals...cells

    (E) Chemicals...cells...organs...tissues...systems...organism

4. All of the following statements about the cell membrane are true, EXCEPT

    (A) It functions as a selective barrier between the intracellular fluid and the extracellular fluid

    (B) The proteins within it are classified as intrinsic (integral) or extrinsic (peripheral)

    (C) The major lipid within it is cholesterol

    (D) The fluid-mosaic model describes the fluidity and mobility of the membrane

    (E) The phospholipids within are amphipathic - that is, they each contain polar and nonpolar regions

5. Which of the following statements is true?

    (A) DNA contains the pentose sugar ribose while RNA contains deoxyribose

    (B) Both DNA and RNA are double-stranded

    (C) Both DNA and RNA contain the bases adenine and thymine

    (D) Only RNA uses the base uracil while only DNA contains thymine

    (E) DNA has a pentose sugar, while RNA has a hexose

6. Semiconservative replication of DNA refers to the fact that

   (A) the daughter DNA is an entirely new duplex and thus the original DNA is intact

   (B) pieces of the old and new DNA duplexes are jumbled together in the daughter generation

   (C) each strand of the parent DNA serves as a template for the synthesis of its new partner strand. Thus one strand is conserved in each new double helix

   (D) DNA replication is modest in its ATP requirements

   (E) DNA replication occurs by base pairing between adenine and thymine and between guanine and cytosine

7. The following statements about the codon are true EXCEPT

   (A) The codon is a triplet of nucleotides on messenger RNA (mRNA)

   (B) The codon is a triplet of bases on transfer RNA (tRNA)

   (C) The codon base pairs with the anticodon

   (D) The codon is degenerate, i.e. most amino acids are represented by more than one codon

   (E) A codon represents an amino acid, or a signal to initiate or terminate protein synthesis

8. Wobble of the anticodon

   (A) refers to the freedom in the pairing of the third base of the codon

   (B) refers to the inaccuracy of base pairing, i.e. if adenine were to pair with guanine, or uracil with cytosine

   (C) is represented, for example, by the codons UUG and CUG, both of which code for leucine

   (D) refers to imprecision in the pairing of the first base of the codon

   (E) occurs only in initiating and terminating codons

9.   In comparing glycolysis under aerobic vs. anaerobic conditions,

(A) pyruvate is converted to lactate under aerobic conditions, while it is converted to acetyl CoA under anaerobic conditions

(B) pyruvate is converted to acetyl CoA under aerobic conditions, while it is converted to lactate under anaerobic conditions

(C) oxygen is used in anaerobic glycolysis, while it is not used in aerobic conditions

(D) humans are only capable of aerobic metabolism

(E) lactate is converted to pyruvate under aerobic conditions, while the reverse occurs under anaerobic conditions

10.   Sickle Cell Anemia is an inherited disorder prominent in the black population of Africa. Which of the following statements is true?

(A) It is caused by a change in one amino acid of the beta chains of hemoglobin

(B) It is transmitted by mosquitoes

(C) Since malaria and sickle cell anemia are positively correlated, this suggests that having malaria increases the chances of contracting sickle cell anemia

(D) Due to wobble, a single nucleotide change cannot cause a disease

(E) Symptoms of sickle cell anemia are expressed equally in those heterozygous and those homozygous for the disease

11.   Nondisjunction, whereby a pair of homologous chromosomes does not separate in the first meiotic anaphase, is responsible for all of the following disorders EXCEPT

(A) Turners Syndrome (X0)

(B) Downs Syndrome (trisomy-21)

(C) Klinefelters Syndrome (XXY)

(D) Hemophilia (X^hY or X^hX^h)

(E) XYY male

12. The following statements about phylum Annelida are true, EXCEPT

(A) the earthworm and the leech are characteristic examples

(B) the nephridium functions as a "kidney," in that it regulates water and solute levels

(C) annelids are segmented worms

(D) a flame cell functions as a "kidney," in that it causes the excretion of excess water

(E) annelids contain bristles called setae

13. The Hardy-Weinberg rule states that, under certain conditions, evolution cannot occur. Which of the following conditions is not required by the Hardy-Weinberg Law?

(A) No mutations

(B) No immigration or emigration

(C) No natural selection

(D) Large population

(E) No isolation

14. In 1953, Stanley Miller and Harold Urey performed an experiment to propose the origin of organic matter from inorganic precursors. Since the prebiotic atmosphere was a reducing one, the experimental reaction chamber contained all of the following molecules, EXCEPT

(A) ammonia

(B) water

(C) oxygen

(D) methane

(E) hydrogen

15.   Evidence of evolution may be based on information obtained from fossils, geological records, studies in comparative bio-chemistry, and comparative embryology. The example below that would be classified as biochemical evidence is:

(A) Similar structure between the flipper of a whale and the arm of a human

(B) Imprint of a footprint

(C) Similarity of the DNA molecules of two mammals

(D) Radioactive dating of a rock sample

(E) Wing of an insect and wing of a bird

16.   The dark reactions of photosynthesis, in which carbon dioxide fixation occurs, are called

(A) Krebs cycle   (D) Cyclic AMP

(B) Calvin cycle   (E) Carbon cycle

(C) Cori cycle

17.   In a food pyramid, most of the energy is concentrated at the level of

(A) the tertiary consumer, since it is a big meat-eater

(B) the producer, since it supports all the consumers at the higher levels

(C) the tertiary consumer, since energy levels increase towards the top of the pyramid

(D) the primary consumer, since less energy is required to digest vegetarian meals

(E) the secondary consumer, since it only has to chase small herbivores, and thus does not waste energy hunting

18.   All of the following processes occur in the nitrogen cycle, EXCEPT

(A) Ammonification

(D) Denitrification

(B) Nitrification

(E) Nitrogen fixation

(C) Deamination

19. Phototropism is a growth response towards light

(A) in which the leaf surface turns away from the light

(B) that is caused by the auxin IAA (Indoleacetic acid)

(C) that is due to the hormone florigen

(D) in which only sunlight but not artificial light causes growth of leaves towards the light source

(E) but it is actually due to the heat emanating from the light source

20. All the statements about circadian rhythms are true, EXCEPT

(A) They cycle over a 24-hour period

(B) They are exemplified in plants by leaf orientations, which change from day to night

(C) They are exemplified in humans by the sleep-wake cycle

(D) They are exemplified in humans by changes in body temperature throughout the day and night

(E) They are entirely controlled by exogenous factors (such as the light-dark cycle)

21. The phloem of vascular plants

(A) contains sieve elements, which function in food conduction

(B) contains tracheids, which function in food conduction

(C) contains sieve elements, which function in water conduction

(D) contains vessel members, which function in water conduction

(E) contains companion cells, which assist in water conduction

22. In comparing photosynthesis to respiration, which of the following statements is true?

(A) Carbohydrate is produced in respiration, but not in photosynthesis

(B) Oxygen is produced in respiration, but not in photosynthesis

(C) Carbon dioxide is produced in photosynthesis, but not in respiration

(D) Water is produced in photosynthesis, but not in respiration

(E) Oxygen is produced in photosynthesis, but not in respiration

23. All of the following fates for sugar produced during photosynthesis are possible in a plant cell, EXCEPT

(A) its polymerization into starch for storage purposes

(B) its decomposition for energy production

(C) its polymerization into glycogen for storage purposes

(D) its use in the synthesis of other organic molecules

(E) its use in the synthesis of sucrose

24. An enzyme functions to increase the rate of a reaction by

(A) increasing the concentration of the substrate

(B) decreasing the E_a (energy of activation)

(C) competing with the substrate

(D) breaking down ATP

(E) hydrolyzing the substrate

25. PKU (phenylketonuria) is an example of an inborn error of metabolism. These "errors" refer to

(A) congenital birth defects

(B) hormonal overproduction

(C) inherited lack of an enzyme

(D) nondisjunction

(E) atrophy of endocrine glands

26. All living organisms are classified as eukaryotes (true nucleus) or prokaryotes (before the nucleus). The only example of a prokaryote listed below is

(A) AIDS virus

(B) *E. coli*

(C) *Homo sapiens*

(D) an oak tree

(E) amoeba

27. The following statements about arthropods are true EXCEPT

(A) snails and slugs are examples

(B) gas exchange occurs via book lungs or tracheae

(C) jointed appendages characterize members of this phylum

(D) crabs and shrimp are examples

(E) some arthropods have wings

28. With respect to the electron transport chain and chemiosmosis, all of the following statements are true EXCEPT

(A) each NADH yields three ATPs

(B) each $FADH_2$ yields two ATPs

(C) the cytochrome enzymes utilize NAD⁺ and FAD as their coenzyme

(D) hydrogen ions are pumped from the mitochondrial matrix into the intermembranal space

(E) the cytochrome enzymes utilize iron as their cofactors

29. All of the following statements concerning ATP are true EXCEPT

(A) it can be formed in anaerobic glycolysis

(B) it can be formed in aerobic respiration

(C) it can be formed in muscle from phosphocreatine

(D) it can be formed from cAMP (cyclic AMP)

(E) it can be formed from ADP

Questions 30 - 31 refer to the diagram below.

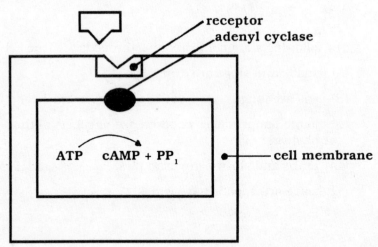

30. The second messenger is

(A) adenyl cyclase

(B) cAMP

(C)

(D)   the cell membrane

(E)   ATP

31.   The symbol [symbol] represents

(A)   a steroid hormone   (D)   glucose

(B)   an operator   (E)   an antibody

(C)   a protein hormone

32.   Lichen, in which an alga and a fungus live in harmony, is an example of

(A)   mutualism   (D)   predation

(B)   commensalism   (E)   competition

(C)   parasitism

33.   Konrad Lorenz researched the phenomenon of imprinting. To test his ideas, he had newly hatched ducks see him first. Subsequently, the ducks were allowed to see their mother. These ducks would tend to follow

(A)   their true mother   (D)   no one in particular

(B)   other ducks   (E)   Konrad Lorenz

(C)   other chickens

34.   The organ that functions as the vascular connection between the embryo and its mother is the

(A)   morula   (D)   blastula

(B)   aorta   (E)   placenta

(C)   amnion

35. The binomial nomenclature for man is *Homo sapiens*. Classify man, in proper order starting with its Kingdom and working through its order.

(A) Chordata, Animalia, Primates, Mammalia, Vertebrata

(B) Animalia, Chordata, Vertebrata, Mammalia, Primates

(C) Animalia, Vertebrata, Chordata, Mammalia, Primates

(D) Primates, Mammalia, Vertabrata, Chordata, Animalia

(E) Animalia, Chordata, Vertabrata, Primates, Mammalia

36. In a longitudinal section of a root, starting from the deepest zone, the four zones occur in the following order:

(A) root cap, zone of cell division, zone of cell enlargement, zone of cell maturation

(B) root cap, zone of cell division, zone of cell maturation, zone of cell enlargement

(C) zone of cell enlargement, zone of cell maturation, zone of cell division, root cap

(D) zone of cell division, zone of cell enlargement, zone of cell maturation, root cap

(E) zone of cell division, zone of cell maturation, zone of cell enlargement, root cap

37. All of the following structures in a leaf may function in photosynthesis, EXCEPT:

(A) cuticle

(B) mesophyll

(C) guard cells

(D) chloroplasts

(E) spongy layer

38. Which of the following statements does not apply to members of Class Aves?

(A) They have feathers

(B) They have compact hollow bones

(C) They are homeothermic

(D) They excrete urea

(E) They use song in mating behavior

39. The kangaroo rat is a desert animal that never needs to drink water. All of the following statements are true EXCEPT

(A) its water requirements are, for the most part, met by metabolic production

(B) its water requirements are, for the most part, met by osomotic influx from the environment

(C) it produces very small amounts of urine

(D) it has long loops of Henle in its kidneys

(E) it has the ability to produce a quite concentrated urine

40. Darwin's theory of natural selection includes all of the following stipulations EXCEPT

(A) every organism produces more organisms than can survive

(B) due to competition, not all organisms survive

(C) some organisms are more fit, i.e., they are able to survive better in the environment

(D) the difference in survivability is due to variations between organisms

(E) variation is due, at least in part, to mutations

41. Hemophilia is a disease caused by a sex-linked recessive gene on the X chromosome; therefore,

(A) females have twice the likelihood of having the disease, since they have two X chromosomes

(B) mothers can pass the gene with equal probability to either a son or daughter

(C) females can never have the disease, they can only be carriers

(D) inbreeding has no effect on the incidence of the disease, since it is purely sex-linked

(E) a hemophiliac son is always produced if his father has the gene and, hence, the disease

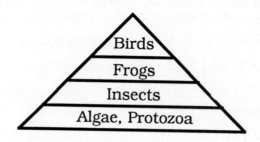

42. If DDT (an insecticide) were sprayed, the organisms shown in the above food pyramid that would suffer the most would be:

(A) frogs   (D) protozoa

(B) insects   (E) algae

(C) birds

43. All of the following statements about vitamin D are correct EXCEPT

(A) a deficiency causes rickets.

(B) it can be produced in the skin in the presence of ultraviolet light.

(C) dairy products are a good source.

(D) night blindness may result from a deficiency.

(E) it helps absorb calcium from the digestive tract and helps incorporate the calcium into bone.

44. In certain flowers, color is inherited by incomplete dominance. A cross between a homozygous red flower (RR) and a homozygous white flower (rr) will always yield pink flowers. When these pink flowers are subsequently crossed, the expected probabilities may include:

(A) 25% pink

(B) 50% red

(C) 0% white

(D) 50% pink

(E) 100% pink

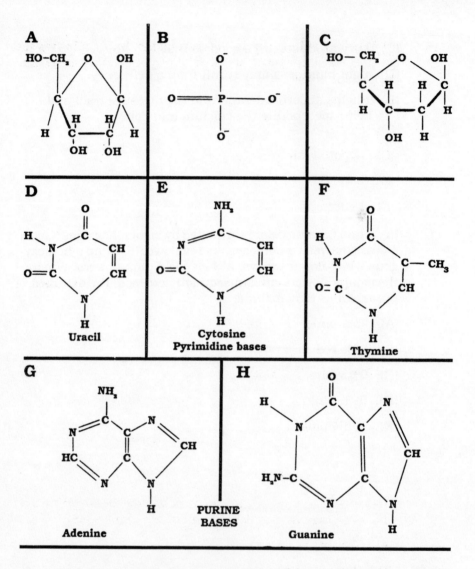

A

HO-CH₂ · O · OH

B

O⁻ — P — O⁻, O=, O⁻

C

HO—CH₂ · O · OH

D

Uracil

E

NH₂

Cytosine
Pyrimidine bases

F

Thymine

G

NH₂

Adenine

H

PURINE
BASES

Guanine

45.   The structures of the components of a nucleotide are illus-
        trated above. The pentose sugars, the phosphate group and
        the five nitrogenous bases are depicted. The two molecules
        that would be found only in DNA, but not RNA are represented
        by the letters

        (A)   A and D

        (B)   B and E

        (C)   C and F

        (D)   C and D

        (E)   A and F

46. The relatively large size of the mammalian brain, allowing for greater learning, association, and memory, is due to the enlargement of the

(A) hindbrain

(B) cerebellum

(C) hypothalamus

(D) cerebrum

(E) midbrain

47. Like eukaryotes, prokaryotes may contain all of the following structures EXCEPT

(A) plasma membrane

(B) cell wall

(C) ribosomes

(D) cytoplasm

(E) mitochondria

48. Angiosperms are classified as monocots or dicots. Which of the following phrases does not pertain to monocots?

(A) Flower parts in groups of three

(B) Netted leaf veins

(C) Scattered vascular bundles in stem

(D) Flower parts in groups of six

(E) Exemplified by grasses and orchids

49. The gametophyte generation in the plant life cycle

(A) is diploid

(B) produces spores

(C)   is haploid

(D)   has become more dominant in the evolution of plants

(E)   is a zygote

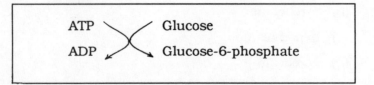

50.   The enzyme-catalyzed sequence depicted above represents

(A)   the dephosphorylation of glucose

(B)   the phosphorylation of ADP

(C)   the first step of glycolysis

(D)   phosphofructokinase activity

(E)   a reaction occurring within the mitochondrial matrix

51.   In humans, there are many anatomical adaptations that function to increase surface area for chemical reactions and transport mechanisms. All of the following are examples EXCEPT

(A) the alveoli of the lungs

(B) the microvilli of the small intestine

(C) the cristae of the inner mitochondrial membrane

(D) the villi of the small intestine

(E) the sensory hairs (cilia) in the cochlea of the inner ear

52.   Concerning the development of the vertebrate brain,

(A) the prosencephalon consists of the pons and cerebellum.

(B) the mesencephalon develops into the myelencephalon and metencephalon.

(C) the rhombencephalon consists of the pons, medulla oblongata, and cerebellum.

(D) the telencephalon develops into the prosencephalon and diencephalon.

(E) the cerebellum is part of the diencephalon.

53.   Adaptations of desert plants to hot, dry environments, may
       include all of the following, EXCEPT

       (A)   wide spacing between plants

       (B)   deep penetrating roots

       (C)   deciduous leaves

       (D)   thick, waxy cuticles

       (E)   superficial stomata

54.   There are various types of plant stems that have different
       functions. Which of the following is not a type of stem?

       (A)   tendrils

       (B)   nodes

       (C)   tubers

       (D)   rhizomes

       (E)   corms

55.   Acquired characteristics

       (A)   refer to traits inherited as genes

       (B)   are not transmitted to the next generation

       (C)   are the basis of Darwin's theory of natural selection

       (D)   are exemplified by the lengthening of the giraffe's neck
             over evolutionary time, due to stretching toward trees

       (E)   can, for instance, explain the lack of pigment in an albino

56.   All of the following statements about embryonic induction are
       true, EXCEPT

       (A)   it is exemplified by the development of the vertebrate lens

       (B)   neurulation is an example

       (C)   it refers to the interaction whereby certain cells can
             stimulate the development of nearby cells

       (D)   it refers to cleavage

       (E)   it most likely occurs due to chemical factors

57. Which of the following statements concerning alternation of generations in plants is true?

   (A) The diploid generation consists of gametophytes

   (B) The haploid generation consists of sporophytes

   (C) Gametes result from mitosis

   (D) Gametophytes result from the fusion of gametes

   (E) Meiosis produces sporophytes

58. Deviation from a Hardy-Weinberg equilibrium may be the result of

   (A) absence of mutation

   (B) large population size

   (C) migration

   (D) no natural selection

   (E) lack of differential reproduction

59. The period of human gestation is divided into three trimesters. The event that is correctly matched to its trimester of occurrence is the following:

   (A) the third trimester is characterized by development and differentiation.

   (B) the greatest growth in size occurs in the first trimester.

   (C) the limb buds develop in the first trimester.

   (D) kicking is felt by the mother in the first trimester.

   (E) organ development begins in the second trimester.

60. Which of the following statements about the gymnosperms is not true?

(A) They are seed plants

(B) They include ginkgo and cycads

(C) They are cone-bearers

(D) They are the flowering plants

(E) They are referred to as the "naked seed" plants

Questions 61 - 63 refer to the diagram below showing DNA replication.

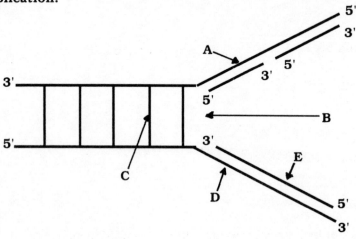

61. The lagging strand

62. A base pair

63. The replication fork

Questions 64 - 67 refer to various transport processes encountered in biology

(A) Transfusion

(B) Translocation

(C) Transcription

(D) Translation

(E) Transpiration

64. This nuclear process refers to the transfer of information from DNA to RNA

65.   The transfer of sucrose in sieve tubes.

66.   Protein synthesis as directed by messenger RNA.

67.   The introduction of blood directly into the bloodstream.

Questions 68 - 70 refer to the stages of the cell cycle.

   (A) Anaphase

   (B) Prophase

   (C) Interphase

   (D) Metaphase

   (E) Telophase

68.   Cytokinesis accompanies this stage of mitosis.

69.   This phase, in which DNA is replicated, is not part of mitosis

70.   This stage of mitosis is characterized by the condensation of the chromosomes.

Questions 71 - 74 deal with anatomical and physiological aspects of the first meiotic prophase (Prophase I).

    (A) Synapsis

    (B) Crossing over

    (C) Sister chromatids

    (D) Chiasma

    (E) Centromeres

71.   The site of crossover between attached homologous chromo-somes.

72.   The constricted point on the chromosome at which sister chromatids are attached.

73.   Close pairing of homologous chromosomes to form tetrads.

74.   Non-sister chromatids of homologous chromosomes exchange segments.

Questions 75 - 76 distinguish between different biomes.

    (A) Tundra

    (B) Taiga

    (C) Desert

    (D) Tropical rainforest

    (E) Deciduous forest

75.   The Arctic exemplifies this biome.

76.   Hot humid weather and no seasonal changes are characteristic of this biome.

Questions 77 - 79 deal with different mechanisms that cause reproductive isolation.

    (A) Mechanical isolation

    (B) Gamete isolation

    (C) Hybrid inviability

    (D) Behavioral isolation

    (E) Temporal isolation

77.   Size, shape, and length of reproductive organs influence mating.

78.   Pollination/mating is a seasonal event.

79.   Courtship rituals are required for mating.

Questions 80 - 82 distinguish five animal phyla.

    (A) Porifera

    (B) Arthropoda

    (C) Echinodermata

    (D) Mollusca

    (E) Platyhelminthes

80.   These freshwater and marine filter feeders are sessile and contain spongin or spicules as supportive structures.

81.   This phylum includes the parasitic flukes and the tapeworms.

82.   As is phylum Chordata, this phylum is classified as Deuterostome.

Questions 83 - 86 show examples of organic nutrients.

    (A) monosaccharide

    (B) disaccharide

    (C) oligosaccharide

    (D) saccharin

    (E) polysaccharide

83.   Fructose

84.   Lactose

85.   Sucrose

86.   Starch

Questions 87 - 90 describe different human cells that are to be matched with the organelle that is most associated with the cell's function.

    (A) Mitochondria

    (B) Cilium

    (C) Rough endoplasmic reticulum

    (D) Lysosome

    (E) Flagella

87.   Especially numerous in skeletal muscle cells

88.   Found only in sperm cells

89.   Leukocyte

90.   Pancreatic acinar cell

Questions 91 - 92 refer to the following blood-typing experiment. Two students type their blood in biology lab using antiserum A and antiserum B. The results of their tests are below:

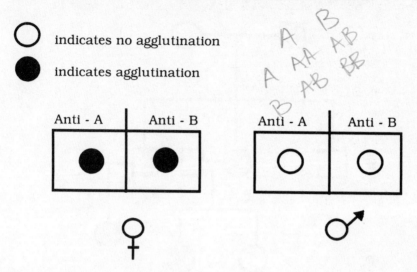

91.   When they marry and have children, the possible phenotypes will be:

(A)   AB only

(B)   Type O only

(C)   Type A and Type B

(D)   Type AB and Type O

(E)   Type B only

92.   The male (♂) depicted above is called a universal donor because

(A)   he has no antigens on his red blood cells

(B)   he has both antibodies in his plasma

(C)   he has no antibodies in his plasma

(D)   he has no antibodies on his red blood cells

(E)   he has no antigens in his plasma

Questions 93 - 94 refer to the pedigree below. A square indicates
a male; a circle indicates a female. A hollow shape indicates the
lack of trait; a darkened shape indicates the presence of trait.

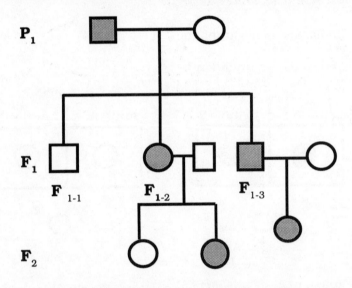

93.   The trait, a preauricular gill remnant, which is a tiny hole in
       the front of the ear, is transmitted by an autosomal dominant
       gene. The probability that the children of F_{1-1}, should he marry
       an unafflicted woman, will have the trait is

       (A)   50%

       (B)   0%

       (C)   100%

       (D)   25%

       (E)   Cannot be determined with the information given.

94. F_{1-3} and his wife just had a baby girl afflicted with this trait.
       The probability that their next son will have the trait is

       (A)   0% - half the kids should have the trait, and one just
             received it so the next one should not.

       (B)   50%

       (C)   100% - all the sons will show the trait

       (D)   25%

       (E)   Cannot be determined with the information given.

96

Questions 95 - 96 refer to the drawings of the two plants. You have two plants in which you wish to stimulate growth. Plant A receives light and responds as indicated below. Plant B receives a chemical and responds as below.

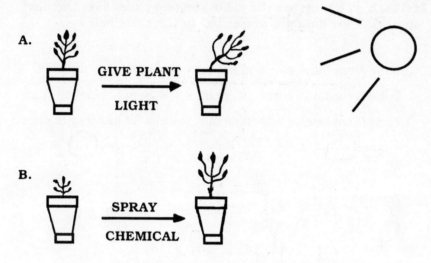

A.

GIVE PLANT
LIGHT

B.

SPRAY
CHEMICAL

95.   The hormone responsible for the growth movements seen in plant A is

   (A)   auxin

   (B)   florigen

   (C)   cytokinin

   (D)   abscisic acid

   (E)   ethylene

96.   The chemical given to plant B, a genetic dwarf, must be

   (A)   ethylene

   (B)   gibberellins

   (C)   cytokinins

   (D)   growth hormone

   (E)   florigen

Questions 97 - 100 refer to an experiment in digestion. Digestions of fat yields glycerol and fatty acids. You wish to determine if and how bile salts and pancreatic juice affect fat digestion. You prepare three tubes as indicated below. While incubating them in a 37°C bath, you measure the pH in the test tubes over the next 45 minutes. Your data are presented in the table below.

Tube A contains 4 mls. of cream + pinch of bile salts

Tube B contains 4 mls. of cream + 4 mls. of pancreatic juice

Tube C contains 4 mls. of cream + 4 mls. of pancreatic juice + pinch of bile salts

Data on pH

Time (mins.)

	0	15	30	45
pH-Tube A	7	7	7	7
pH-Tube B	7	6.9	6.7	6.3
pH-Tube C	7	6.3	6.0	5.4

97. Digestion of the cream occurs most rapidly in

(A) Tube A

(B) Tube B

(C) Tube C

(D) No digestion occurred

(E) Cannot be determined from the data

98. If you only have two test tubes with which to do your experiment and you wish to figure out the role of bile salts, it is necessary to prepare

(A) Tubes A and B

(B) Tubes B and C

(C) Tubes A and C

98

(D) Just Tube C

(E) Just Tube A

99. It is clear that the ingredient necessary for digestion of fats is

(A) cream

(B) water

(C) bile salts

(D) pancreatic juice

(E) not in any of the tubes

100. The function of bile salts is to

(A) digest the fat

(B) chemically degrade the fat

(C) emulsify the fat into smaller globules

(D) activate the enzymes in pancreatic juice

(E) acidify the solution which it is in

Questions 101 - 104 refer to a urinalysis report. A lab technician forgets to label three patients' recent urine samples and a sample of distilled water. He performs various tests on these samples to try to determine which urine belongs to which patient. One patient was on a high-salt diet; one was on a high-protein diet; and one had uncontrolled diabetes mellitus. The results of the tests on the three urine samples and the distilled water tube are shown below:

	A	B	C	D
Specify gravity	1.030	1.029	1.010	1.000
Glucose	negative	299 mg/dl	negative	negative
pH	6.3	5.2	5.0	7
Odor	aromatic	sweet	aromatic	none
Volume	130 mls.	160 mls	96 mls	100 mls

101. The urine sample from the diabetic patient is in which test tube?

(A) Tube A

(B) Tube B

(C) Tube C

(D) Tube D

(E) Cannot be determined from data available

102. Tube D must contain the distilled water because

(A) by definition, the specific gravity of distilled water = 1.000

(B) by definition, the pH of distilled water is 7

(C) process of elimination leaves only Tube D for water

(D) specific gravity is the weight of a volume of water divided by the weight of an equal volume of substance

(E) there is no glucose in it

103. The effect of a diuretic drug would most directly affect which of the following parameters?

(A) specific gravity

(B) glucose

(C) pH

(D) odor

(E) volume

104. The tube with the highest concentration of hydrogen ions is

(A) Tube A

(B) Tube B

(C) Tube C

(D) Tube D

(E) Cannot be determined by the data available

The graph on the following page indicates the interaction between a predator (solid) and prey (dotted) population over the years 1976 to 1980. Their population sizes are indicated on either side of the graph.

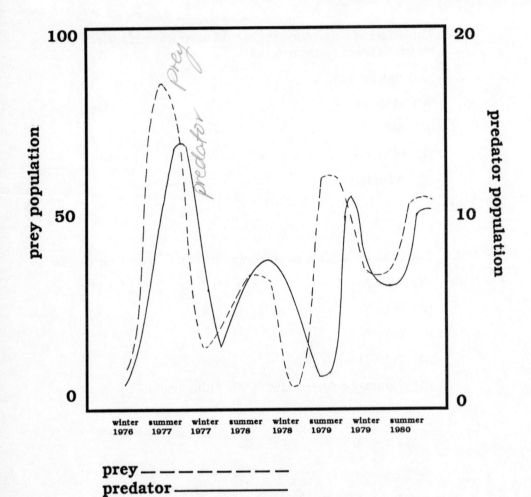

prey — — — — — — — —

predator ————————————

105. The oscillations in the predator/prey population

(A) occur because an increase in prey population will increase predator population indefinitely.

(B) occur because an increase in prey population will allow an increase in predator population until the predators eat too many prey.

(C) occur because an increase in predator population causes a direct and immediate increase in prey population.

(D) occur because of seasonal changes

(E) are unique in this case and not seen in most predator/prey interactions.

106. The drop in predator population during the winter months is probably due to

(A) cold

(B) snow

(C) drop in prey population

(D) hibernation

(E) coincidence

107. The most mild winter probably occurred in the year

(A) 1976   (D) 1979

(B) 1977   (E) 1980

(C) 1978

108. The peak prey population was achieved

(A) in the summer of 1977 with a population of 100

(B) in the summer of 1977 with a population of 75

(C) in the summer of 1977 with a population of 20

(D) in the summer of 1977 with a population of 15

(E) in the summer of 1977, population unknown

109. In the winter of 1980, a chemical, toxic and fatal to the prey but not to the predator, is introduced into the environment; the subsequent effect to the predator will be

(A) no effect, since it is non-toxic to him

(B) a drop in population due to decreased food resources

(C) a rise in population

(D) independent of any changes in prey population

(E) a drop in population due to the yearly winter drop

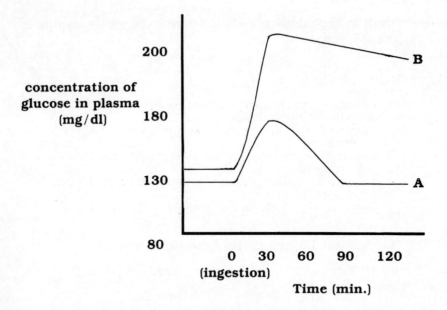

In a glucose tolerance test, two fasting subjects drink glucose solutions (in amounts that are proportional to their body weight). Plasma glucose levels are measured just before ingestion, and at 30-minute intervals for the next two hours.

110. The plasma glucose levels of the subjects, while still fasting, are

(A) 0 minutes

(B) 80 mg/dl

(C) 130 mg/dl

(D) 180 mg/dl

(E) not depicted in the graph above

111. Peak plasma glucose levels are reached

(A) at the time of ingestion

(B) at 30 minutes

(C) at 60 minutes

(D) at different times for the two subjects

(E) at 30 mg/dl

112.   Graph A indicates

(A) a normal response to glucose ingestion

(B) a diabetic response to glucose ingestion

(C) a maximum plasma glucose level of 180 mg/dl

(D) a return to fasting glucose levels within a half hour

(E) a fasting glucose level of 130 mg/dl

113.   Graph B indicates

(A) excess insulin has been secreted

(B) high tolerance to glucose

(C) a diabetic condition

(D) a maximum glucose level of 180 mg/dl

(E) rapid return to fasting glucose levels

114.   Urinary excretion of glucose may occur

(A) in subject A

(B) in subject B

(C) in both subjects

(D) in neither subject

(E) but it is unrelated to plasma glucose levels

Questions 115 - 117 refer to transport processes through the cell membrane.

The following table indicates the effects of two cell poisons on the transport processes of the cell membrane. One group of cells was given ouabain, and another group was given cyanide. A "✓" indicates that the process is still functional, while an "**X**" indicates that the process cannot occur in the presence of the poison.

	Diffusion	Na⁺-K⁺ Pump	Endocytosis
Ouabain	✓	X	✓
Cyanide	✓	X	X

115. Which of the following statements is true?

(A) Ouabain did not affect any active processes

(B) Cyanide did not affect any active processes

(C) Cyanide prevented ouabain from acting on endocytosis

(D) Passive processes were unaffected by either poison

(E) Only processes that do not require ATP were affected by cyanide

116. The probable mechanisms of action are that

    (A) cyanide inhibits the Na⁺-K⁺ pump directly

    (B) ouabain inhibits the production of ATP

    (C) cyanide inhibits the production of ATP and, hence, the activity of the Na⁺-K⁺ pump

    (D) cyanide inhibits the activity of the Na⁺-K⁺ pump and, hence, the production of ATP

    (E) ouabain inhibits the production of ATP directly and, hence, the activity of the Na⁺-K⁺ pump

117. In the presence of ouabain,

    (A) Na⁺ will accumulate outside the cell

    (B) K⁺ will accumulate inside the cell

    (C) K⁺ will accumulate outside the cell due to diffusion

    (D) K⁺ will accumulate outside the cell due to active transport

    (E) Na⁺ will accumulate inside the cell due to endocytosis

Questions 118- 120 refer to observations on simple plant tissues.

In a lab examination in your botany class, you are to identify the following three cell types A, B, and C. Your determinations are based on the following key observations.

	A	B	C
Capacity to divide (Mitosis)	Yes	No	No
High concentrations in mesophyll of leaves	Yes	No	No
High numbers in young stems	Yes	Yes	No

118.	To identify a parenchyma cell, the key observation that you would look at would be

(A)	its presence in leaves, since leaves are alive at maturity

(B)	its ability to divide, since it is the only live cell

(C)	its presence in stems, since only stems function in food storage

(D)	its presence in leaves, since leaves contain a lot of chloroplasts

(E)	air spaces in the cells when viewed with a microscope

119.	According to the chart, the inability of cell C to divide

(A)	indicates that it must be a parenchyma cell, which is the least specialized and thus cannot undergo mitosis, a specialized function

(B)	indicates that it must be a parenchyma cell, since the air spaces preclude the capacity to divide

(C)	indicates that it must be sclerenchyma cell, since dead cells cannot divide

(D)	indicates that it must be a parenchyma cell, since photosynthesizing cells cannot divide

(E)	is not a clue to the identification of any cell type

120.	Which of the following is true?

(A)	A is a sclerenchyma cell

(B)	B is a collenchyma cell

(C)	C is a parenchyma cell

(D)	A is a collenchyma cell

(E)	C is a collenchyma cell

SECTION II

DIRECTIONS: Answer each of the following four questions in essay format. Each answer should be clear, organized, and well-balanced. Diagrams may be used in addition to the discussion, but a diagram alone will not suffice. Suggested writing time per essay is 22 minutes.

1. Discuss the complete oxidation of glucose by the processes of glycolysis, the Krebs cycle, and the electron transport chain. Be sure to include in which part of the cell each process occurs. Do not dwell on details and numerical balance of the reactions, but focus on the major reactants, products, and destinations of these products.

2. Different patterns of genetic inheritance cause variation and sometimes speciation. Discuss how dominant/recessive alleles, incomplete dominance, codominance, and sex-linked inheritance can affect two populations of the same organism separated by geographic barriers, giving an example.

3. Negative feedback plays an important role in the counteraction of stress.
    A) Explain the mechanism's role in a stressful environment. Include how at least two bodily systems are affected.
    B) Design a controlled experiment to test the postulate that adrenaline prepares the body for "Flight or Fight."

4. Explain the role of plant hormones, including their effects.
    A) Growth hormones
    B) Postulated flowering hormone
    C) Aging and ripening hormone

ADVANCED PLACEMENT
BIOLOGY EXAM II

ANSWER KEY

1.	A	31.	C	61.	A	91.	C
2.	A	32.	A	62.	C	92.	A
3.	B	33.	E	63.	B	93.	B
4.	C	34.	E	64.	C	94.	B
5.	D	35.	B	65.	B	95.	A
6.	C	36.	A	66.	D	96.	B
7.	B	37.	A	67.	A	97.	C
8.	A	38.	D	68.	E	98.	B
9.	B	39.	B	69.	C	99.	D
10.	A	40.	E	70.	B	100.	C
11.	D	41.	B	71.	D	101.	B
12.	D	42.	C	72.	E	102.	A
13.	E	43.	D	73.	A	103.	E
14.	C	44.	D	74.	B	104.	C
15.	C	45.	C	75.	A	105.	B
16.	B	46.	D	76.	D	106.	C
17.	B	47.	E	77.	A	107.	D
18.	C	48.	B	78.	E	108.	A
19.	B	49.	C	79.	D	109.	B
20.	E	50.	C	80.	A	110.	B
21.	A	51.	E	81.	E	111.	B
22.	E	52.	C	82.	C	112.	A
23.	C	53.	E	83.	A	113.	C
24.	B	54.	B	84.	B	114.	B
25.	C	55.	B	85.	B	115.	D
26.	B	56.	D	86.	E	116.	C
27.	A	57.	C	87.	A	117.	C
28.	C	58.	C	88.	E	118.	D
29.	D	59.	C	89.	D	119.	C
30.	B	60.	D	90.	C	120.	B

ADVANCED PLACEMENT
BIOLOGY EXAM II

DETAILED EXPLANATIONS
OF ANSWERS

SECTION I

1.   (A)

The complete oxidation of glucose yields 38 ATPs. Three cellular processes are required. In glycolysis, which occurs in the cytoplasm, only 2 ATPs are formed. Another two are formed in the Krebs cycle in the mitochondrial matrix. However, the Krebs cycle is also the major site of production of the reducing equivalents NADH and $FADH_2$. Since these coenzymes are in the reduced state, they carry many electrons (and hydrogen ions). They pass their electrons via the electron transfer chain of the inner mitochondrial membrane ultimately to oxygen. The formation of the other 34 ATPs is concomitant with electron transfer. While there are many electron transfers in this respiratory chain, oxygen is the ultimate oxidizing agent, and thus, the ultimate electron acceptor, since it must be reduced.

The last step of the respiratory chain is catalyzed by the enzyme cytochrome c oxidase. As its copper cofactor is oxidized (from the Cu^+ to the Cu^{+2} state), oxygen $(1/2\ O_2)$ is reduced by accepting two hydrogen ions and two electrons to form water (H_2O).

2.   (A)

Photosynthesis involves reactions that are light-dependent and reactions that are light-independent. The light reactions entail the use of two photosystems (pigment complexes) that capture the energy of sunlight. These light reactions can be cyclic or noncyclic, depending on the pathway taken by the released electrons upon being charged with energy from sunlight.

In noncyclic photophosphorylation, water is split into its component oxygen and hydrogen atoms, releasing electrons to replace electrons lost by the pigment complex. This liberation of oxygen is vital to animal life. Noncyclic photophosphorylation is also responsible for the reduction of the coenzyme NADP to $NADPH_2$. Both cyclic and noncyclic processes result in the generation of ATP (adenosine triphosphate, the energy currency of the cell).

The dark reactions of photosynthesis are responsible for the reduction of carbon dioxide to carbohydrate, using the $NADPH_2$ and ATP produced in the light reactions.

3.   (B)

The fundamental anatomical and physiological unit of life is the cell, which is composed of all the chemical substances essential to sustain life. While cells have basic similarities, each cell type has a unique structure and performs a specific function. Thus there are blood cells, muscle cells, and nerve cells, to name a few.

Similar cell types are grouped together to form a tissue, which subserves a special function. Muscle tissue, nerve tissue, epithelial tissue and connective tissue are the basic subdivisions of tissue.

An organ is a distinct structure composed of two or more different tissue types. The organ also carries out a specific function. Examples of organs are stomach, liver, lung, brain, ureter, ovary, etc.

An organ system consists of a group of organs that work together towards a specific function. The urinary system consists of four organs: kidney, ureter, bladder and urethra.

Finally, the organism is the sum of all the organ systems functioning in harmony. Man is an example of this highest level of organization.

4.   (C)

The cell membrane, or plasma membrane, is composed primarily of phospholipids and proteins. The phospholipids form a lipid bilayer with their fatty acid chains oriented toward the center and the phosphate heads oriented towards the extracellular and intracellular fluids. Since the fatty acid tail, a long hydrocarbon chain, is nonpolar, and the charged phosphate head is polar, the phospholipid is said to be amphipathic.

Proteins also form an important part of the plasma membrane. The intrinsic proteins span all or half of the lipid bilayer, while the extrinsic proteins lie on the inner or outer surfaces, exposed to the intracellular or extracellular fluids. Any protein on the extracellular side may have a sugar moiety attached which functions as a recognition site; the protein is now called a glycoprotein.

The fluid mosaic model of the plasma membrane is a reminder that the membrane is very dynamic. The proteins form a mosaic upon a fluid phospholipid background; thus the membrane has been likened to "icebergs (proteins) floating on a sea (of lipid)."

The primary function of the plasma membrane is to separate the intracellular and extracellular fluids, forming a selective barrier between the two and maintaining quite a different composition in the two fluid compartments.

Cholesterol, though present in cell membranes, is a steroid. It is not a major component of the cell membrane.

5.   (D)
DNA (deoxyribonucleic acid) and RNA (ribonucleic acid) are both nucleic acids and are, therefore, composed of a sequence of nucleotides. A nucleotide consists of a nitrogenous base, a pentose sugar (five-carbon sugar), and a phosphate group. The distinctions between the two nucleic acids are in the sugar and in the bases.

The sugar in DNA is deoxyribose, while in RNA, it is ribose. Deoxyribose has one less oxygen in its structure than ribose. Ribose and deoxyribose are both pentose sugars. Hexoses (six-carbon sugars) are found in carbohydrates. Examples are the monosaccharides glucose, galactose and fructose.

The nitrogenous bases are classified as pyrimidines (cytosine, uracil and thymine) or purines (adenine and guanine), as determined by their single-ring or double-ring heterocyclic structure, respectively. Thymine is unique to DNA, while uracil is unique to RNA. The other bases are found in both nucleic acids.

DNA is a double-stranded molecule in which the bases from separate strands pair. Cytosine pairs with guanine; thymine pairs with adenine. RNA is single-stranded.

6. (C)

DNA replicates by forming two daughter molecules from the original parent. There are three possible ways in which DNA can replicate by base pairing. In semiconservative replication, each strand of the double helix serves as a template for the synthesis of its new partner. Thus there are two identical molecules formed from the original one; the daughter molecules are also identical to the parent molecule. It is named semiconservative replication since only one original strand is conserved in each daughter molecule. This is the way in which DNA replicates.

The other proposed mechanisms by which DNA might replicate, are conservative and dispersive replication, neither of which occur. In conservative replication, the parent duplex is left intact and an entirely new double-stranded molecule is formed. In dispersive replication, pieces of parent and daughter DNA are mixed together in the new generation.

The three models are shown below. A solid line indicates a parent strand, while a dotted line indicates a newly synthesized strand.

	Semiconservative	Conservative	Dispersive
Parent	‖	‖	‖
Daughter	¦¦¦¦	‖ ¦¦	¦¦¦¦

(D) is incorrect, as semiconservative replication is not concerned with what the energy requirements may be. Likewise, while (E) is a true statement, it does not describe semiconservative replication.

7. (B)

The codon is a triplet of nucleotides on messenger RNA (mRNA). Since there are four nucleotides in RNA, a triplet sequence would yield 64 possible codons (4^3), 61 of which code for an amino acid. There are only twenty amino acids; therefore most of the amino acids are coded for by more than one codon. This is referred to as the degeneracy of the genetic code. Three of the triplet codons are a signal to stop, while the one that codes for the amino acid methionine, is also a signal to initiate protein synthesis.

The anticodon is a triplet of nucleotides on a loop of transfer RNA (tRNA). Since the tRNA is specific for an amino acid, when the mRNA codon base pairs with the tRNA anticodon, the tRNA will bring the appropriate amino acid to the elongating protein chain on mRNA.

8.   (A)

The codon is degenerate, meaning that most amino acids are represented by more than one codon. The difference in two or more different codons representing the same amino acid is usually the base of the third nucleotide in the sequence. This relative freedom in base-pairing at the third position is referred to as wobble. Thus UUC and UUU both code for the amino acid phenylalanine. While UUG and CUG do both code for leucine, this is not an example of wobble since the mismatch occurs at the first position.

Base pairing occurs between one purine and one pyrimidine base. In RNAs, adenine pairs with uracil while guanine pairs with cytosine.

While the stop codons show some degeneracy (UAA and UAG), most of the amino acids are represented by degenerate codons, and thus tRNA molecules bond to mRNA molecules allowing wobble at the third base.

9.   (B)

Glycolysis is a sequence of chemical reactions that occur in the cell cytoplasm and initiate glucose oxidation. In the tenth step of glycolysis, pyruvate is formed. While glycolysis itself does not require oxygen, the fate of pyruvate is determined by the presence or absence of oxygen (aerobic and anaerobic conditions, respectively).

Under aerobic conditions, pyruvate enters the mitochondrial matrix and is converted to acetyl CoA. Acetyl CoA will then enter the Krebs Cycle and the complete oxidation of glucose will ensue; much ATP will be formed.

Under anaerobic conditions, glycolysis continues one step farther in the cytoplasm. Pyruvate is reduced to lactate. This is important because the reduction step is coupled to the oxidation of NADH to form NAD^+. Since NAD^+ is necessary for an earlier oxidative step of glycolysis, the reduction of pyruvate allows glycolysis to continue to produce minimal amounts of ATP, despite the lack of oxygen.

Note that in either case, aerobic or anaerobic, oxygen is not directly used in glycolysis. Also note that while humans normally metabolize aerobically, anaerobic metabolism can be maintained for brief intermittent periods. For instance, a sprinter metabolizes glucose anaerobically. But lactate will diffuse into the blood and cause pain. This would cause the sprinter to slow down or stop and thus begin to consume oxygen once again.

10.   (A)
Sickle cell anemia is an inherited disease in which one base substitution, a uracil for an adenine, codes for a new amino acid, valine (instead of glutamic acid).This single amino acid substitution occurs in the sixth position of each B chain of hemoglobin.

Valine is nonpolar, while glutamic acid is polar. Thus the properties of hemoglobin are altered and this defective hemoglobin, designated Hemoglobin S to distinguish it from normal Hemoglobin A, has a lessened ability to transport oxygen, due to the fact that it will bind to other Hemoglobin S molecules.

A single nucleotide change does not always cause such a disease, since, due to wobble, the amino acid coded for may not be different, or, even if a new amino acid is called for, the properties of that new amino acid may not differ much from the original one. However, in this case, the substituted base occurs in the second position of the codon sequence and thus wobble cannot compensate for it. GUA and GUG specify valine, while GAA and GAG specify glutamic acid. Thus a new amino acid is called for, and it has quite different properties from the original one.

Sickle cell anemia is prevalent in the Black population of Africa. It has been maintained in the population because it serves as a protective mechanism against malaria — the parasite causing malaria uses blood cells during part of its life cycle, and the sickled shape of the red blood cell hinders the parasite. People may die of sickle cell anemia or malaria, but the heterozygotes are protected from malaria and are relatively free of anemia. (Those who are homozygous for the sickle-cell gene are highly anemic.) Thus malaria is endemic in regions in which resistance is high.

The protozoan parasite that causes malaria is transmitted to humans by the bite of a mosquito; mosquitoes cannot cause sickle cell anemia, because it is an inherited disease. The link between the coincidence of malaria and sickle cell anemia is due to the protection against malaria afforded by the sickle cells.

11.   (D)

Nondisjunction refers to the lack of separation of homologous chromosomes in Anaphase I of meiosis (or lack of separation of a pair of sister chromatids in Anaphase II of meiosis). While this may occur in the ovaries or the testes, it appears to be more common in the ovaries, and is often associated with being middle-aged. Nondisjunction results in one gamete (egg or sperm cell) with an extra chromosome and one lacking a chromosome. If the gamete participates in fertilization, the result may be Down's Syndrome or a sex-chromosomal anomaly.

Down's Syndrome is caused by an extra chromosome #21 (trisomy-21) and is more common in babies born to women who conceived past age 40.

When the extra or lost chromosome is a sex chromosome (X or Y chromosome), various anomalies result. A normal female has the genotype XX, while a normal male has the genotype XY. A Turner's Syndrome patient is a female with 45 chromosomes, of which one, not two, is an X-chromosome. Klinefelter's Syndrome refers to a male with 47 chromosomes. He is an XXY male. An XYY male, referred to as a "supermale" is often very tall and aggressive.

It is important to note that nondisjunction may occur in other chromosomes as well, yet the embryos, like many of the cases reported above, often result in a spontaneous abortion.

Hemophilia is caused by a recessive gene carried on the X chromosome and thus is a sex-linked disease and is not a result of nondisjunction.

12.   (D)

Phylum Annelida includes the segmented worms, best characterized by earthworms and leeches. Characteristics of annelids include chitin-containing bristles called setae, which may function in anchoring or swimming. The nephridium functions as a "kidney" in that it regulates water and solute levels of the body, removing wastes from the coelom. A flame cell (or proto-nephridium) is the water-regulating mechanism characteristic of certain flatworms (phylum Platyhelminthes).

13.   (E)

The Hardy-Weinberg rule states that allele frequencies and genotype frequencies will be infinitely stable. The frequencies are expressed by the formula:

$$p + q = 1$$

where p is the frequency of one allele at a given locus and q is the frequency of the alternative allele.

Implicit in the statement is that no evolution is occurring. The four factors that bring about change (evolution) are (1) mutation; (2) genetic drift (random changes in allele frequency often seen in small populations); (3) gene flow (a change in allele frequency due to emigration or immigration); (4) natural selection, due to differential fertility and survivorship. Hence, the requirements for a Hardy-Weinberg equilibrium are merely the opposition of the factors above: (1) No mutations; (2) large population; (3) isolation (no migration); and (4) equal viability and fertility of all genotypes, i.e. random reproduction.

14.   (C)

By 1953, it was believed that the prebiotic atmosphere (i.e., the atmosphere before life arose) was filled with hydrogen-containing molecules. Such an atmosphere can be referred to as a reducing atmosphere. Examples of such molecules are ammonia (NH_3), water vapor (H_2O), molecular hydrogen (H_2), and methane (CH_4). These simple gases can react in the presence of energy (such as from sunlight or lightning) to form the simple organic compounds that are necessary for life.

In Miller and Urey's experiment, a chamber was prepared with the prerequisite molecules. Using electric sparks to mimic an energy source, they successfully formed amino acids and other basic building blocks of life in their chamber. Of critical importance was the lack of oxygen, which would have oxidized and destroyed any building blocks that were formed.

15.   (C)

Evidence of evolution comes from many scientific disciplines. Evidence from comparative biochemistry refers to the analysis of enzymes, proteins, nucleic acids, (i.e. DNA), etc. to determine similarities in amino acid sequences, nucleotide sequences, etc. A similar molecular structure suggests a recent (by evolutionary standards) bifurcation of the original ancestral stock.

Comparative anatomy distinguishes between homologous structures and analogous structures. Homologous structures are those structures that share an anatomical similarity due to a common evolutionary origin; this is exemplified by the basic bone structure in the forelimb of all vertebrates such as the flipper of a whale and the arm of a human. Analogous structures are those that subserve similar functions despite differences in their anatomy, and thus have no common ancestry. The wing of an insect and the wing of a bird exemplify analogous structures.

The fossil record is an important clue to evolutionary origins. While fossils can actually be the preserved remnants of life (shells, skeletons), they also include imprints and molds, such as an animal footprint.

Fossils can be dated by the use of carbon-dating. The ratio of radioactive carbon (C^{14}) to non-radioactive carbon (C^{12}) is determined. This is based on the half-life of the element in question. The half-life of a radioactive element is the time span in which half the atoms in a sample will have decayed. The half-life for C^{14} (to decay to C^{12}) is 5730 years. This spontaneous rate of decay is a constant.

16.   (B)

In 1961, Melvin Calvin elucidated the series of reactions in which six carbon dioxide molecules are converted to one glucose molecule (a six-carbon sugar) using NADPH and ATP during the dark reactions of photosynthesis. The Calvin cycle occurs in the stroma (the semifluid matrix) of the chloroplast and in many ways is analogous to the Krebs cycle occurring in the mitrochondrial matrix.

In the Calvin cycle, a five-carbon sugar, ribulose bisphosphate is regenerated in each turn of the cycle, whereas in the Krebs cycle, oxaloacetic acid, a four-carbon molecule, is regenerated every turn. Sir Hans Krebs described the sequence in 1937. Synonyms for the Krebs cycle are Citric Acid cycle and Tricarboxylic acid cycle, both names referring to the intermediates of the cycle.

The Cori cycle refers to the biochemical sequence whereby glucose is converted into lactate in skeletal muscle. The lactate then moves into the liver, in which it can be reconverted to glucose, which then becomes available to the muscle.

Cyclic AMP (adenosine monophosphate) is a ubiquitous substance synthesized from the catalysis of ATP by the enzyme adenyl cyclase. It functions as the "second messenger" in the action of many hormones, such as glucagon and epinephrine.

The carbon cycle refers to all the interrelationships whereby carbon is cycled between different inorganic and organic molecules. Photosynthesis and cellular respiration are the two major aspects of this cycle: in photosynthesis, carbohydrate is produced from carbon dioxide, whereas the reverse occurs in cellular respiration, maintaining a steady equilibrium. Photosynthesis occurs in plant cells, but both plant and animal cells undergo cellular respiration. Other aspects to be considered in the carbon cycle are the burning of fossil fuels, and environmental conditions. Note that the carbon cycle does not merely refer to the fixation of carbon dioxide.

17.   (B)

A food pyramid is depicted below by the producer organisms at the base, supporting the consumers above:

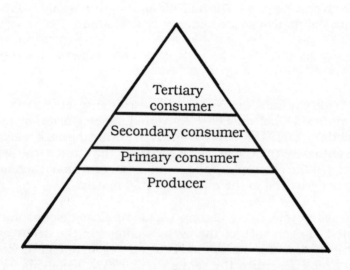

When one ascends the pyramid, the levels decrease in size regardless of whether one is concerned with number of individuals, biomass, or energy content. Therefore the producer, being at the basal level, is characterized by showing the greatest numbers, highest total biomass, and greatest energy content. It is generally accepted that 90% of the energy present in one level is lost as one ascends to the next trophic level; in other words, only 10% of the energy is transferred to the next level.

18.   (C)
The nitrogen cycle involves several distinct types of bacteria that metabolize the various forms of nitrogen. Nitrification refers to the conversion of ammonia to nitrate in two oxidation steps: the nitrite bacteria convert ammonia (NH_3) to nitrite (NO_2^-) and then nitrate bacteria convert the nitrite to nitrate (NO_3^-).

In the absence of oxygen, denitrifying bacteria reduce nitrate or nitrite to atmospheric nitrogen (N_2). Nitrogen-fixing bacteria live in the soil and convert atmospheric nitrogen to organic nitrogen molecules. Ammonification refers to the process whereby nitrogenous organic remains are decomposed by soil bacteria and fungi, which utilize the proteins and amino acids for themselves and release the excess nitrogen as ammonia.

Deamination is the initial step in the degradation of amino acids. The amino group (NH_2) is removed and converted to ammonia, which will be eliminated in the form of urea (H_2N-$\overset{\displaystyle ||}{\underset{\displaystyle O}{C}}$-$NH_2$).

19.   (B)
A tropism is a growth response. A positive phototropism is a growth towards while a negative tropism is a growth away from. Therefore, a positive phototropism is a growth response towards light; specifically, a plant bends towards a light source (either natural or artificial). It is due to a plant hormone called auxin, in this case, IAA (indoleacetic acid), the first of the plant hormones to be discovered. In the presence of light, auxin migrates to the dark side of the stem and causes elongation of the cells there, which causes the plant to bend towards the light. The stem curves toward the light and the flat leaf blades become perpendicular to the light to maximally expose their surfaces.

Florigen is a postulated plant hormone that may function in the flowering process. While temperature certainly does affect plant growth, it does not have specific effects on stem bending and does not explain the phototropic response.

20.   (E)

Circadian literally means "about a day," hence a circadian rhythm is a rhythm that cycles over a twenty-four hour period. Examples of these rhythms in humans can be as obvious as the sleep-wake cycle, or as obscure as the urinary excretion of ions. Other rhythms include fluctuations in body temperature and secretion of certain hormones. Metabolism of drugs may even follow a circadian rhythm.

Examples of circadian rhythms in the plant kingdom include the secretion of nectar from flowers. In certain leguminous plants, the orientation of the leaves, which lie horizontally in the daytime and yet more vertically at night, exemplifies a circadian rhythm.

21.   (A)

Vascular tissue in plants consists of xylem and phloem. The former transports water and minerals from the roots to the leaves, while the latter transports organic food substances.

The cell types of the xylem that conduct water are the vessel members and tracheids, both elongated and dead cells. In phloem, the sieve element cells function in food conduction with the aid of companion cells. Both these cell types are alive - sieve elements are anucleate, but do have a cytoplasm. It is proposed that the nucleus of the companion cell has control over the sieve element cell as well.

22.   (E)

The complete oxidation of glucose is basically the opposite process of photosynthesis. The reactions can be written as below:

$$6CO_2 + 6\ H_2O \overset{energy}{\Longleftrightarrow} 6\ O_2 + C_6H_{12}O_6 \text{ (glucose)}$$

in which photosynthesis is represented by the forward reaction and respiration by the reverse reaction.

Photosynthesis occurs in plants. In the presence of sunlight (energy), atmospheric CO_2, and water, plants can evolve oxygen and form glucose, which is stored as the polysaccharide carbohydrate starch.

Animals eat plant foods (starch) and respire: in the presence of oxygen, animals oxidize glucose completely to CO_2, and H_2O and form ATP (energy).

Thus, animals depend on plants for oxygen and plants depend on animals for carbon dioxide. Note that while animals do not photosynthesize, plants must respire.

23.   (C)
In photosynthesis, plant chloroplasts produce glucose and oxygen from carbon dioxide and water, in the presence of sunlight.

Glucose is a monosaccharide (simple sugar) that has many possible fates, depending upon the needs of the plant cell. Within the chloroplast, glucose can be polymerized into the polysaccharide starch. Alternatively, plant cells can respire: glucose can enter the cytoplasm and be oxidized/degraded to generate ATP. Cytoplasmic glucose can also participate in other chemical reactions, including the formation of the disaccharide sucrose (glucose + fructose) or other organic molecules.

Glycogen is the storage form of glucose in animal cells (liver and muscle cells). It is the analogue of plant starch, but is not found in plant cells.

24.   (B)
Enzymes are proteins that interact with specific substrates, and increase the rate of the chemical reaction which the substrate undergoes. While heat and increased concentration of substrate can increase the reaction rate by increasing the collision between molecules, the mechanism of action of enzymes is different.

All chemical reactions have an energy barrier that they must overcome in order for the reaction to occur. This is true of endothermic (energy-requiring) and exothermic (energy-releasing) reactions. This energy barrier is called the activation energy (E_a). It is analogous to lighting a fire: once lit, the fire will produce a lot of heat (energy), but you must first put energy in, i.e. light the match. An enzyme functions by lowering the energy of activation, and thus making it more likely that the colliding molecules will react, overcome the barrier, and form products. The mechanism of action of an enzyme is indicated below for both endothermic and exothermic reactions. Note that whether the reaction consumes or produces energy (ATP) will not affect the way an enzyme works. The decrease in the energy barrier in the presence of the enzyme is indicated by the dotted line.

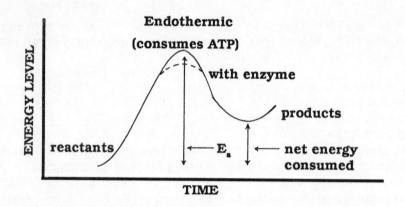

Endothermic
(consumes ATP)

ENERGY LEVEL

with enzyme

products

reactants

E_a

net energy
consumed

TIME

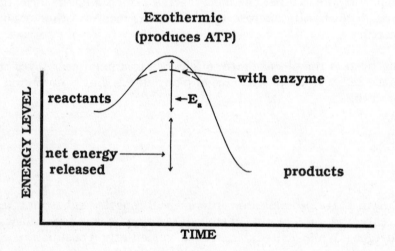

Exothermic
(produces ATP)

ENERGY LEVEL

with enzyme

reactants

E_a

net energy
released

products

TIME

25.   (C)

Phenylketonuria is an inherited disease in which the enzyme phenylalanine hydroxylase, which converts the amino acid phenylalanine to the amino acid tyrosine, is missing. Phenylalanine therefore is diverted to other usually insignificant metabolic pathways. These alternate metabolites develop to high levels in body fluids. Mental retardation and early death by age 20 are characteristic of the disease.

Like other inborn errors of metabolism, this disease is caused by an inherited lack of an enzyme (or synthesis of a deficient form of the enzyme), which leads to the accumulation of abnormal, or excessive levels of normal metabolites (metabolic intermediates). Symptoms may appear shortly after birth. There is no obvious "defect" at birth,

124

since the disease is only manifest upon ingestion of the amino acid. Furthermore, birth defects may be environmentally or medically induced, rather than genetically induced.

Disorders of endocrine glands often include overproduction or lowered production of a hormone. For instance, a tumor of a gland may cause excess secretion. Underproduction may be a result of atrophy of the gland.

Nondisjunction refers to the event whereby the chromosomes do not separate during meiosis: the resultant gamete therefore carries an extra chromosome, or lacks a chromosome.

26. (B)

Cells are classified as eukaryotic or prokaryotic. The former are characterized by a membrane-bound nucleus, while the latter do not have an organized nucleus. All living things are grouped into one of five kingdoms. Only kingdom Monera consists of prokaryotic organisms. Monera include the cyanobacteria (blue-green algae) and the bacteria. *E. coli (Escherichia coli)* is a bacterium.

The other four kingdoms include only eukaryotic organisms. Kingdom Animalia, the animals, include the human being, or *Homo sapiens*. An oak tree falls into kingdom Plantae. *Amoeba* is a member of the kingdom Protista.

The AIDS (Acquired Immune Deficiency Syndrome) virus is not classified here because viruses are not truly living organisms; they depend on a living host (plant, animal, bacterium) for their metabolic and reproductive mechanisms.

27. (A)

Arthropods are characterized by a lightweight yet protective chitinous external skeleton (exoskeleton) and jointed appendages. There are many classes of arthropods. For instance, class Arachnida includes spiders and scorpions. Class Crustacea includes crabs, lobsters, crayfish and shrimp. In Class Insecta, the organism typically has two pairs of wings on its thorax. The ability to fly is a key to the success of the insects.

The respiratory system of arthropods varies with the class. While the marine arthropods have gills, these would not function out of the water. Book lungs, especially prominent in spiders, function in gas exchange in terrestrial arthropods. However, the result of further evolution is displayed by tracheae, a branching network of tubes that brings air throughout the body. Tracheae can be noted particularly in insects.

Snails and slugs are not arthropods; they are in the Class Gastropoda of Phylum Mollusca.

28.   (C)
(See diagram). There are three sites along the electron transfer chain at which pairs of hydrogen ions are extruded (pumped) from the mitochondrial matrix into the intermembranal space. The hydrogen ions flow back into the matrix via special channel proteins that penetrate the inner mitochondrial membrane. ATP synthetases are on the ends of the channels facing the matrix; they sit like heads on the channel stalk. For each pair of hydrogen ions flowing through a protein channel due to the H^+ concentration gradient from the intermembrane space into the matrix, one ATP is produced.

Since NADH releases its electrons (and hydrogen ions) at the beginning of the electron transport chain, it causes three pairs of hydrogen ions to be extruded - one at each aforementioned site. Therefore, three pairs of hydrogen ions will re-enter the matrix and cause the formation of three ATP's. $FADH_2$ releases its electrons (and hydrogen ions) at a later site, bypassing the first site. Therefore only two pairs of electrons will re-enter the matrix and thus two ATPs will be formed.

Many of the enzymes that participate in electron transfer are cytochromes. Their active portions are iron-sulfur groups in which inorganic iron (Fe) is either in the ferrous, reduced (Fe^{+2}), or ferric, oxidized (Fe^{+3}), state.

NAD+ and FAD are coenzymes in the dehydrogenase enzymes, which predominate in the Krebs cycle. These coenzymes are reduced and pass their electrons to the electron transfer chain.

The term cofactor as opposed to coenzyme is merely a function of its chemical nature. A cofactor is inorganic, e.g. Fe, while a coenzyme is organic, e.g. NAD^+.

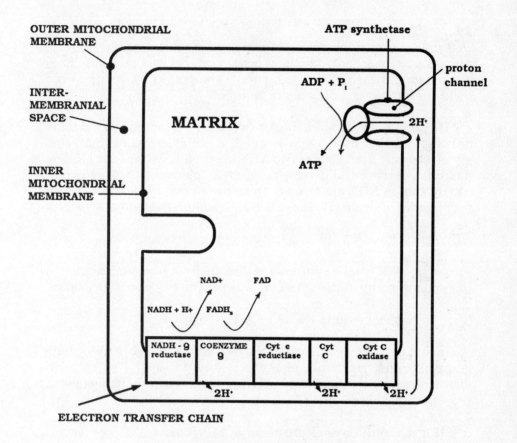

MITOCHONDRION
(SCHEMATIC REPRESENTATION)

29.   (D)

ATP (adenosine triphosphate) is a nucleotide consisting of the purine nitrogenous base adenine, the pentose sugar ribose, and three phosphate groups. It is a high-energy compound by virtue of the bonds of the last two phosphate groups.

ATP can be synthesized under aerobic or anaerobic conditions although aerobic production is much more efficient. In anaerobic glycolysis, only a net yield of two ATPs are produced from one glucose molecule. However, in aerobic respiration, glucose is oxidized completely and 38 ATPs are formed. Oxidative phosphorylation refers to the coupling of electron transfer to the phosphorylation of ADP to form

ATP (ADP + P_i → ATP, in which P_i is inorganic phosphate).

However, ATP can be synthesized directly from phosphocreatine, a high-energy compound in skeletal muscle. The reaction is as follows:

Phosphocreatine + ADP ↔ ATP + creatine.

Note that in this reaction, catalyzed by creatine kinase, a phosphate group is transferred from phosphocreatine to ADP. This reaction allows rapid, albeit short-lived, ATP production before the metabolic processes discussed above kick in.

cAMP (cyclic AMP) is not a precursor of ATP. Rather, ATP is converted to cAMP in certain cells.

30.   (B)

Protein hormones utilize the same mechanism of action as the amino-acid-derivative hormone epinephrine. The mechanism is described by the second messenger theory of hormone action. The hormone, illustrated by ⌐⌐⌐ , binds to a membrane receptor, illustrated by ⌐⌐⌐ . This hormone-receptor complex activates an enzyme located on the inner surface of the plasma membrane. This enzyme, adenyl cyclase, represented by ●, catalyzes the conversion of ATP into cyclic adenosine-3'-5'-monophosphate abbreviated cAMP. cAMP is the second messenger (the hormone was the first messenger) that will activate a cascade of intracellular reactions, which vary with the hormone.

31.   (C)

Chemically, hormones are divided into three groups: proteins, steroids, and derivatives of the amino acid tyrosine. The steroid hormones include those of the adrenal cortex (i.e. aldosterone, corticosterone) and the sex steroids produced primarily in the gonads (i.e. testosterone, progesterone). The amino acid derivatives include epinephrine and the thyroid hormones. All other hormones are proteins. This includes all the hormones of the pituitary gland (i.e. growth hormone, prolactin, oxytocin, etc.), pancreas (glucagon and insulin), and many more.

Steroid hormones are lipid-soluble and diffuse into the cell. They bind to a cytoplasmic receptor. The hormone-receptor complex moves into the nucleus, where it can activate particular genes. Steroid hormones have no need for a second-messenger system.

Protein hormones cannot diffuse through the lipid membrane and thus must activate a membrane receptor facing the extracellular fluid. This receptor ultimately stimulates the cAMP system represented in the diagram.

While glucose and antibodies each have membrane receptors, they do not activate adenyl cyclase and ultimately the cAMP system. Receptors are quite specific, as are their mechanisms of activation.

An operator is a gene that regulates the transcription of particular genes.

32.   (A)

Symbiosis, which literally means "living together," refers to an association between organisms of distinct species. When both organisms benefit from the association, the symbiosis is called mutualism. An extreme example of this is a lichen, which, through evolutionary time, is now a single organism. A lichen is a combination of a fungus and an alga. The alga supplies the photosynthetically produced food for the fungus while the fungus provides water and minerals, and a mechanical support for the alga.

In a commensalistic symbiosis, one species benefits, while the other is neither benefitted nor harmed. Epiphytes grow in the branches of trees to maximize light exposure, with no ill effect on the tree.

In parasitism, one species benefits at the expense of the other species, designated the host. Parasites may live on or within the host's body. A hookworm (phylum Nematoda) bores through human skin and eventually comes to live in the intestine, causing diarrhea and anemia in the host.

Members of a predatory species kill and consume other organisms in order to survive.

Competition occurs between organisms that share a limited resource in the environment. It can be intraspecific as well as interspecific.

33.   (E)
Konrad Lorenz is a noted ethologist (one who studied behavior). He described the phenomenon of imprinting, a type of learning behavior pattern in which birds (e.g.: ducks, geese, chickens) form a strong attachment with whomever they are first exposed to in a critical time period shortly after hatching.

Under normal conditions, the bird first sees its mother and thus follows her around; the bird also thus learns to associate and mate with its own species. However, in his experiment, Lorenz had newly hatched ducks exposed to himself. The ducklings followed him as if he were their mother.

34.   (E)
The placenta is an organ formed from maternal and fetal tissue (endometrium and chorion, respectively), each retaining its own blood supply. The placenta is continuous with the umbilical cord. Nutrients, including oxygen, diffuse from the maternal bloodstream into the fetal bloodstream via the placenta; wastes, including carbon dioxide, diffuse in the opposite direction. Note that at no time do the maternal and fetal bloodstreams actually mix.

The aorta is the largest artery in the body. It carries oxygenated blood away from the left ventricle of the heart.

The amnion, one of the extraembryonic membranes, is a fluid-filled sac that surrounds the embryo. The fluid within keeps the embryo hydrated, decreases friction, and acts as a shock absorber. Genetic information can be determined by sampling this amniotic fluid.

The morula and blastula both refer to stages of cleavage. A morula is a solid cluster of cells that the egg has become by the time it reaches the uterus. The morula develops into a hollow ball of cells called the blastula.

35.   (B)

Taxonomy is the classification of organisms and is based on categorizing similar organisms. The highest or broadest classification scheme is the kingdom. All living organisms are placed within one of the five kingdoms - Monera, Protista, Fungi, Plantae or Animalia. The subsequent subdivisions are phylum, class, order, family, genus and species. There may also be subdivisions (e.g. subphyla) and super-divisions (e.g. superclasses).

The last two names of the classification form the binomial nomen-clature (a two-worded Latin name) that is unique for each species. The binomial nomenclature is underlined or italicized; the genus name is capitalized, while the species name is not.

The complete classification for man is as follows:

Kingdom	Animalia
Phylum	Chordata
Subphylum	Vertebrata
Class	Mammalia
Order	Primate
Family	Hominidae
Genus	Homo
Species	*Homo sapiens*

Thus *Homo sapiens* is the binomial nomenclature for the human being.

36.   (A)

A longitudinal section of a root shows four zones. The root cap forms the tip of the root. The cells here protect the next layer above as the root penetrates the soil. That next layer is the zone of cell division, or the meristematic region, where the cells undergo mitosis. Above this is the zone of cell elongation, which is the primary provision of root growth. Finally, in the zone of maturation, the cells differentiate into specialized cells such as the sieve-tube members and the vessel members (conducting cells of the phloem and xylem, respectively).

37. (A)

Since chloroplasts are the photosynthetic organelles, it is important to identify those cells/structures in the leaf that contain chloroplasts. The middle section of the leaf, the mesophyll, contains the photosynthetic cells. These parenchymal cells are divided into two layers - the upper palisade layer and the lower spongy layer. The former is responsible for most of the photosynthesis although the latter contribute. The upper and lower layers of the leaf consists of epidermal cells. Stomata, or openings, penetrate the epidermis, in particular the lower epidermis. The pore sizes are regulated by two guard cells, which are specialized epidermal cells. The guard cells contain chloroplasts.

The epidermal cells secrete a waxy substance called cutin, which forms the cuticle. The cuticle forms the outer coating of the leaf. While it allows light to penetrate to the photosynthetic cells within the leaf, it does not participate in photosynthesis, since it contains no chloroplasts.

38. (D)

Class Aves, within the phylum Chordata, refers to birds. Birds' unique feature is that they have feathers. They are homeotherms, or warm-blooded animals; i.e. like mammals, they maintain a constant and high body temperature. This is in contrast to reptiles, which are poikilothermic: these cold blooded animals have a body temperature that varies with the ambient temperature. As an adaption for flying, birds have air sacs and hollow bones, and thus do not weigh very much. Birds use song as a behavioral mechanism to identify their own species and territory, and for courtship practices.

Like terrestrial reptiles and insects, birds excrete uric acid crystals as the final waste product of nitrogen metabolism. In contrast, mammals excrete their nitrogen wastes as urea. Uric acid is less soluble than urea and thus little water is lost by its excretion.

39. (B)

The nephron is the anatomical and functional unit of the kidney. It produces the urine. A nephron is a renal tubule and its vascular component (see figure 1). The renal tubule consists of Bowman's capsule, the proximal convoluted tubule, the loop of Henle (with descending and ascending limbs), the distal convoluted tubule, and the collecting duct. The glomerular capillaries filter fluid (derived from the plasma) into Bowman's capsule.

The longer the loop of Henle, the greater the capacity for water reabsorption and hence elimination of a concentrated urine. The mechanism that allows this is called the countercurrent multiplier. Chloride ion is actively pumped from the thick ascending limb of the loop into the medullary interstitium. Sodium ion passively follows. However, this limb is impermeable to water; hence the urinary fluid becomes diluted as it ascends. When urine travels through the collecting duct (which is permeable to water), water will diffuse osmotically into the interstitium (the permeability of the collecting tubule to water is under the influence of a hormone called ADH - antidiuretic hormone). Urea thus becomes concentrated in the lower collecting duct as water is reabsorbed, and thus urea diffuses into the medullary interstitium. This increases the concentration of urea in the medullary interstitium.

Furthermore, the descending limb of the loop is selectively permeable to water and thus water is reabsorbed at this site too. Since a longer loop can create a greater concentration of sodium ion in the deep medulla, which then causes the movement of water out of the collecting tubule, it favors osmotic water reabsorption.

One adaption of desert animals is that they have long loops of Henle, specialized for water reabsorption and hence conservation. By the reabsorption of water, they can produce very little but very concentrated urine - far more concentrated than the plasma originally filtered through the glomerulus. A kangaroo rat is so well adapted, that all its water needs are met by that produced in its metabolism. No terrestrial animals show osmotic water influx from the environment.

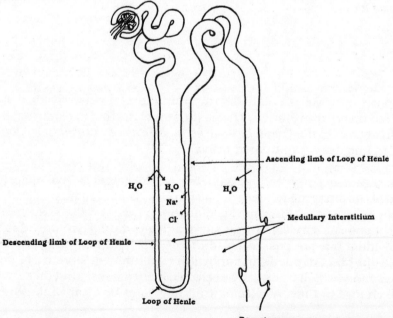

Ascending limb of Loop of Henle

H_2O H_2O H_2O
Na⁺
Cl⁻

Medullary Interstitium

Descending limb of Loop of Henle

Loop of Henle

Countercurrent mechanism of the nephron

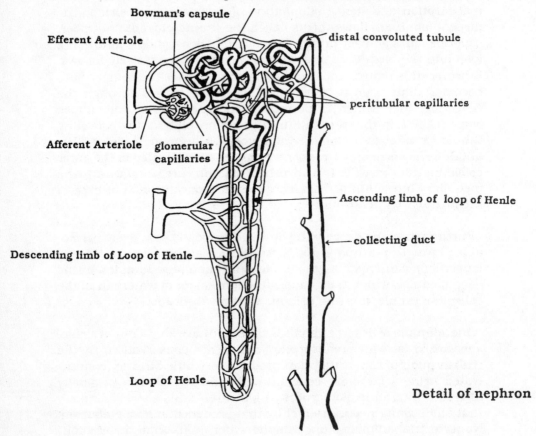

proximal convoluted tubule

Bowman's capsule

Efferent Arteriole

distal convoluted tubule

peritubular capillaries

Afferent Arteriole

glomerular capillaries

Ascending limb of loop of Henle

collecting duct

Descending limb of Loop of Henle

Loop of Henle

Detail of nephron

40.   (E)

Charles Darwin is credited with formulating the most widely supported theory of evolution. The postulates of his theory came together in his book *On the Origin of Species by Means of Natural Selection* in 1859, and are recapitulated below.

All organisms overproduce gametes. Not all gametes form offspring, and of the offspring formed, not all survive. Those organisms that are most competitive (in various different aspects) will have greater likelihoods of survival. These survival traits vary from individual to individual but are passed on to the next generation, and thus over time, the best adaptions for survival are maintained. The environment determines which traits will be selected for or against; and these traits will change in time. A selected trait may later be disadvantageous.

The key drawback to Darwin's theory is that he did not suggest the key to variation in traits. It is now known that variation may be due to genetic mutations, gene flow due to migration, genetic drift, especially in small populations, and natural selection of genotypes, i.e. a differential ability to survive and/or reproduce.

41.   (B)

Hemophilia is a sex-linked recessive disease. Like color-blindness, the gene for hemophilia, h, is carried on the X-chromosome. If a male inherits the gene, he will have the genotype X^hY and will be a hemophiliac (a normal male is X^HY) since the recessive gene will be expressed. If a female inherits the gene, she will have the genotype X^HX^h and will carry the trait since her other X chromosome has the normal dominant gene, H.

The common pattern of transmittal is from carrier mothers to their sons. Note that a carrier mother X^HX^h has an equal (50%) chance of passing the gene on to either a son (X^hY) or a daughter (X^HX^h); however, the daughter will not express the disease. It is unlikely for a female to be a hemophiliac, X^hX^h, since she must have acquired the recessive gene from both her carrier mother and her hemophiliac father. However, this is possible, and as expected, the incidence increases when there is marriage between relatives.

If the gene were Y-linked, then a diseased father would always produce a hemophiliac son. However, a son inherits the gene only from his mother, since the mother contributes his sole X-chromosome.

42.   (C)

In the food pyramid presented, the producers are the algae and protozoa. The herbivorous insects consume the algae. DDT is a potent insecticide. DDT is a nondegradable compound that accumulates in fatty tissue. As any nondegradable substance ascends the trophic levels, its concentration in each organism of a higher trophic level will increase. Frogs may feed on the poisoned insects; one insect would carry very little DDT, but an entire meal of insects would contain a rather large amount of DDT - - all of which would remain in the frog. If enough frogs consumed the affected insects, and if a bird were to consume enough affected frogs, then the amount of DDT present in the bird could be enough to harm it, perhaps fatally.

43.   (D)

Vitamin D is a fat-soluble vitamin that is chemically similar to steroid compounds. Cholesterol is the precursor to the steroid hormones and vitamin D. Vitamin D is produced in the skin in the presence of ultraviolet (UV) light. Its synthesis requires enzymes in the kidneys and the lungs. Good sources of vitamin D include dairy products, such as milk, and fish liver oils. Among the functions attributed to vitamin D are an increase in calcium absorption from the gastrointestinal tract and the incorporation of calcium and phosphate into bones. Thus, a vitamin D deficiency manifests itself in bone problems. Children show rickets, malformed bones; adults show osteomalacia, a softening of the bones.

Night blindness is due to vitamin A deficiency, since vitamin A is a constituent of rhodopsin, a visual pigment.

44.   (D)

In incomplete dominance, the "dominant" allele cannot completely mask the expression of the recessive allele. While the result of a cross between a dominant allele and a recessive allele may appear to be a blended trait, in future generations the dominant and recessive allele can each be independently expressed (i.e., the original traits will re-emerge); hence, no blending has occurred.

In certain flowers, color is inherited by incomplete dominance. The Punnett square for the cross between the red and white flower is shown below:

RED x WHITE
RR x   rr

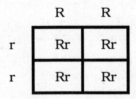

One hundred percent pink flowers (Rr) are produced. The Punnett square for a cross between two pink flowers is shown as follows:

PINK   x   PINK
Rr   x   Rr

	R	r
R	RR	Rr
r	Rr	rr

The expected probabilities of phenotypic expression are 25% red (RR), 50% pink (Rr), and 25% white (rr).

45.   (C)

Nucleic acids are composed of polymers of nucleotides, each nucleotide consisting of a pentose sugar (five-carbon sugar), a phosphate group, and a nitrogenous base. DNA and RNA both contain the phosphate group (Figure B in the diagram). The only difference between DNA and RNA is the pentose sugar and the types of nitrogenous bases. The pentose sugars are deoxyribose (Figure C) in DNA, and ribose (Figure A) in RNA. DNA and RNA utilize both of the purine bases (double-ringed heterocyclic bases), adenine and guanine. They both also use the pyrimidine base (single-ringed heterocyclic base) cytosine. However, thymine (Figure F) is specific to DNA, while uracil (Figure D) is specific to RNA.

46.   (D)

The brains of all vertebrates are divided into the hindbrain, midbrain, and forebrain. In lower vertebrates, such as fish, the hindbrain is the dominant portion of the brain. It consists of the medulla oblongata and pons. The former deals with vital reflexes, such as cardiac activity, and it links the spinal cord to the rest of the brain. The latter contains the respiratory center, and like the former, it is the origin of many cranial nerves. The cerebellum is also part of the hindbrain: it functions in equilibrium and proprioception (awareness of body/limb position and movement).

The midbrains of fish process visual information and their forebrains function in olfactory (smell) sensation. These sensory functions are attributed to the cerebrum in higher vertebrates.

The midbrain is relatively small in humans, and, as the origin of several cranial nerves, it controls eye movements and pupillary size.

In all vertebrates, the forebrain consists of the diencephalon and the cerebrum. The diencephalon consists of the thalamus, a relay center for sensory input en route to the cerebrum, and the hypothalamus, which regulates many activities, including circadian rhythms, body temperature, emotions, food intake, and some hormone secretion. The cerebrum is highly developed in mammals: it constitutes 7/8 of the human brain. The cerebrum functions in learning, association, and memory. Birds also show great development of the cerebrum.

47.   (E)

Cells are broadly classified as prokaryotic ("before the nucleus") or eukaryotic ("true nucleus"). Kingdom Monera consists of organisms with prokaryotic cells, while the other four kingdoms (plant, animal, protista, and fungi) are characterized by eukaryotic organisms. The primary distinction between the two cell types is that prokaryotic cells do not have a membrane-bound nucleus, hence their name. Their DNA is contained within a nuclear region called a nucleoid.

Like eukaryotes, prokaryotes have a plasma membrane, which separates the cell from its external environment. All prokaryotic cells are further surrounded by a cell wall, while this is true of only certain eukaryotic cells (plant and fungal). They also have cytoplasmic ribosomes, which function in protein synthesis.

However, the eukaryotic cell has many other organelles, forming subcompartments in the cytoplasm. These organelles carry out specific functions. Two examples are the mitrocondrion, which functions in oxygen utilization and energy production, and the lysosome containing hydrolytic enzymes, which function in degradation and disposal of wastes.

48.   (B)

Angiosperms are flowering plants. There are two large subdivisions of them: The monocots (class Monocotyledoneae) and the dicots (class Dicotyledonae). Characteristics of the monocots (versus those of the dicots) follow: They have (1) one cotyledon (seed leaf) per seed (versus two); (2) flower parts occur in groups of three or multiples thereof (versus groups of four and five and multiples thereof); (3) parallel leaf veins (versus a netted pattern); and (4) "scattered" vascular bundles in the stem (versus forming a ring around the central pith). Examples of monocots are grasses, lilies, irises, and orchids. Herbs and many shrubs are dicots.

49.   (C)

In plants and some green algae, the organism's life cycle is characterized by an alternation of generations in which a diploid, spore-producing generation alternates with a haploid gamete-producing generation. The former is called the sporophyte generation while the latter is the gametophyte generation. Thus a gametophyte plant is composed of haploid cells. It undergoes mitosis to produce haploid gametes. Two gametes fuse to form the zygote, the first diploid cell of the new generation. The zygote develops into the sporophyte plant. The sporophyte consists of diploid cells. It undergoes meiosis to produce haploid spores, which may disperse. A spore need not fuse with another to develop into a new organism.

Depending upon the species, the two generations may appear similar or quite dissimilar, yet one generation is often the conspicuous one. In the evolution of plants, the sporophyte generation has become more dominant.

50.   (C)
The two reactions depicted occur simultaneously. ATP is converted (dephosphorylated) to ADP, while glucose is phosphorylated to glucose-6-phosphate. The phosphate group of ATP is transferred to glucose. This is the first step of glycolysis, which occurs in the cytoplasm. Since the phosphorylation of glucose requires energy, it is coupled to the breakdown of ATP. It is catalyzed by a hexokinase.

Glucose-6-phosphate is then isomerized to fructose-6-phosphate. Then phosphofructokinase (PFK) catalyzes the next energy requiring reaction of glycolysis in which the fructose-6-phosphate is phosphorylated to fructose-1,6-diphosphate. Phosphofructokinase is an important enzyme, as it is a chief site of regulation of glycolytic activity.

51.   (E)
Alveoli are tiny thin-walled air sacs within the lungs. There are widespread sheets of capillaries that cover the alveoli. The millions of alveoli in each lung increase the surface area through which gases can diffuse between air and blood.

The small intestine has many adaptions that increase its surface area for both digestion and absorption (transport of nutrients from the lumen into the blood or lymph). Firstly, the small intestine is very long (about 12 feet); secondly, the inner layer, or mucosa, is folded into villi that project into the lumen. Microvilli, which are folds of the epithelial cells composing the villi, further increase the surface area. Microvilli are also referred to as the brushborder. There are brushborder enzymes that complete chemical hydrolysis of nutrient molecules. Then the small monomers formed are absorbed.

The inner mitochondrial membrane has enzymes that function in the electron transport chain or as ATP synthetases. This inner membrane is folded into projections called cristae, which increase the surface area for these chemical reactions that sustain life.

Cilia are often considered to be cell projections that push substances past the cell. The hair cells in the cochlea of the inner ear are actually cilia. However, they are mechanoreceptors which, when distorted, will signal an auditory sensation. They thus have a neural function and do not function in chemical or transport reactions.

52. (C)

All vertebrate brains have three primary divisions: the prosencephalon is the forebrain; the mesencephalon is the midbrain; and the rhombencephalon is the hindbrain.

The rhombencephalon develops into the myelencephalon (medulla oblongata) and the metencephalon (pons and cerebellum). The medulla oblongata and pons cooperate in the regulation of breathing. They are also the origin of many of the cranial nerves. The cerebellum deals with reflexes and equilibrium.

The mesencephalon is simply the midbrain; while it is a major association site in lower vertebrates such as fish, it is reduced in size and function in higher vertebrates. However, it maintains communication with the eyeball.

The prosencephalon develops into the diencephalon (thalamus and hypothalamus) and telencephalon (cerebrum). The thalamus functions as a relay center for sensory stimuli. The hypothalamus controls the pituitary gland; it also controls visceral activity via the autonomic nervous system. The cerebrum is highly developed in higher vertebrates such as birds and mammals; it functions in learning and other associative functions.

53. (E)

Desert plants show many adaptations to their hot and arid environment. Plant life is relatively sparse, minimizing competition for water for widely spaced plants. The plants often have roots that penetrate deep and reach the groundwater. The leaves may be deciduous (able to fall off) during particularly stressful periods. The coating, or cuticle, of the plant leaf is often thick and waxy, thus functioning in water retention. Finally, water loss through the stomata (the openings between guard cells) is minimized by keeping the stomata hidden deep within the leaves.

54.   (B)

There are many types of stems that have specialized functions. The functions of stems generally may include transport of water and food between root and leaves, leaf support, and food storage.

Tendrils are modified stems that assist climbing plants such as the grapevine. Underground stems include tubers, rhizomes and corms. Tubers, found in the potato, function to store starch. Rhizomes, in ferns, function in vegetative propagation. Corms are found in gladiolus and are actually fleshy leaves that store food.

Nodes are not stems, but rather are the site on the stem at which the leaves attach. Internodes are thus the region between nodes.

55.   (B)

In the 1800's, two major theories of evolution were proposed. In 1800, Jean Baptiste Lamarck proposed a theory based on the inheritance of acquired characteristics. In other words, organs, and therefore animals, evolve through use. His classic example was his explanation for why the giraffe had such a long neck: He said that giraffes stretch their necks to reach the leaves high in trees. He assumed that an animal that stretched its neck could pass "a stretched and hence lengthened neck" on to its offspring.

Lamarck's theory was incorrect, since characteristics acquired in life cannot be passed on to the next generation, since the information is not in the genes. For instance, if a man develops his muscles by lifting weights, his offspring would not therefore be muscular as well.

In the 1830's, Charles Darwin started collecting information that would lead him to propose his theory of evolution based on natural selection. The basic premise of his theory is that individuals have differing capacities to cope with their environment. Some individuals have characteristics which are advantageous in the environment and hence, those individuals will tend to survive and reproduce.

According to Darwin's theory, a giraffe has a long neck because those giraffes that by chance had the selective advantage (long necks) would eat food in the trees, and hence survive and reproduce. Their offspring would be like the parents and hence have long necks. Giraffes with short necks would be selected against and not be able to survive and reproduce other short-necked giraffes.

Despite all of Darwin's insight, he never could explain the mechanism whereby traits were passed on. At about the same time, although unknown to Darwin, Gregor Mendel was experimenting with garden peas and was the first to introduce the concept of genes (the heritable factors), although he did not use that term.

Albinism is a genetic disease in which there is no production of melanin. Note that while many genetic diseases may be obvious at birth, others become manifest later in life. This does not mean that the disease is an acquired characteristic, for the genetic predisposition to develop the disease is present at birth, regardless of the age of manifestation of an inherited disease.

56.   (D)
Embryonic induction refers to the interaction between adjacent cells whereby one group of cells can determine the morphological development of a neighboring group. It is believed that this induction is due to the production and release of chemicals.

The classic example of induction is the formation of the vertebrate lens. When the optic vesicle contacts the overlying epidermis, it induces the formation of a lens from the epidermal tissue. The formation of the neural tube (neurulation) is yet another example of embryonic induction. The interaction occurs between the dorsal ectoderm and dorsal mesoderm.

Cleavage is the series of cell divisions whereby a zygote (fertilized egg) divides mitotically into two cells: the cells continue to divide into two cells until a hollow ball of cells called a blastula is formed.

57.   (C)
All plant life cycles exhibit an alternation of generations. The diploid generation is the sporophyte generation. It is characterized by a diploid (2n) number of chromosomes. The sporophyte plant divides by meiosis to produce spores that are haploid and hence the first cells of the new haploid generation. The spore divides mitotically to form the haploid gametophyte plant, which only has one of each pair of chromosomes. The gametophyte divides mitotically to produce gametes (egg and sperm cells). Haploid gametes of opposite sex fuse to produce a diploid zygote. This marks the beginning of the new diploid generation. This zygote grows into a sporophyte plant and the cycle begins anew.

58.   (C)

The Hardy-Weinberg equilibrium refers to a condition in which the proportion of alleles at a given locus remain constant. Four factors are required to maintain the equilibrium: large population size, no mutation, no migration, and no natural selection.

The best way to see why these factors are required is to study how evolution (and hence deviation from the equilibrium) may occur when the criteria are not met.

In a small population, chance alone may cause a shift in allele frequency. This random fluctuation is called genetic drift.

A mutation is a change in the DNA and as such is passed on to future generations.

Migration can be immigration (entrance) or emigration (exit) of individuals (and their genes!) to or from a population. This change in allele frequency due to migration is called gene flow.

Natural selection and differential reproduction are basically synonymous. This is the key factor of evolutionary change. Each individual of the population has a unique genotype (except identical twins, which have a common genotype) responsible for determining its phenotype. The environmental circumstances at the time determine whether that phenotype is at a survival advantage. Those individuals with the phenotypes that can cope best in the environment will differentially reproduce. Their genes will increase in frequency in the next generations.

59.   (C)

Human embryonic development takes nine months; these months are divided into three trimesters of three months each.

In the first trimester, cleavage and implantation occur within the first week. The embryonic membranes begin to develop, followed by gastrulation (the differentiation of the three primary cell layers: ectoderm, mesoderm, and endoderm) and neurulation (formation of the neural tube). By the end of the first month, organ development has begun - these organs include the eyes, heart, limb buds, and most other organs. In the second month, morphogenesis occurs. Morphogenesis refers to the development of form or structure. In short, the first trimester is a critical period of differentiation and development, but growth is not pronounced.

The second trimester is a period of rapid growth in size and weight. The mother may become aware of kicking in the baby. Growth continues and is most prominent in the final trimester.

60.   (D)
Seed-producing plants are evolutionarily advanced because their reproductive mechanisms do not depend on an aqueous medium in which their gametes must travel. The seed-producing plants include the gymnosperms (cone-bearers) and the angiosperms (the flowering plants). Seeds of gymnosperms are exposed, although embedded in cones. Thus the gymnosperms are referred to as "naked seed" plants, due to lack of protection afforded the seed. However, angiosperms have seeds that are well-protected within fruits.

All gymnosperms bear cones, as the cones are the reproductive structure - the site of seed production. Cones can be either male or female, and may or may not be found on the same plant.

61. (A)   62. (C)   63. (B)
The diagram illustrates semiconservative replication of DNA. The 3' and 5' illustrate the location of the phosphate group. The hydrogen bonds between the bases (labelled C in diagram) are broken, and the double-stranded DNA molecule begins to unwind. The replication fork (labelled B in the diagram) marks the area from which the single strands emerge. Each strand then serves as a template for DNA synthesis. The DNA polymerase enzyme reads the template in the 3' to 5' direction and hence synthesizes the new chain only in the 5' to 3' direction, since the two strands of the double helix are antiparallel (i.e. they run in opposite directions). Therefore, only one strand, designated the leading strand (labelled D in the diagram) will show continuous synthesis as the replication fork moves towards the DNA molecule's 5' end. The lagging strand (labelled A in the diagram) shows discontinuous synthesis. Nucleotides are added in short bursts in the appropriate 5' to 3' direction. Later these short polymers of nucleotides will be connected to form one strand.

64. (C)   65. (B)   66. (D)   67. (A)
Transcription is the nuclear process in which one strand of DNA dictates a complementary sequence of nucleotides in RNA. DNA unwinds in a particular region to expose particular genes and transcribe the necessary RNA. The DNA strand that is at this site is the template strand, while the other DNA strand is the anti-template strand. Note that the RNA transcript formed will contain the same sequence of bases as the anti-template strand, since it, too, had complementary base-pairing.

In plants, the term translocation refers to the movement of the organic products (such as sucrose) of photosynthesis from a leaf to other parts of the plant. This food is conducted through the phloem, the vascular tissue that consists of sieve tubes and companion cells.

Translation is the cytosolic process whereby a strand of messenger RNA dictates the amino acids that will be incorporated into protein. The messenger RNA has codons (triplets of nucleotides), which form base pairs with the anticodons on transfer RNA. Transfer RNAs carry the amino acids to the messenger RNA strand. Protein synthesis occurs on ribosomes, which are cytoplasmic organelles consisting of ribosomal RNA and protein. Hence all three types of RNA (messenger, transfer, and ribosomal) are involved in translation.

Transpiration is the loss of water vapor from plant stems and leaves. It occurs primarily through the stomata. This evaporated water loss may create a tension that will pull water upward from the root through the stem.

Transfusion is the direct infusion of blood into the bloodstream.

68. (E)   69. (C)   70. (B)

The cell cycle consists of, in continuous sequence: mitosis, in which the nucleus divides, cytokinesis, in which the cytoplasm divides, and interphase. Interphase is the longest period of the three and it is in this period that DNA is replicated. It consists of three distinct stages: G_1, a gap before DNA replicates, S, the stage of actual synthesis of DNA, and G_2, the gap after replication and prior to mitosis. It is the variable time spent in G_1 that is responsible for the widely variable times for the cell cycles of different cells.

Mitosis only occupies a small time segment of the total cell cycle. It consists of four stages. In prophase, the chromosomes condense into distinct rodlike structures. By metaphase, the nuclear membrane and nucleolus have disappeared completely. The chromosomes move to and align themselves along the spindle equator. In anaphase, sister chromatids of each replicated chromosome separate to each form a complete chromosome. Their migration to opposite poles of the cell seems to be based on the action of microtubules. Telophase is basically the opposite of prophase - the chromosomes decondense into a threadlike form. A new nuclear membrane is formed.

In the strict sense, mitosis only refers to division of the nucleus. Cytokinesis, or cytoplasmic division, usually accompanies mitosis. It begins in late anaphase or in telophase and is marked by the formation of a depression called a cleavage furrow. At this site, the plasma membrane is pulled inward by a ring of microfilaments. The continual contraction of the microfilaments cuts the cytoplasm into usually equal halves. In plants, cytokinesis occurs via the formation of a cell plate. Now interphase will start anew.

71. (D)   72. (E)   73. (A)   74. (B)

While mitosis maintains the chromosome number, meiosis functions to halve the chromosome number. In meiosis, there are two sequential nuclear divisions (Meiosis I and Meiosis II). DNA duplication only precedes the first meiotic division; the interphase between the two divisions is short or absent; no duplication occurs.

Due to replication, there are two sister chromatids/chromosome. These chromatids are linked at the centromere, a constricted portion of the chromosome. In the first meiotic division (Anaphase I), the homologous chromosomes separate. These are the paired chromosomes that code for the same type of genetic trait. In the second meiotic division (Anaphase II), the sister chromatids of each chromosome separate.

Prophase I includes several important events that ultimately contribute to the uniqueness of each individual. Homologous chromosomes pair closely in a process called synapsis. The resultant four linked chromatids are called a tetrad. Non-sister chromatids in the tetrad can undergo a process called crossing over: they exchange segments of the homologous chromosomes at a site called the chiasma. Thus an original paternal and original maternal gene will end up in the same chromosome and hence in the same gamete. Recombination refers to this new combination of alleles. Note that there is no reason for sister chromatids to undergo crossing over, since they carry the same genetic material.

75. (A)   76. (D)

A biome is a broad yet distinct vegetational formation. Each biome is associated with particular fauna as well. A classification of biomes usually includes the following:

The tundra is a treeless region with long cold winters such as those that occur in the Arctic. A taiga is a coniferous forest, characterized by a preponderance of evergreens such as spruce and fir trees. The

146

winters are long and harsh. A desert is a region of little rainfall, hot days, and cool nights. Plants, such as cacti, have adaptations to obtain and retain water. A tropical rainforest has much rainfall, high humidity, and high temperatures. There is little variation in this pattern throughout the year. The trees are tall and the vegetation is thick. A deciduous forest has alternating warm and cold seasons. The leaves fall (hence the term deciduous) in the winter. Examples of deciduous trees are oak, maple, and beech trees. The final biome, the grasslands, shows irregular rainfall. Grazing animals are numerous.

77. (A)   78. (E)   79. (D)

A mechanism that causes reproductive isolation is one that prevents interbreeding (breeding between different species). There are many types of these mechanisms, whose effects may manifest themselves before or after mating. That is, they may prevent the formation of the zygote or they may prevent the development of the zygote or the fertility of the organism.

Mechanical isolation refers to differences in the anatomy (i.e., size and length) or physiology of the reproductive organs such that mating is impossible.

Gamete isolation refers to an incompatibility (perhaps biochemically based) between the gametes of two different species.

Hybrid inviability is displayed if a hybrid embryo fails to develop, or, in the best of circumstances, develops normally, except for the fact that its reproductive system is nonfunctional. This is exemplified best by the sterile mule — a cross between a female horse and a male donkey.

Temporal isolation refers to a mechanism concerned with time. The mating seasons of two different species may not correspond.

Behavioral isolation involves courtship rituals, such as song and dance, which may be specific to each species.

80. (A)   81. (E)   82. (C)

Phylum Porifera, the sponges, consists of freshwater and marine animals. They are sessile filter feeders, and acquire food through their many pores. Needlelike spicules of calcium or fibers of spongin protein serve as supportive structures.

147

Members of phylum Arthropoda are characterized by jointed appendages and chitinous exoskeletons. Arthropods undergo molting to reform their exoskeletons. Examples include spiders, lobsters, and all insects.

Echinoderms are marine animals such as sea stars and sea cucumbers. They have radially symmetrical bodies. Echinoderms, as well as chordates, are classified as deuterostomes; this term refers to the fact that in the ontogeny of echinoderms and of chordates, the blastopore becomes the anus of the adult.

Mollusks have muscular or membranous mantles that protect the inner soft organs, called the visceral mass. In many cases, the mantle secretes a shell. Mollusks also have a muscular foot, which may function in locomotion or burrowing. Examples of mollusks include snails, clams, squid, scallops and oysters.

Phylum Platyhelminthes, the flatworms, include planarians, and parasitic flukes and tapeworms. The two parasitic types show reduced or absent nervous, digestive, respiratory and circulatory systems when compared to the planarians.

83. (A)   84. (B)   85. (B)   86. (E)

The organic nutrients that contribute calories to the diet are carbohydrates, lipids, and proteins. Carbohydrates have the empirical formula $C_nH_{2n}O_n$. Saccharum is Latin for sugar; hence the simple sugars are called monosaccharides. These include glucose, fructose and galactose, all six-carbon sugars. Disaccharides are the chemical bonding of two monosaccharides. Glucose and fructose yield sucrose (table sugar); glucose and galactose yield lactose (milk sugar); and two glucoses combine to form maltose. An oligosaccharide is a molecule composed of a few monosaccharides - it typically contains around three to seven sugars in its structure, although the number is rather arbitrary. Polysaccharides are molecules composed of many monosaccharides; they are long polymers. The three major polysaccharides are all polymers of glucose. Plants store glucose in the form of starch which is edible. Animals store glucose as glycogen in the liver and muscle. Another polysaccharide in plants is cellulose - it is a component of plant cell walls and thus serves a structural role. However, unlike starch, cellulose found in wood and vegetable fibers, is not digestible by man due to lack of the enzyme that hydrolyzes it.

Saccharin is an artificial sweetener that was banned by the FDA (Food and Drug Administration) due to its possible carcinogenic (cancer-causing) properties.

87. (A)   88. (E)   89. (D)   90. (C)

Eukaryotic cells have many organelles that subserve specific functions. While most cells contain most of these organelles, some cells may have an abundance or a complete lack of certain organelles, depending on the function of the cell. Skeletal muscle cells are multinucleated and have an abundance of mitochondria. Skeletal muscle cells use a lot of energy for muscular contraction and hence need many mitochondria to supply the energy.

Sperm cells must swim to the egg and hence also have many mitochondria. However, a sperm cell is the only cell in the human body that has a flagellum. Flagella and cilia are cellular projections both based on a microtubular core and function in movement. The main differences are that a flagellum is a single, long projection which can propel the entire sperm cell, while cilia are short, numerous projections which move substances past the cell. The epithelial cells of the respiratory tract contain cilia to move mucus containing entrapped dust and debris up and out of the tract.

All secretory cells, such as the pancreatic acinar cell, which secretes pancreatic juice, have a preponderance of rough endoplasmic reticula. Ribosomes on the rough endoplasmic reticulum synthesize proteins that are destined for secretion by the cell (whereas free ribosomes synthesize proteins that will not be exported). The synthesized protein enters the rough endoplasmic reticulum as it is being synthesized on the ribosome; then the rough endoplasmic reticulum transports the protein, often to the Golgi apparatus for modifications before release.

Lysosomes contain hydrolytic enzymes that digest phagocytosed material such as bacteria and cell debris. Leukocytes (white blood cells) specialize in phagocytosis. In the process of endocytosis, a vesicle is formed around the bacterium, and this phagocytic vesicle fuses with the lysosomes, whose enzymes proceed to destroy the bacterium.

91. (C)   92. (A)

Red blood cells may contain antigens (markers) on their surfaces. Blood plasma contains the antibodies. A person only has antibodies against those antigens that he does not have. If he had antibodies against his own antigens, the antibodies and antigens would bind and agglutination (clumping) of the red blood cells would occur. In contrast to most antibodies that develop in response to exposure to an antigen, these plasma antibodies exist despite lack of exposure to the antigen.

Blood Type	A	B	AB	O
Antigens	A	B	A,B	---
Antibodies	anti-B	anti-A	---	anti-A anti-B

The female in the question showed agglutination with both antisera. The anti-A antiserum agglutinated with her A antigens, while the anti-B antiserum agglutinated with her B antigens. She has Type AB blood.

The male in the question showed no agglutination and therefore has no antigens on his red blood cells. He has Type O blood.

Type AB and Type O are the phenotypes of the students. Their genotypes are $I^A I^B$ and ii, respectively. Alleles I^A and I^B are codominant with each other, but each is dominant to the recessive allele i. A Punnett square shows the results of a cross between these two students:

Type AB x Type O

	I^A	I^B
i	I^Ai	I^Bi
i	I^Ai	I^Bi

The probabilities of their children's blood types are: Type A (I^Ai), 50%; Type B (I^Bi), 50%. Note that the phenotypes of the children are completely different from those of the parents.

The concern with blood transfusions is that a donor's red blood cells (by virtue of the antigens on them) will be agglutinated by the recipient's antibodies. There is not the opposite fear that a donor's plasma will agglutinate the recipient's red blood cells because the donor's plasma is diluted by the recipient's plasma, and thus the effects of the antibodies are minimized.

A subject with Type O blood, such as the male in the question, is said to be a universal donor. This is because he has no antigens on his red blood cells that could be agglutinated by the recipient's antibodies. Although he does have both plasma antibodies, his antibodies will be diluted by the recipient's plasma and thus will not cause harm. He can thus donate blood to anybody.

93. (B)   94. (B)

Since the trait is due to an autosomal dominant gene, if a family member does not express the trait, he does not carry it either. Suppose P stands for the preauricular gene and p represents the recessive normal gene. Thus, F_{1-1} (pp) does not contain the gene in his chromosomes and hence if he marries an unafflicted woman (pp), none of their children will have the trait.

For F_{1-2} (Pp) and F_{1-3} (Pp) both of whom show the trait, half the children should inherit the gene P while half should inherit the normal allele, p. This is evident in the children of F_{1-2}. For F_{1-3}, his next son (or daughter) has a 50% chance of inheriting the trait. The trait is not sex-linked. Each baby is independent of any other and thus, even though F_{1-3} already has an afflicted daughter, this will not affect the next outcome. Probability simply means that in the long run a 50-50 ratio is expected.

95. (A)   96. (B)

Auxin was the first of the plant hormones to be discovered. Auxins are responsible for the phototropic effect. They stimulate stem elongation. The auxin, probably indoleacetic acid (IAA) migrates and concentrates in the cells on the side of the plant away from the light and causes those cells to elongate. This causes the stem to bend in the opposite direction, towards the light, as depicted in experiment A.

Gibberellins also promote stem elongation. One particular gibberellin, gibberellic acid$_3$, (GA$_3$) is widely used in growth experiments. Dwarf plants treated with gibberellic acid will approach normal size, as depicted in experiment B.

Florigen is a postulated hormone that may play a role in the flowering process.

Cytokinins function in cytoplasmic division (cytokinesis), enhance leaf size, and slow the yellowing of leaves.

Abscisic acid was named according to its presumed, but now questioned role in abscission - the dropping of flowers, leaves and fruits. However, its more important function is inhibitory in nature: it induces stomatal closure and seed dormancy.

Ethylene is an alkene with the structure $H_2C=CH_2$. It stimulates fruit ripening and promotes abscission.

Growth hormone is an animal hormone. In man, it increases bone and muscle growth, but also stimulates mitosis and protein synthesis in most other cells.

97. (C)   98. (B)   99. (D)   100. (C)

A triglyceride (neutral fat molecule) is one glycerol bound by ester bonds to three fatty acids. Digestion of fat thus yields free glycerol and fatty acids. The free fatty acids will decrease the pH of a solution. The pH drops the most and the most rapidly in tube C, which contains the fat source (cream), pancreatic juice, and bile salts.

To determine the function of bile salts, you must compare two tubes with exactly the same contents, except one will have bile salts as well, tube C has the same contents as tube B, but it also has bile salts. You would compare tubes A and C if you wished to elucidate the function of pancreatic juice. No real comparison can be made between tubes A and B since they only have cream in common. Of course, the most information would be obtained by looking at the data for tubes A, B and C.

Each tube has the same fat source, cream. From tube A, it is apparent that bile salts have no enzymatic activity. It is clear that tube B contains the necessary ingredient for fat digestion, since fatty acids must have been produced in order for the pH to have dropped. That ingredient is pancreatic juice, which contains the enzyme lipase. Tube C contains bile salts in addition. Apparently bile salts are able to aid the pancreatic enzymes in their work. But bile salts alone have no enzymatic activity and therefore do not function in digestion. Digestion can be mechanical (e.g. churning) or chemical (e.g. enzymes).

The liver synthesizes bile, a fluid composed primarily of bile salts. The bile is then stored in the gallbladder until released as needed, such as after a fatty meal. The function of bile is to emulsify fats; it breaks fats up into smaller globules. This increases the surface area on which pancreatic enzymes can work. Thus tube C shows greatest digestion because the pancreatic enzymes have many smaller globules to digest. Note that the bile has no direct role in activating pancreatic enzymes. It simply increase the available surface area.

101. (B)   102. (A)   103. (E)   104. (C)

Tube A contained the urine of the subject on a high-salt diet. Tube B contained the diabetic's urine. Tube C contained the urine of the subject on a high-protein diet. Tube D contained the distilled water.

There are several clues that tube B contains the urine of the diabetic. The most obvious is that there is excess glucose in the urine (200 mg/dl). Excess glucose in the urine often draws water with it and a diabetic shows an osmotic diuresis. There is then a high volume of urine. Volume is a function of many factors, but compared to the other tubes, tube B contains the highest volume. Since a diabetic cannot utilize the blood sugar (the sugar will not be transported into the cells), he has excessive fatty acid oxidation for energy. This leads to the build-up of ketone bodies in the blood and urine. The ketone bodies are acetoacetic acid, B-hydroxybutyric acid and acetone. Acetone is the substituent in urine that casts off a sweet odor. The formation of these acids will lower the pH of the urine. Urinary pH is normally between 5 and 8. Diet influences urinary pH. A high protein diet often results in acidic urine, while a vegetarian diet results in more alkaline urine. The high concentration of solutes in a diabetic's urine increases the specific gravity. Specific gravity is defined as the weight of a volume of substance divided by the weight of an equal volume of distilled water. Therefore, by definition, the specific gravity of water is 1.000. Since urine is primarily water with solutes dissolved in it, it will always weigh more than an equal volume of pure water. Therefore, the specific gravity of urine is always greater than one. The higher the solute concentration, the higher the specific gravity, since the numerator is a higher number than the denominator. Urinary specific gravity varies between approximately 1.010 and 1.030. The man on the high-salt diet, as well as the diabetic, will also show relatively high specific gravities. The concentration of salt in the urine is due in part to diet. Although excess salt may be excreted with water, the specific gravity is still usually high. A high protein diet will not have much effect on specific gravity or volume.

Diuresis refers to an increased output of water. A diuretic drug is one that primarily affects urinary volume. Some diuretics also affect the output of specific urinary constituents. Thus the effect of diuretics on specific gravity is variable. pH is a measure of the concentration of hydrogen ions ($[H^+]$), or acidity of a solution. Mathematically, pH is defined as follows:

$$pH = \log [H^+]$$

However, it is an inverse relationship, in which an increase in $[H^+]$ causes a drop in pH. The pH scale varies from O to 14 in which pH 7 is neutral. At neutrality, there is a balance of H^+ and OH^- ions. Water is an example of a neutral compound. When the pH is less than 7, there is an excess of H^+ (over OH^- ions).

Tube C has the lowest pH (5.0) or highest $[H^+]$. It is 100 times more acidic that tube D with a pH of 7. Urinary pH normally varies between pH 5 - 8 depending on diet, drugs, and other factors.

105. (B)   106. (C)   107. (D)   108. (A)   109. (B)

Sizes of predator and prey populations fluctuate regularly and are interdependent. If the prey population is allowed to increase, this provides more food for the predator population, which then flourishes. However, more predators will eat more prey and the prey population will then decrease. Now the predator population will of necessity decrease as well, due to fewer food sources. The decrease in predator population allows the prey population to flourish and the cycle begins anew.

Note that while these oscillations occur independently of other factors, in the graph depicted, the oscillations coincide with seasonal changes. It appears that the prey succumb to the winter environment and hence the predator population will shrink as well. Note that the size of the predator population lags behind that of the prey population. There is no evidence that the winter poses any direct danger to the predator population, only that the population drop is due to the drop in prey population. The mildest winter was that of 1979 - although the normal oscillation occurred, the drop in prey population was not as severe as in other years.

The peak prey population was achieved in the summer of 1977, with a population of 100. At this time, the predator population was 15. Note the axes for population - the prey population size is read on the left; that of the predator is on the right.

The data for the graph ends in the summer of 1980. If in the following winter, a chemical were introduced into the environment which proved fatal to the prey, the predator population would decrease severely due to decreased food resources. Of course, if the predators had other sources of food, this drop need not occur; however, there is no evidence that this is the case, as their population size always lags behind that of the prey.

110. (B)   111. (B)   112. (A)   113. (C)   114. (B)

A glucose tolerance test is often administered to patients suspected of having diabetes. In this test, the patient fasts overnight to bring his plasma glucose to a stable basal level, often around 80 mg/dl. Both subjects A and B show the same fasting glucose level: 80 mg/dl. The subjects then ingest a sweet syrupy concoction. The volume ingested is proportional to body weight usually 1 gm of glucose per kg. of body weight. Normally, after a glucose load, the body responds by releasing insulin, a pancreatic hormone that stimulates glucose uptake by cells and hence lowers plasma glucose levels back to normal.

In the graphs, the rate of rise (slope) of plasma glucose level is the same for both subjects, although subject B shows a much higher glucose level for the given stress. In either case, the maximum plasma glucose levels are reached at 30 minutes, the time of the first of many blood tests used to measure the current glucose status. Thus, there is little latency between glucose ingestion and peak glucose levels.

Subject A exhibits a normal response to glucose ingestion: the plasma glucose reaches its maximum at about 130 mg/dl and returns to baseline within two hours. This is due to the release of insulin, which lowers plasma glucose.

Graph B indicates an abnormal response to glucose ingestion: the response given is indicative of diabetes. The peak plasma glucose level is close to 200 mg/dl and stays high. In two hours, the plasma glucose level is still excessively high. This indicates diminished tolerance to the glucose stress. The probable cause is diabetes, a disease in which the pancreas does not release insulin, and hence the body cannot counteract the glucose stress.

Not only do diabetics have high plasma glucose concentrations, but glucose is often excreted in the urine. When the plasma glucose level rises above 180 mg/dl (as it does in subject B), the kidneys cannot reabsorb all the glucose, and it begins to "spill over" into the urine.

115. (D)   116. (C)   117. (C)

It is evident that the two poisons have no effect on passive transport processes such as diffusion. They both affect active processes (processes that require the expenditure of ATP), either active transport and/or endocytosis. To determine the differences in their action, it would be most informative to first look at endocytosis, since it is blocked by only one of the poisons, cyanide. Endocytosis requires the presence of ATP, thus it is suggested that cyanide may inhibit ATP production. Furthermore, cyanide also blocks active transport, which also requires ATP. Since ouabain does not inhibit endocytosis, it presumably cannot affect the production of ATP. In comparing the effects of ouabain on active transport and on endocytosis, it is clear that the inhibition by ouabain on active transport is independent of the production of ATP. It is probable that ouabain directly inhibits the activity of the sodium-potassium pump, which would thus inhibit active transport, but not endocytosis, since the latter is independent of the pump. Cyanide and ouabain were not placed into the same cell, so one cannot determine cyanide's effect on ouabain.

Cyanide binds to the last enzyme of the electron transport chain and inhibits utilization of oxygen. Since oxidation is coupled to the phosphorylation of ADP to form ATP, cyanide inhibits ATP production.

Cyanide inhibits the ultimate production of ATP. There is a secondary inhibition of the activity of the pump (active transport) which requires the hydrolysis of ATP. In contrast, ouabain has a direct inhibitory effect on the pump with no effect on other active cell processes that do not utilize the pump, but do require ATP.

In the presence of ouabain, sodium cannot be pumped out of the cell into the extracellular fluid and potassium cannot be pumped in the opposite direction. However, these ions can still diffuse passively down their concentration gradients. Thus sodium, which is normally in higher concentration extracellularly (due to the pump), will diffuse into the cell, while potassium will diffuse out of the cell. Without active transport to maintain the gradient, the concentration gradient will dissipate by passive diffusion, and sodium will accumulate intracellularly and potassium will accumulate extracellularly relative to the standard conditions of the cell.

118. (D)   119. (C)   120. (B)

The three observations made were chosen because they can identify each of the cell types. A parenchyma cell contains chloroplasts, which are present in leaves. There are "air spaces" between the parenchyma cells composing the tissue; this does not mean that there are air spaces within any one cell. The upper and lower epidermis sandwich the palisade and spongy mesophyll layers. The latter layer, in particular, is laden with air spaces, which aid in nutrient and gas exchange. The chloroplasts are located in cells throughout the mesophyll layer. As the site of food production, leaves also store food. Food storage also occurs in stems and roots. While leaves are alive at maturity, so are the stems, roots, etc.

Both parenchyma and collenchyma cells are alive. While it can not be stated that every live cell will undergo mitosis, it is certainly true that a dead cell, such as a sclerenchyma cell, cannot divide. Note that the more specialized a cell is, the more likely that it will lose its capacity for division. Hence, parenchyma cells are the least specialized and are able to divide.

Since they serve a supportive role, one might correctly presume that collenchyma is present particularly in the stems. It is also present in leaf stalks and along the veins of leaves. Thus cell B is a collenchyma cell, cell A is a parenchyma cell, and cell C is a sclerenchyma cell.

SECTION II

ESSAY I

The complete oxidation of glucose is represented by the following equation:

$$\text{glucose} + 6\ O_2 \rightarrow 6\ CO_2 + 6\ H_2O = 38\ ATP$$

Glucose, in the presence of oxygen forms carbon dioxide, water, and ATP. ATP, or adenosine triphosphate, is the energy currency of the cell. Three major series of reactions are involved in the above reaction, with a total of about 30 individual steps. These reaction series are called glycolysis, Krebs cycle, and the electron transport chain. Each series occurs in a certain part of the cell, where the necessary enzymes are compartmentalized. A cell, with only the plasma membrane and one mitochondrion is depicted schematically below. The mitochondrion is the sole organelle required for energy synthesis. Of course, there are many mitochondria in cells, particularly in active cells such as skeletal muscle cells.

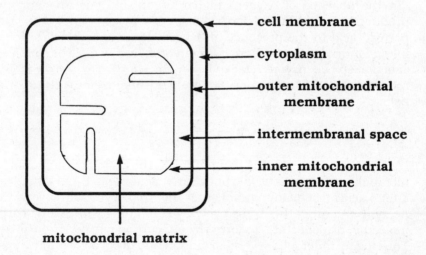

cell membrane

cytoplasm

outer mitochondrial membrane

intermembranal space

inner mitochondrial membrane

mitochondrial matrix

Note that the mitochondrion (plural, mitochondria) has two membranes. The inner membrane is folded and these folds are called cristae. The folding increases the surface area available for enzymatic reactions.

Glycolysis occurs in the cytoplasm. The Krebs cycle occurs in the mitochondrial matrix. The electron transport chain occurs on the inner mitochondrial membrane, including the cristae.

The process begins when glucose enters the cell as stimulated by the action of insulin. The first glycolytic enzyme is hexokinase. It transfers a phosphate group from ATP to form glucose-6-phosphate:

$$\text{Glucose} + \text{ATP} \rightarrow \text{Glucose-6-phosphate} + \text{ADP}$$

Note that although ultimately many ATPs will be formed, one ATP is required initially. In the next step, glucose-6-phosphate is isomerized to fructose-6-phosphate. The third step also requires an ATP molecule, since a phosphate group is added to fructose-6-phosphate to form fructose-1,6-diphosphate. In the next step, the six-carbon molecule is split into two three-carbon molecules. Several more steps are required to produce the final product of glycolysis, two molecules of pyruvic acid (or pyruvate, the salt form), a three-carbon molecule. Four molecules of ATP are produced and two molecules of NAD$^+$ (nicotinamide adenine dinucleotide) are reduced in those same steps.

Pyruvic acid has several fates in the cell. In yeast and other microorganisms, pyruvic acid is converted into ethanol in a process called alcoholic fermentation. This occurs in the absence of oxygen.

In the absence of oxygen, higher organisms such as ourselves produce lactic acid from pyruvic acid. Of interest here is the fate of pyruvic acid in the presence of oxygen. Pyruvic acid will enter the mitochondrial matrix, where it is converted to a two-carbon molecule called acetyl coenzyme A (acetyl CoA, for short). The third carbon is released as CO_2. The reaction catalyzed by this enzyme is summarized below:

$$\text{pyruvic acid} + \text{NAD}^+ + \text{CoA} \rightarrow \text{acetyl CoA} + CO_2 + \text{NADH}$$

Acetyl CoA is the starting point of the Krebs cycle. It is an important metabolic intermediate, as both fatty acids (via B-oxidation) and amino acids (via deamination) enter the Krebs cycle with acetyl CoA. Acetyl CoA enters the Krebs cycle by combining with a four-carbon molecule, oxaloacetate, to form the six-carbon citrate, or citric acid. The citrate undergoes several oxidation and decarboxylation steps to eventually become, once again, the four-carbon oxaloacetate. Each molecule of pyruvic acid that enters the Krebs cycle produces one molecule of GTP (guanosine triphosphate), and causes the reduction of three molecules of NAD^+ and one molecule of FAD, forming NADH and $FADH_2$, respectively.

The NADHs and $FADH_2$s formed in glycolysis and the Krebs cycle pass to the inner mitochondrial membrane to enter a series of oxidation/reduction reactions there. The oxidation of NADH yields 3 ATPs,whereas the oxidation of $FADH_2$ yields only 2 ATPs. The oxidation steps are coupled to the phosphorylation of ADP to form ATP, hence oxidative phosphorylation refers to the sequence of reactions of the electron transport, or respiratory, chain.

In the respiratory chain, the major group of electron carriers are cytochrome enzymes. The cytochromes can exist in an oxidized or a reduced form. Yet the cytochromes are pure electron carriers and do not pick up hydrogens. Like hemoglobin, cytochromes contain a heme group. Heme is an iron-containing pigment. Iron (Fe) can exist in the oxidized form Fe^{3+} or the reduced form Fe^{2+}. The electrons of the reducing equivalents are transferred to the cytochromes in a series of steps. The final step is catalyzed by cytochrome oxidase: oxygen is reduced to water as it oxidizes the cytochrome.

In 1961, Peter Mitchell proposed the chemiosmotic hypothesis to explain how oxidation is coupled to phosphorylation. The basic premise is that during electron transport, a proton (H^+) gradient develops across the inner mitochondrial membrane, with protons accumulating in the intermembranal space. This is due to the pumping of a pair of protons into this space at three sites along each respiratory chain. A "lollipop"-like enzyme traverses the inner mitochondrial membrane. The stalk penetrates the membrane, while the head faces the matrix. Protons flow back into the matrix through the stalk. The ATP synthetase portion is the head of the stalk. As protons flow back to the matrix down their gradient enough energy is generated to phosphorylate ADP to form ATP. Thus most of the ATP produced in the cell is due to oxidative phosphorylation. By comparison, glycolysis provides only a little ATP.

A diagram is shown below. Note that the original reaction is proven, where the reactants are glucose and oxygen and the products are water, carbon dioxide and ATP. These reactants and products are circled.

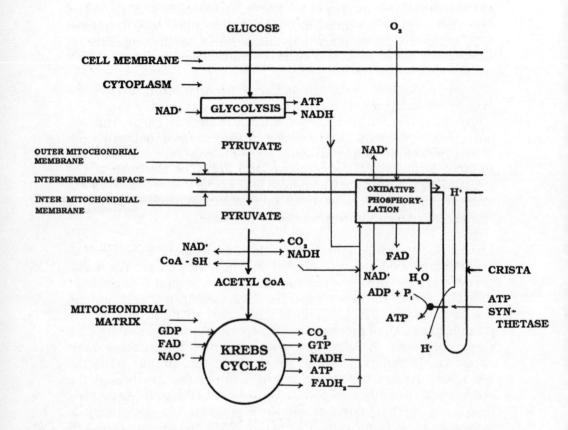

The energy that is required for the reaction that is catalyzed by the ATP synthetase, namely ADP + Pi ⟶ ATP is provided by the movement of protons (H⁺) along their concentration gradient.

ESSAY II

Genes are the units of heredity, located on chromosomes. Genes occur in various forms, or alleles, the combinations of which code for the specific expression of traits. Each individual inherits one allele of each gene from his mother and one from his father. If an individual inherits two identical alleles of a gene, he is said to be homozygous for that trait; he is a homozygote. If the alleles are different, the individual is heterozygous for the trait; he is a heterozygote. The genotype is the individual's genetic constitution, whereas the phenotype refers to the expression of the genotype.

There are many different types of genetic inheritance patterns. The simplest is that of dominant and recessive inheritance. This is best exemplified by the inheritance of eye color in man. Let B represent the dominant allele for brown eyes and b represent the recessive allele for blue eyes. If an individual has the genotype BB, he is homozygous dominant and hence has the phenotype of brown eyes. An individual with the genotype bb is homozygous recessive and has the phenotype of blue eyes. A heterozygote, Bb, has brown eyes because B is dominant to b. Note that there are two genotypes (BB and Bb) that represent the phenotype brown eyes.

A Punnett square below shows a cross between two heterozygotes.

	Bb	X	Bb
		B	b
B	BB	Bb
b	Bb	bb

The parents are brown-eyed. Phenotypically, 75% of their offspring will have brown eyes and 25% will have blue eyes. The genotypes are 25% BB (homozygous dominant), 50% Bb (heterozygous) and 25% bb (homozygous recessive).

Some traits show incomplete dominance, whereby a dominant allele cannot fully mask the recessive allele. This is best exemplified in flower color. Let R be the dominant allele for red flower color and r be the recessive allele for white flower color. When a red flower (RR) is crossed with a white one (rr), the first generation plants are all pink (Rr). The Punnett square shows this cross:

161

	RR	X	rr
		R	R
r		Rr	Rr
r		Rr	Rr

If the allele for red were fully dominant, the heterozygote plants would all be red.

In codominance, a heterozygote with two dominant alleles expresses them both equally. This is best exemplified by the inheritance of blood antigens. There are multiple alleles, I^A, I^B, and i, possible at the locus that codes for blood type. Of course, any one individual inherits only two alleles. I^A and I^B are dominant to i, yet are codominant with each other.

Type A blood is expressed by the genotypes I^AI^A and I^Ai. Recall that I^A is dominant to i. This person has only A-antigens on his red blood cells.

A phenotypically type B person has the genotype I^BI^B or I^Bi. This person has only B antigens on his red blood cells.

The AB phenotype is expressed by the genotype I^AI^B. These alleles are codominant and the person has both A and B antigens on his red blood cells.

The homozygous recessive genotype ii is expressed phenotypically by Type O blood. Type O blood only occurs when there is no dominant allele. This person has no antigens on his red blood cells.

Sex-linked inheritance is a little more complex. There are two sex chromosomes: X and Y. A female has XX, while a male has XY. Unlike other pairs of chromosomes, these chromosomes do not look alike. The X-chromosome is much larger than the Y. While the Y codes only for male sex, the X-chromosome contains other genes on it. The genes for color-blindness and hemophilia are carried on the X-chromosome. These two disorders are coded by recessive genes. Some types of baldness may be inherited in the same manner.

Since hemophilia and color-blindness are carried on the X-chromosome, these disorders are seen primarily in males, who have only one X-chromosome. The reason is that because males have only one X chromosome, when a male inherits an X-chromosome with the recessive gene, he will express the disease, since his Y-chromosome has no dominant allele to mask it. However, if a female inherits an X-chromosome with the recessive gene, she most likely will have the normal dominant allele on her other X-chromosome; thus she will carry the trait, but will not express it. For a female to express it, her father must have the disease while her mother must at least be a carrier. The likelihood of that combination is low, unless there is marriage between relatives.

The example provided deals with color blindness, although hemophilia is transmitted in an analogous manner. Suppose C is the dominant allele for normal color vision, and c is the allele for color blindness. A male is either normal (X^CY) or color blind (X^cY); a female is either normal (X^CX^C), a carrier (X^CX^c), or color blind (X^cX^c). A cross between "a carrier" mother and a normal father is shown below:

$$X^CX^c \times X^CY$$

	X^C	X^c
X^C	X^CX^C	X^CX^c
Y	X^CY	X^cY

Note that the offspring are 25% of each of the following: normal female (X^CX^C), carrier female (X^CX^c), normal male (X^CY), and color blind male (X^cY). The normal mode of transmission is in this manner, i.e., from a carrier mother to her son. The disorder is not passed from fathers to their sons, because sons inherit the Y chromosome from their father and the Y chromosome does not contain these genes.

It is important to note that the gene for color blindness or hemophilia is passed equally to male and female offspring; however, it is expressed more often in the male due to lack of a dominant allele to mask it.

Also, when two populations of one organism are separated geographically, natural selection becomes a factor in the resulting phenotypical differences. As variance occurs through genetic inheritance, different environmental circumstances affect selection of characteristics. For example, if you separated a species of birds raised in the one environment, so that no interbreeding could occur, one could observe the differences. For example, if one group was kept in the same environment in which it was raised, and the other was placed in an environment where the bugs lived deep in the woods, one would expect that over generations, the second group would exhibit longer beaks. The different phenotype is expressed because it became favorable to survive in a different niche. Speciation is caused when the differences become incompatible with reproductive success.

ESSAY III

Homeostasis refers to the stability of the internal environment, which is composed of the extracellular fluid compartment (plasma and interstitial fluids). The concept was first introduced by Claude Bernard, who, around 1870, discussed the importance of the "milieu interieur." In the 1930's, Walter Cannon coined the term *homeostasis*. In short, all organ systems, but primarily the nervous and endocrine systems, strive to keep parameters of the internal environment constant, for this is the fluid in which our cells live. Some of the parameters monitored include the composition, temperature, osmotic pressure, and pH of the blood.

Stress causes an imbalance in the internal environment, be it external stress such as heat, noise, or surgery, or internal stress, such as infection or poison. The body compensates and counteracts the stress by homeostatic mechanisms. Thus, for instance, body temperature is normally regulated at 98.6 degrees F. (37^0 C), despite changes in the ambient temperature. If homeostasis were not functional, disease, and ultimately death, would ensue. Fever occurs when homeostasis of body temperature is non-functional; hypertension is a result of unregulated blood pressure; dehydration is due to a loss of blood and fluid volume; acidosis is evident when the pH of the blood drops.

The major mechanism that functions to maintain homeostasis is negative feedback. In feedback, the information about the status of a parameter (such as blood pressure, blood glucose) is monitored and reported to a central control region, which has a setpoint. The input (the status) is compared to the setpoint for the parameter. The control region then initiates a response if the status of the monitored parameter is not what it should be (the setpoint). In negative feedback, response of the body counteracts the effects of stress. Thus, a drop in blood glucose concentration elicits the release of glucagon to bring the concentration back up; the opposite input, a rise in the blood glucose level would elicit the release of insulin, which lowers the blood glucose level by transporting glucose into the cells.

While all the organ systems function in homeostasis (the respiratory system exchanges gases that are dissolved in the blood with gases that are present in alveoli; the kidneys regulate the volume and composition of the plasma), the major systems that control homeostasis are the nervous and endocrine systems. The nervous system has receptors that may detect an imbalance and send messages to counteract the imbalance or elicit hormonal secretion from endocrine glands. The nervous system sends action potentials that induce action immediately, albeit the signals are of short duration. In contrast, the endocrine system sends chemical signals, whose effects are longer-lasting, though there may be a latency to the effect.

A simplistic view of the homeostatic control of blood pressure and volume follows. Suppose the stress of hemorrhage occurs. The nervous system receives input concerning decreased blood pressure and blood volume. The nervous system responds by decreasing arterial baroreceptor activity. This causes vasoconstriction and an increase in heart rate. The vasoconstriction is an important way to increase blood pressure back to normal, while the increase in heart rate will increase cardiac output. The drop in blood volume is a signal to the hypothalamus to stimulate the posterior pituitary gland to release ADH (antidiuretic hormone). This hormone acts on the kidneys to reabsorb water into the blood and thus raise blood volume and ultimately blood pressure. Thus the interplay of nervous and endocrine activity functions in homeostasis.

To test if adrenaline and noradrenaline are important in preparing the body for "flight or fight," take four sets of mice, two test groups — one with their adrenal medulla destroyed, one group with normal adrenal medullas — and one control group for each. Artificially stimulate the adrenal glands for one group in each set. Monitor their heart rate, blood glucose level and respiration rate. Under a stressful condition, one would expect in a normal animal for all of the rates to be increased, as epinephrine and norepinephrine are known to exert those effects. The gas exchange will be increased because the hormone relaxes smooth muscle in the air passages. Compare the rates of the stimulated groups to each other and to the control groups. One would expect to find the normal mice with higher rates, above all of the other groups. To test the hypothesis further, inject the normal and mutant rats with epinephrine, and follow the same experiment. In this scenario, one would expect to find the test groups to have significantly different levels than the control group.

ESSAY IV

Just as an animal hormone is secreted by endocrine glands and released into the blood, a plant hormone is produced in one tissue and transported to another, exerting a specific effect. Only small amounts are needed to have the desired effect. Like animal hormones, plant hormones function in growth, development and metabolism. The major groups of plant hormones are the auxins, cytokinins, gibberellins, ethylene, abscisic acid, and perhaps florigen. Unlike animal hormones, a plant hormone may exert an effect on the tissue or cells in which it was produced.

Auxin was the first of the plant hormones to be discovered. The primary auxin is called IAA (indoleacetic acid). Auxin causes cell elongation and is responsible for the phototropic effect, whereby a plant bends toward a light source. This is due to cell elongation on the side of the plant that is away from the light, to which the auxin has migrated in response to the light. Auxin also plays a role in abscission, apical dominance, and the growth of fruit. Diminished concentrations of auxin promote fruit and leaf dropping.

Cytokinins function in cytoplasmic division, hence their name. They are especially important in tissues that are growing, such as germinating seeds. Yet they have other effects as well. They prevent the aging and abscission of leaves. It is of interest that cytokinins resemble the purine adenine, one of the nitrogenous bases in nucleic acids. Gibberellins were named after a fungus, *Gibberella fujikuroi*, which infected rice plants and caused hyperelongation of the stems, causing these rice plants to fall over. Gibberellins are thus useful in the treatment of genetically dwarfed plants, which can grow to normal size after being treated with gibberellins. Gibberellins also have secondary functions. They can induce cellular differentiation and seed germination by causing enzymes to convert food from the edospore into a usable form for seedlings.

Ethylene is a hydrocarbon gas ($H_2C = CH_2$) released from the burning of kerosene. It was discovered that ripening occurred more rapidly in the presence of kerosene combustion, although it was erroneously believed that it was the heat produced that caused the ripening. In fact, plants themselves produce ethylene just prior to fruit ripening. It also causes the aging and subsequent abscission, or dropping, of leaves. Other effects include a role in seed germination, and may determine sex in some flowers.

Abscisic acid is the primary inhibitory hormone, albeit named incorrectly for its postulated role in abscission. Abscisic acid in general does not cause abscission, but rather causes dormancy. It protects against unfavorable environmental conditions. For example it causes the stomata to close to prevent water loss. It encourages downward growth of roots if seedlings are in an unfavorable position.

Florigen is a postulated hormone that induces flowering. However, no specific hormone has been identified, and perhaps flowering is due to a combination of hormones. For instance, both auxin and gibberellin may induce flowering. There are probably also hormones that inhibit flowering. Florigen, the postulated flowering hormone, has not been isolated. Florigen is not a specific hormone, but probably a combination of plant hormones already identified. Auxin and gibberellin are two hormones which are likely to be identified as "florigen." It is possible that those hormones may inhibit flowering as well. The amount or level of each hormone present is likely to be a major determining factor.

THE ADVANCED PLACEMENT EXAMINATION IN

BIOLOGY

TEST III

ADVANCED PLACEMENT BIOLOGY EXAM III

SECTION I

120 Questions
90 Minutes

DIRECTIONS: For each question, there are five possible choices. Select the best choice for each question. Blacken the correct space on the answer sheet.

1.   Isomers of a given nutrient molecule have:

   (A)   atomic numbers that vary

   (B)   different kinds of atoms

   (C)   different molecular weights

   (D)   the same molecular formula

   (E)   the same structural formula

2.   Select the most abundant element of protoplasm from the following choices:

   (A)   calcium

   (B)   carbon

   (C)   phosphorus

   (D)   sulfur

   (E)   zinc

3. Water's ability to regulate environmental temperatures within a small range conducive to life is partially due to its

(A) function as a universal solvent.

(B) high heat capacity.

(C) nonpolarity.

(D) viscosity.

(E) maximum density at zero degrees Celsius.

4. Separation of homologous chromosomes occurs during what phase of mitosis?

(A) Interphase   (D) Telophase

(B) Anaphase   (E) Metaphase

(C) Prophase

5. A factor that contributed greatly to the prolonged existences of simple organic molecules in Earth's prebiotic oceans was

(A) the presence of simple amino acids.

(B) the lack of high concentrations of atmospheric ammonia.

(C) the presence of rudimentary enzymes.

(D) the extremely low concentrations of atmospheric methane.

(E) the virtual absence of atmospheric oxygen.

6. The metabolization of one molecule of glucose will produce how many molecules of reduced NAD in the Krebs cycle?

(A) 1   (D) 8

(B) 3   (E) 9

(C) 6

7. *Canis lupus* (wolf), *Canis latrans* (coyote) and *Canis dingo* (dog) all share the same:

    (A) class   (D) phylum

    (B) family   (E) species

    (C) genus

8. Among angiosperms, dicots are distinguished by their:

    (A) absence of xylem and phloem

    (B) leaf net veination

    (C) presence of cell centrioles

    (D) scattered vascular bundles in the stem

    (E) single embryonic leaf

9. A cell deprived of its series of Golgi complexes has difficulty:

    (A) maintaining its shape

    (B) synthesizing DNA

    (C) synthesizing mRNA

    (D) storing molecules

    (E) synthesizing protein

10. An evolutionary trend of adaptive radiation is:

    (A) absence of many new kinds of organisms produced

    (B) convergence of separate evolutionary lines

    (C) divergence of many new species into different habitats

    (D) extinction of most members in a species

    (E) geographic isolation of all new-formed species involved

11. Name the bone that does not articulate with the humerus.

(A) clavicle

(D) shoulder blade

(B) radius

(E) ulna

(C) scapula

12. A cell's nucleolus is found in its:

(A) cytoplasm

(D) nucleus

(B) E.R.

(E) plasma membrane

(C) mitochondrion

13. Removal of a cell's ribosomes would result in a cell's inability to utilize which of the following molecules?

(A) carbon dioxide

(D) oxygen

(B) carbon monoxide

(E) phosphorus

(C) lysine

14. Which hormone opposes the effect of the other four?

(A) cortisol

(D) growth hormone

(B) epinephrine

(E) insulin

(C) glucagon

15. Smoking cigarettes over a long period harms the upper respiratory tract's

(A) alveoli

(D) goblet cells

(B) cilia

(E) villi

(C) smooth muscle

16. The first living cells to appear on earth probably resembled today's:

(A) bacteria

(D) protozoa

(B) eukaryotes

(E) viruses

(C) plant cells

17. A product of anaerobic plant metabolism is:

(A) carbon dioxide

(D) lactic acid

(B) ethyl alcohol

(E) pyruvate

(C) hydrochloric acid

18. A plant cell accepting water will eventually show:

(A) active transport

(D) weight loss

(B) flaccidity

(E) vacuoles

(C) turgor

19. Synapsis of homologous chromosomes in prophase of meiosis-one means:

(A) attraction and pairing

(B) building and synthesis

(C) division and separation

(D) duplicating and splitting

(E) repulsion and separation

20. Select the protease:

(A) amylase

(D) ptyalin

(B) maltose

(E) sucrase

(C) pepsin

21. The region of the chromosome to which spindle fibers attach is composed of

(A) centromeres.

(D) stomates.

(B) vacuoles.

(E) centrioles.

(C) gametes.

22. Tetraploidy that occurs in an organism whose species normally has 80 chromosomes per cell will cause the organism's somatic cells to contain how many chromosomes?

(A) 20

(D) 80

(B) 40

(E) 160

(C) 60

23. A primary sex cell experiences one nondisjunction during meiosis. If the normal diploid number of chromosomes in the cell is 42, how many chromosomes are present in one of the defective gametes?

(A) 20

(D) 41

(B) 21

(E) 43

(C) 24

24. Another name for a fertilized egg cell is:

(A) blastula

(D) secondary oocyte

(B) polar body

(E) zygote

(C) primary oocyte

25. The term metastasis refers to the fact that cancer cells tend to:

(A) destroy

(D) shrink

(B) divide

(E) wander

(C) reproduce

26. Muscles pull on bones from their:

(A) antagonists to prime movers

(B) insertions to origins

(C) origins to insertions

(D) prime movers to synergists

(E) synergists to antagonists

27. A cell is inhibited during the S phase of its cycle. It will not reproduce due to lack of:

(A) ATP availability

(B) centriole migration

(C) centromere formation

(D) DNA synthesis

(E) plasma membrane structure

28. One of hemoglobin's subunits, about 150 amino acids long, requires a coded DNA sequence that contains how many nucleotides?

(A) 3

(D) 150

(B) 50

(E) 450

(C) 100

29. Eye receptors and their function can best be summarized as:

(A) cones - color discrimination, rods - twilight vision

(B) cones - twilight vision, rods - color discrimination

(C) ganglia - color discrimination, rods - twilight vision

(D) lens - light refraction, cornea - light refraction

(E) neurons - twilight vision, ganglia - color discrimination

30. Certain oncogenic viruses violate the central dogma of biochemical genetics by transferring genetic information in which of the following sequences?

(A) DNA, protein, RNA

(B) DNA, RNA, protein

(C) RNA, DNA, protein

(D) protein, DNA, RNA

(E) RNA, protein, DNA

31. Which stimulus will activate the lactose operon in a bacterial cell?

(A) Absence of lactose

(B) Availability of an inducer

(C) Cistron repression

(D) Regulator gene dominance.

(E) Repressor molecule binding to the operator gene.

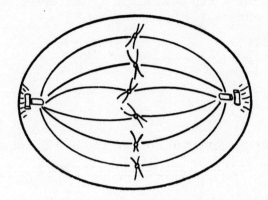

32.   This illustration shows a cell in the mitotic stage of:

(A) anaphase   (D) prophase

(B) interphase   (E) telophase

(C) metaphase

33.   Four different DNA bases, transcribed in groups of three with possible base repetition, yield a number of codons, or three-letter combinations equaling:

(A) 1   (D) 64

(B) 4   (E) 128

(C) 16

34.   During transcription, an RNA polymerase adds a uracil-containing nucleotide to the RNA molecule. The base on the strand of DNA that is not being transcribed is:

(A) Adenine   (D) Thymine

(B) Cytosine   (E) Uracil

(C) Guanine

35. After chromosomal mapping, it is found that genes A and B are ten units apart, genes B and C are five units apart, and genes A and C are fifteen units apart. Their sequence on the chromosome is:

(A) A C B

(B) B A C

(C) B C A

(D) C A B

(E) C B A

36. The sequence of stem layers from the inside out is:

(A) cambium-xylem-phloem-pith-cortex

(B) cortex-phloem-cambium-xylem-pith

(C) phloem-xylem-pith-cambium-cortex

(D) pith-xylem-cambium-phloem-cortex

(E) xylem-phloem-cortex-pith-cambium

37. The pattern of xylem growth is:

(A) cambium lays it down to the inside year by year

(B) cambium lays it down to the outside year by year

(C) it alternates with phloem bands year by year

(D) it is internal to the pith as a thin layer

(E) the oldest layers are closest to the cambium

38. The monosaccharide products of lactase-controlled hydrolysis are:

(A) fructose and maltose

(B) galactose and glucose

(C) glucose and fructose

(D) glucose and glucose

(E) sucrose and fructose

39. A mother of blood type B-negative gives birth to an infant of blood type O-positive. Which of the following blood types could be that of the father?

(A) AB-negative

(D) A-positive

(B) AB-positive

(E) B-negative

(C) O-negative

B^-

40. Which of the following organisms has the largest surface area-to-volume ratio and is listed with its proper relative metabolic rate?

(A) dog - varying metabolism

(B) elephant - rapid metabolism

(C) horse - slow metabolism

(D) human - varying metabolism

(E) shrew - rapid metabolism

41. Assuming a 1/2 probability of reproducing a male offspring, the chance for five successive male births in a family is:

(A) 1/4

(D) 1/32

(B) 1/8

(E) 1/64

(C) 1/16

42. Which of the following is a key component of Darwin's principles?

(A) Abundance of resources is available for organisms.

(B) More population members are produced than can survive.

(C) Organisms in a population are uniform.

(D) Population members contribute equally to a gene pool.

(E) Survival rates do not vary among a population's members.

43. Two parents of genotype AaBb mate. Assuming independent assortment and random recombination, the chance for an offspring to phenotypically express the dominant allele of the first gene and the recessive allele of the second gene is:

(A) 9/16

(D) 2/16

(B) 6/16

(E) 1/16

(C) 3/16

44. Among humans, a universal recipient is a person that has which blood type?

(A) A+

(D) O+

(B) AB+

(E) O-

(C) AB-

45. An organism with three independently sorting gene pairs, AaBbCc, can produce a number of genetically different sex cells equaling:

(A) 2

(D) 8

(B) 4

(E) 16

(C) 6

46. Select the bacterial genus that is composed of many pathogens.

(A) *Azobacteria*

(D) *Paramecium*

(B) *Clostridium*

(E) *Bacillus*

(C) *Escherichia*

47. All of the following are terms that describe viruses except:

(A) free-living

(B) host-dependent

(C) noncellular

(D) protein and nucleic acid makeup

(E) ultramicroscopic

48. Humans, great apes, and monkeys are all members of which of the following taxonomic categories?

(A) genus

(B) family

(C) subfamily

(D) order

(E) species

49. A zygote will produce a 32-cell blastula after dividing mitotically by a number of divisions equaling:

(A) 2

(B) 4

(C) 5

(D) 7

(E) 8

32

Question 50 refers to the drawing of the two pairs of chromatids below.

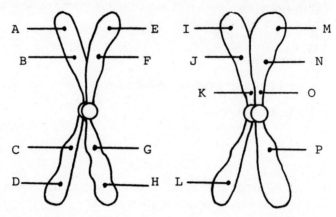

50. The two pairs of chromosomes are in a diploid cell that is in Prophase I of meiosis. If crossing-over of the medial chromosomes were to occur, one of the new chromosomes could have which of the following gene sequences?

(A) ABGH

(D) IJEF

(B) EFCD

(E) IJKO

(C) EFKL

51. Greatest humidity is found in which region of a leaf ?

(A) cuticle

(D) parenchyma

(B) lower epidermis

(E) upper epidermis

(C) mesophyll

52. Select the proper pair of phrases stating the condition and resulting plant growth of the chrysanthemum:

(A) lateral leaf buds removed, terminal bud grows

(B) lateral leaf buds remain, terminal bud inhibited

(C) terminal bud remains, lateral buds grow

(D) terminal bud removed, lateral buds unaffected

(E) terminal bud removed, lateral buds grow

53. One thousand offspring are counted in a genetic cross. Five hundred and two appear dominant in phenotype while four hundred and ninety-eight appear recessive. The genotypes of the parents are most likely:

(A) AA, AA

(D) Aa, aa

(B) AA, Aa

(E) aa, aa

(C) Aa, Aa

54. Following menses, the initiation of the menstrual cycle follows an increase in the secretion of the hormone:

(A) ACTH

(D) GnRH

(B) FSH

(E) GH

(C) LH

55. A pair of mammalian kidneys reabsorbs 188 liters of water, eliminating 1.5 liters during that same period. The volume of water filtered during that time was:

(A) 189.5 l

(D) 141 l

(B) 188 l

(E) 1.5 l

(C) 186.5 l

56. The two criteria used most often in taxonomic classifications are:

(A) color and height

(B) evolution and lifespan

(C) lifespan and morphology

(D) morphology and phylogeny

(E) phylogeny and evolution

57. Select the organism whose metabolism would be most affected by a sudden temperature decrease in its ecosystem.

(A) bird

(B) frog

(C) raccoon

(D) snake

(E) turtle

58. A heterozygous blood-type A person who is also Rh negative mates with a heterozygous blood-type B person who is Rh positive (heterozygous). The probability of producing an off-spring of blood-type O-negative is:

(A) 0

(B) 1/8

(C) 3/8

(D) 1/2

(E) 5/8

59. Oak seeds are allowed to germinate in a Petri dish whose medium is fortified with high concentrations of auxin. As the seeds germinate into seedlings, the observed growth pattern is:

(A) lengthening of shoots and roots

(B) lengthening of shoots, inhibition of root growth

(C) inhibition of shoot and root growth

(D) inhibition of shoot growth, lengthening of the root

(E) no major effect on the roots or shoots

60. Tracheophytes:

(A) cannot conduct photosynthesis

(B) conduct and transport materials

(C) function as heterotrophs

(D) include fungi and mosses

(E) lack sexual reproduction

61. A protein currently synthesized by bacterial cells that have been altered by gene splicing is:

(A) actin
(D) insulin

(B) AMP
(E) myosin

(C) hemoglobin

62. The role of RNA polymerase is to:

(A) bind DNA nucleotides together during translation

(B) bind ribonucleotides together during transcription

(C) break down RNA during digestion

(D) destroy ribosomes during translation

(E) digest RNA nucleotides in an organism's diet

63. The mineral necessary for thyroxin synthesis is:

(A) Ca
(D) P

(B) Fe
(E) S

(C) I

64. A person's blood pressure is taken, revealing a diastolic pressure of 90 mm Hg. The pulse pressure is 30 mm Hg. Systolic pressure is:

(A) 30 mm Hg
(D) 120 mm Hg

(B) 60 mm Hg
(E) 270 mm Hg

(C) 90 mm Hg

65. An invertebrate is collected from a freshwater pond. Dissection of it shows three developmental body layers. It has a cuticle as its outer body covering. It belongs to the phylum:

(A) Arthropoda

(D) Echinodermata

(B) Annelida

(E) Porifera

(C) Coelenterata

66. A micro-organism suspected of causing a disease in a host is discovered. It is isolated and grown in a culture. Micro-organisms from the culture are then injected into a healthy host. What must now be done in order to determine whether or not the micro-organism is a pathogen?

(A) The micro-organism's karyotype must be determined.

(B) The toxin must be extracted from the micro-organism.

(C) The micro-organism must be isolated and grown in culture; the new culture must then be compared to the original culture.

(D) A gram stain and motility test must be performed on the organism.

(E) The size, shape, motility, and nutritional requirement of the organism must be determined.

67. Rejection of a tissue graft in an organism is associated with heightened activity of:

(A) B lymphocytes

(D) T monocytes

(B) B monocytes

(E) neutrophils

(C) T lymphocytes

68. A drug induces paralysis by blocking the function of acetylcholine. This drug specifically inhibits activity at which of the following structures?

(A) cell body

(D) neurolemma

(B) dendrite

(E) synapse

(C) myelin sheath

69. Cells that sense sound are found in the ear's:

(A) cochlea

(D) semicircular canals

(B) eardrum

(E) vestibule

(C) pinna

70. Blood normally circulates through vessels in which sequence?

(A) artery - arteriole - capillary - venule - vein

(B) arteriole - artery - capillary - vein - venule

(C) capillary - arteriole - artery - vein - venule

(D) vein - capillary - venule - artery - arteriole

(E) vein - venule - arteriole - artery - capillary

71. Primates are characterized by each of the following traits except:

(A) relatively large brains

(B) opposable thumbs

(C) rotating shoulder joints

(D) stereoscopic vision

(E) three-chambered hearts

72. A randomly-mating population has an established frequency of 36% for organisms homozygous recessive for a given trait. The frequency of this recessive allele in the gene pool is:

(A) .24

(D) .6

(B) .36

(E) .64

(C) .5

73. The abiotic source of nitrogen in the nitrogen cycle is in the:

(A) atmosphere   (D) minerals

(B) biomass   (E) water

(C) ground

74. Primary consumers in a food chain are:

(A) animals   (D) decomposers

(B) bacteria   (E) herbivores

(C) carnivores

75. A gas that causes asphyxiation by binding to hemoglobin thus preventing oxygen from doing so, is:

(A) carbon dioxide

(B) carbon monoxide

(C) nitrous oxide

(D) sulfur dioxide

(E) water vapor

76. The chromosomal mutation by which a chromosome fragment attaches to a nonhomologous chromosome is termed a:

(A) deletion   (D) inversion

(B) diversion   (E) translocation

(C) duplication

77.   The order in which glands add seminal fluid to migrating semen is:

(A) Cowper's-seminal vesicle

(B) prostate-seminal vesicle-Cowper's

(C) prostate-Cowper's-seminal vesicle

(D) seminal vesicle-prostate-Cowper's

(E) seminal vesicle-Cowper's-prostate

78.   Most energy in a food chain will flow to:

(A) decomposers   (D) secondary consumers

(B) producers   (E) tertiary consumers

(C) primary consumers

DIRECTIONS: The following groups of questions have five lettered choices followed by a list of diagrams, numbered phrases, sentences, or words. For each numbered diagram, phrase, sentence, or word choose the heading which most directly applies. Blacken the correct space on the answer sheet. Each heading may be used once, more than once, or not at all.

Questions 79 - 82 refer to the drawings of functional groups below.

$$-N\begin{smallmatrix}H\\ \\H\end{smallmatrix}$$

A

$$-\overset{\overset{\textstyle H}{|}}{C}=O$$

C

$$-\overset{\overset{\textstyle O}{||}}{\underset{\underset{\textstyle OH}{|}}{P}}-OH$$

B

$$\overset{|}{\underset{|}{N}}-O-H$$

D

E $\quad >C=O$

79. amino acid

80. fructose

81. glucose

82. nucleotide

Questions 83 - 85 refer to the diagram of the organelle below.

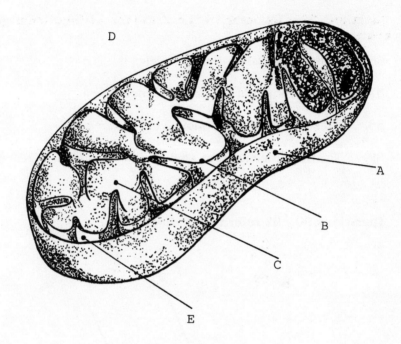

83. The Krebs Cycle (citric acid cycle) occurs here.

84. The majority of this organelle's energy harvest occurs here.

85. Anaerobic metabolism of glucose occurs here.

Questions 86 - 89.

(A) Water

(D) Nitrogen

(B) Carbon dioxide

(E) Oxygen

(C) Glucose

86. Enters the Calvin-Benson cycle.

87. Final product of the Calvin-Benson cycle.

88. Gaseous product of photosynthesis.

89. Provides the electrons that are used in the light reactions.

Questions 90 - 92 refer to the diagram below.

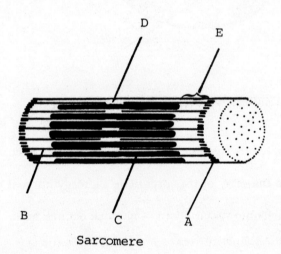

Sarcomere

90. An I band.

91. Crossbridges are specifically found on this molecule.

92. The proteins to which the crossbridges temporarily bind.

Questions 93 - 94 refer to the diagram below.

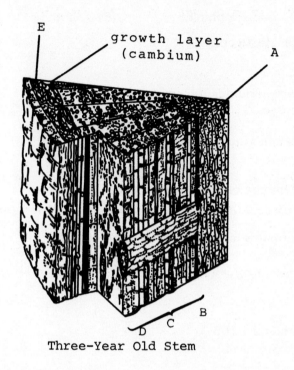

E

growth layer
(cambium)

A

B

D C

Three-Year Old Stem

93.   A non-vascular layer is designated by this label.

94.   The third-year growth layer is designated as this region.

Questions 95 - 97

(A) Ctenophora   (D) Annelida

(B) Aschelminthes   (E) Mollusca

(C) Arthropoda

95.   Small, wormlike, and have flame cells, but lack circulatory
       and respiratory systems.

96.   Radially symmetrical organisms with saclike bodies and
       mesogleal layers.

97.   Organisms with a visceral mass and muscular foot.

Questions 98 - 101

        (A) Sarcodina   (D) Sporozoa

        (B) Bacteriophage   (E) Ciliata

        (C) Mastigophora

98.   *Paramecium* is a member.

99.   Members move by flagella.

100.   Members move by pseudopodia.

101.   Parasites immobile in the adult stage.

Questions 102 - 105

        (A) Porifera   (D) Platyhelminthes

        (B) Annelida   (E) Coelenterata

        (C) Chordata

102.   Organisms with some form of a dorsal supportive structure.

103.   Organisms which are radially symmetrical and possess nematocysts.

104.   These organisms are dorsoventrally flattened and have gastrovascular cavities.

105.   Organisms which retain many primitive animal characteristics.

Directions: The following questions refer to experimental or laboratory situations or data. Read the description of each situation. Then choose the best answer to each question. Blacken the correct space on the answer sheet.

Questions 106 - 108 refer to the diagram below.

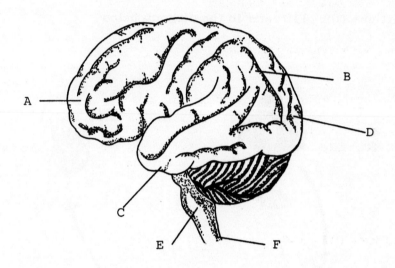

106.   A function of region F is:

(A) abstract reasoning

(B) control of blood pressure

(C) hearing and vision processes

(D) memory

(E) vision

107.   Select the function not performed by region C.

(A) gustation   (D) taste

(B) hearing   (E) vision

(C) smell

108. Region E is the:

   (A) cerebellum   (D) medulla

   (B) cerebrum   (E) pons

   (C) hypothalamus

Questions 109 - 110 refer to the diagram below.

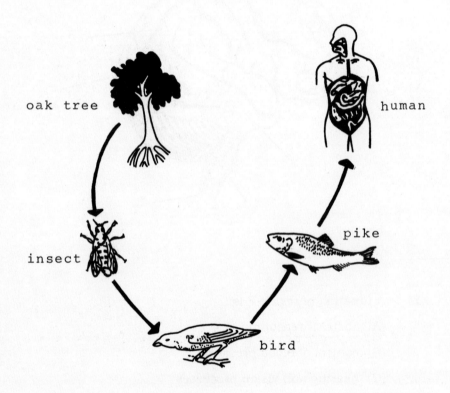

Food Chain

109. In this food chain, which heterotroph receives the largest amount of useful energy?

   (A) bird   (D) oak tree

   (B) human   (E) pike

   (C) insect

110. Which of the following members of the food chain is an omnivore?

(A) bird

(D) oak tree

(B) human

(E) pike

(C) insect

Questions 111 - 113 refer to the drawing below.

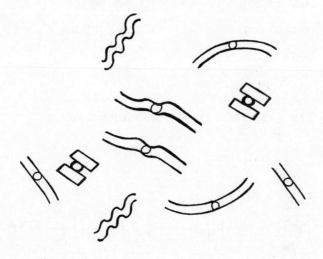

```
Chromosomes extracted from
a cell
```

111. Select the correct statement about this chromosome allotment.

(A) Chromosomes here are from a haploid cell.

(B) Five homologous chromosome pairs are shown.

(C) The chromosomes are single-stranded.

(D) The chromosomes are triple-stranded.

(E) There are ten chromatids shown.

112. Select the incorrect statement about this chromosome allotment.

(A) A cell produced from these meiotically has five chromosomes.

(B) A cell produced from these mitotically has twelve chromosomes.

(C) Each chromosome has a homologue.

(D) The cell containing these chromosomes is diploid.

(E) The chromosomes contain chromatids.

113. An organized chromosome alignment made from this group is
a:

(A) autosome   (D) karyotype

(B) complement   (E) sex linkage

(C) homologue

Questions 114 - 116 refer to the drawing below.

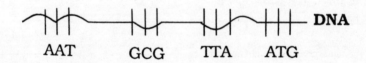

AAT   GCG   TTA   ATG

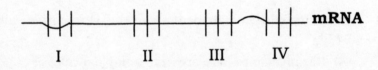

I   II   III   IV

114. RNA codons III and IV contain which of the following sequences?

(A) CCG and GCA

(B) AAU and UAC

(C) GGU and UGT

(D) AAT and TAC

(E) UUC and CUA

115. The function of the ribosome is to:

(A) allow transcription between DNA and mRNA

(B) guide DNA along mRNA for duplication

(C) provide a meeting site for mRNA and tRNA

(D) transfer amino acids to DNA for ordering

(E) transport tRNA to DNA for translation

116. The tRNA molecules that will bind to mRNA codons I and II will have which anticodons?

(A) UUA and CGC

(B) TTA and CGC

(C) CCG and TAT

(D) AAU and GCG

(E) GGC and ATA

Questions 117 - 118 refer to the following figure.

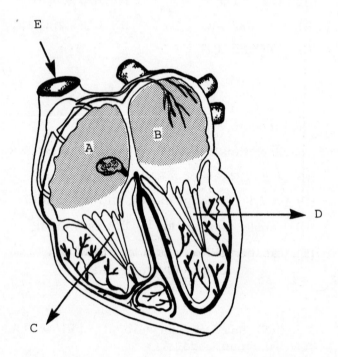

117. According to the figure, blood flows in the sequence:

    (A) A-B-C-D-E   (D) B-C-A-B-E

    (B) A-C-B-D-E   (E) E-B-D-C-A

    (C) B-D-E-C-A

118. The region that directly sends blood to the lungs is:

    (A) A   (D) D

    (B) B   (E) E

    (C) C

Questions 119 - 120 refer to the graph below.

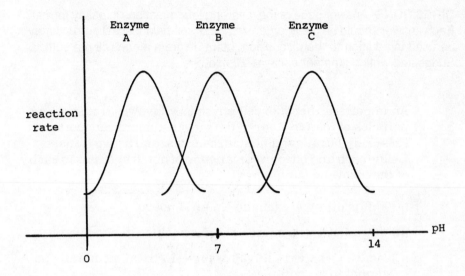

119. Select the correct statement about the graph of enzyme activities.

    (A) All three enzymes work best at the same pH.

    (B) Enzyme C works best at the most acid pH.

    (C) Enzyme C works best at the most alkaline pH.

    (D) Enzyme A works best at the most basic pH.

    (E) Each enzyme works over pH values 0 to 14.

120. Enzyme B could be which human enzyme?

    (A) amylase   (D) trypsin

    (B) pepsin   (E) maltase

    (C) sucrase

DIRECTIONS: Answer each of the following four questions in essay format. Each answer should be clear, organized, and well-balanced. Diagrams may be used in addition to the discussion, but a diagram alone will not suffice. Suggested writing time per essay is 22 minutes.

1. An important theme of cell physiology involves transporting particles of matter. Four of the most common transport processes are filtration, diffusion, osmosis, and active transport. Define each and offer an explanation of how it operates in each of the following situations.

    (A) diffusion – gas exchange in an *Amoeba*

    (B) filtration and osmosis – fluid exchange in a capillary

    (C) active transport and diffusion – electrical membrane potential in a neuron

2. How does Darwin's explanation of natural selection complement modern genetic knowledge to produce the current model of evolution? Offer an example of how this model could explain the change of a population characteristic through time: e.g., eye color, limb length, etc.

3. Plant and animal cells reproduce by two patterns: mitosis and meiosis. Compare and contrast the two by:

    (A) contribution to plant and animal life cycles

    (B) manipulation of chromosome number in cells

    (C) roles in growth and development

    (D) relationship to each other in sexually reproducing organisms

4. Compare water balance problems dealt with by a saltwater and freshwater fish. What adaptions have they evolved to solve these problems? How do these mechanisms compare to solutions evolved by terrestrial mammals, particularly to their kidney structure and function?

ADVANCED PLACEMENT
BIOLOGY EXAM III

ANSWER KEY

1.	D	31.	B	61.	D	91.	C
2.	B	32.	C	62.	B	92.	B
3.	B	33.	D	63.	C	93.	A
4.	B	34.	D	64.	D	94.	D
5.	E	35.	E	65.	A	95.	B
6.	C	36.	D	66.	C	96.	A
7.	C	37.	A	67.	A	97.	E
8.	B	38.	B	68.	E	98.	E
9.	D	39.	D	69.	A	99.	C
10.	C	40.	E	70.	A	100.	A
11.	A	41.	D	71.	E	101.	D
12.	D	42.	B	72.	D	102.	C
13.	C	43.	C	73.	A	103.	E
14.	E	44.	B	74.	E	104.	D
15.	B	45.	D	75.	B	105.	A
16.	A	46.	B	76.	E	106.	B
17.	B	47.	A	77.	D	107.	E
18.	C	48.	D	78.	B	108.	E
19.	A	49.	C	79.	A	109.	C
20.	C	50.	C	80.	E	110.	B
21.	A	51.	C	81.	C	111.	B
22.	E	52.	E	82.	B	112.	B
23.	A	53.	D	83.	C	113.	D
24.	E	54.	B	84.	B	114.	B
25.	E	55.	A	85.	D	115.	C
26.	B	56.	D	86.	B	116.	D
27.	D	57.	B	87.	C	117.	B
28.	E	58.	B	88.	E	118.	C
29.	A	59.	B	89.	A	119.	C
30.	C	60.	B	90.	E	120.	A

ADVANCED PLACEMENT
BIOLOGY EXAM III

DETAILED EXPLANATIONS OF ANSWERS

SECTION I

1.   (D)

Isomers of a molecule have the same molecular formula, but different structural formulas. For example, the monosaccharide $C_6H_{12}O_6$ can exist in forms of glucose, fructose or galactose depending on the arrangement of the very same atoms in space. The numbers and weights of each atom in the isomer remain the same as well.

2.   (B)

Of the choices given, carbon is the most abundant element found in protoplasm. Together with oxygen, hydrogen, and nitrogen, it composes over 90% of cellular structure. Calcium, phosphorus, and sulfur are found in varying amounts, depending upon the nature of the cell. Zinc is found in trace amounts.

3.   (B)

Heat capacity is the amount of heat energy required to separate a given amount of molecules. The higher the heat capacity, the more stable the set of molecules is. Water has a high heat capacity (1 cal/gm/^{0}C), therefore, it is highly stable and relatively unaffected by large changes in environmental temperature. Water is not a universal solvent because it is a polar molecule and will, therefore, only dissolve other polar (mostly inorganic) molecules. Viscosity refers to the rate at which a liquid flows and does not pertain to this question.

4.   (B)

Separation of homologous chromosomes occurs during anaphase. During interphase, the cell performs metabolic activities and DNA synthesis. Condensation of chromosomes, separation of the centrioles, and dissolution of the nuclear membrane are some of the major events of prophase. During telophase, the chromosomes begin to decondense, the nucleolus reappears and cytokinesis occurs. During metaphase, all chromosomes are oriented in the central plane of the cell.

5.   (E)
   The virtual absence of atmospheric oxygen allowed organic molecules in Earth's prebiotic areas to exist without undergoing oxidation. Had oxygen been present in appreciable concentrations, early organic molecules would have been oxidized and would therefore have been rendered unable to form macromolecules, the formation of which represented the next step towards "life." Simple amino acids were present, but they did not contribute to their own survival. Atmospheric ammonia was present in high concentrations, as was atmospheric methane. Rudimentary enzymes were not present at first, and could not have contributed to the prolongation of the molecules' existences.

6.   (C)
   After the completion of glycolysis, the original molecule of glucose will have been broken down into two molecules of pyruvic acid. Each molecule of pyruvic acid is then converted to one of Acetyl-coenzyme A. Each of the two molecules of Acetyl-coenzyme A will then enter the Krebs cycle, producing, among other products, three molecules of reduced NAD, for a total of six reduced NAD molecules per molecule of glucose.

7.   (C)
   In the scientific binomial name, the genus name is first and capitalized. The species name follows, beginning with a lower case letter.

8.   (B)
   Dicots and monocots are two main classes of angiosperms, within the division Anthophyta. They are flowering plants. As vascular plants, they have the conducting tissues of xylem and phloem. The name dicot refers to the appearance of two cotyledons (embryonic leaves) in early development. A dicot's leaf surface has a central internal conducting vessel with continuous side vessels extending from it - - net venation. The xylem and phloem that can be observed

in a dicot stem's cross-section are laid down in organized concentric rings or bundles. Monocots develop only one cotyledon and have a parallel arrangement of vessels in the leaves. Vascular tissue in their stems is randomly scattered. Centrioles are absent in higher plant cells.

9.   (D)

The role of the Golgi body or complex is to store and modify synthesized material for secretion from the cell. Modification is accomplished by the terminal glycosylation of the molecules, forming glycoproteins and glycolipids. Secretory vesicles bud from the end of the Golgi body. The cell's shape is maintained by microtubules, which form a supportive cytoskeleton. They are of the 9 + 2 variety, found in cilia, hair cells of the ear, rod and cone cells of the eye, as well as in spindles during cellular division. DNA replication is carried out in the nucleus during the S stage of the cell cycle. RNA is replicated using a DNA template, under the direction of the nucleolus. Proteins are synthesized in the ribosomes upon mRNA translation.

10.   (C)

The best-known example of adaptive radiation is that of Darwin's finches on the Galapagos Islands. Originating from the same species, each adapted to a slightly different niche, in order to avoid competition for limited island resources. Eventually they diverged into new species, each with its own niche. Current-day orders of mammals arose from a common, shrew-like originator. Evolved forms diverged from it, exploring uniquely different habitats from arboreal to aquatic to land-based.

11.   (A)

The humerus, or the upper arm bone, articulates with the radius and ulna (forearm bones) at its distal end. Its other end fits into the lateral socket of the scapula, or shoulder blade. It does not articulate with the clavicle, or collarbone.

12.   (D)

The nucleolus is a dark-staining spherical body inside the cell's nucleus. It is not associated with an organelle nor is it in the cytoplasm.

13.   (C)
Lysine is an amino acid, a subunit of proteins. Ribosomes are the sites of protein synthesis in a cell. By their loss, a cell would lose its ability to incorporate incoming amino acids into the proteins that it builds.

14.   (E)
Insulin lowers blood sugar while the other four tend to elevate it. Cortisol raises blood glucose levels by stimulating the mobilization of noncarbohydrate energy sources. Epinephrine (adrenalin) increases the conversion of glycogen into glucose and the release of glucose into the blood stream. Glucagon promotes the release of glucose from the liver. Growth hormone elevates blood glucose levels by apparently inhibiting the muscle's utilization of glucose and stimulating the release of fatty acids from adipose tissue. This release is subsequently converted to glucose. It must be noted that the anabolic character to promote protein synthesis makes it simultaneously similar to insulin as well.

15.   (B)
Cells that line the upper respiratory tract have cilia. These cilia trap and filter foreign debris. Smooth muscle around the respiratory tract is not in contact with air. Alveoli, which would also be harmed by smoking, are in the lower respiratory tract. Villi and goblet cells are located in the small intestine.

16.   (A)
Fossil imprints dated at well over three billion years old show the cell characteristics of the simple, prokaryotic cell type that bacteria still display.

17.   (B)
Lactic acid is the product of anaerobic animal metabolism. In the absence of sufficient oxygen for the cell, pyruvic acid (pyruvate) will not enter the aerobic reactions of the Krebs cycle. Instead, this product of glycolysis is shunted to an alternative chain of reaction whose final product is lactic acid. This toxic substance is converted back to pyruvic acid for further metabolic processing as oxygen becomes available to the cell. Ethyl alcohol is the anaerobic side shunt product of plants and is also produced by certain microorganisms.

Synthesis of ethyl alcohol via anaerobic respiration is known as fermentation. The remaining choices are not products of anaerobic respiration.

18.   (C)
Turgor is a term for the rigidity of a plant cell due to the uptake of incompressible water into that cell. The plant cell will not burst due to the outer, restraining effect of the rigid cell wall. A cell gains weight by water uptake through osmosis, resulting in turgor. Flaccidness is the consistency of a cell when water is lost.

19.   (A)
During synapsis, homologous chromosomes are attracted to each other and pair up. This occurs during the prophase stage of meiosis one.

20.   (C)
Pepsin is a protease. This enzyme catalyzes chemical breakdown of proteins. The other four choices belong to the enzyme class of carbohydrases. They catalyze carbohydrate digestion. Amylase, also known as ptyalin, catalyzes the digestion of the polysaccharide starch into the sugar maltose. Sucrase and maltase break down the carbo-hydrate disaccharides sucrose and maltose, respectively.

21.   (A)
Centromeres are structures that hold chromatids together in double-stranded chromosomes. The centromere is the region of the chromosome to which spindle fibers attach. Vacuoles (B) are cell organelles that contain storage materials. Gametes (C) are specialized reproductive cells. Stomates (D) are small openings in the epidermis of leaves through which gases diffuse. Centrioles (E) are cell or-ganelles found in animal cells that function in the process of cell division.

22.   (E)
Somatic cells are diploid. In this species the diploid (di = 2) chromosome number is 80. A tetraploid (tetra = 4) chromosomal would contain 160 chromosomes, because there would be four, instead of two, choices of each chromosome.

23. (A)

A diploid cell proceeding through meiosis yields haploid daughter cells with a halved chromosome number. A cell with a diploid chromosome number of 42 has 21 pairs of homologous chromosomes. If all twenty-one pairs undergo normal disjunctions, then each daughter cell will contain twenty-one chromosomes. If only twenty pairs do so, then one daughter cell will have twenty chromosomes and one will have twenty-two chromosomes.

24. (E)

This is a fact. The blastula is a hollow ball of cells and early embryo, produced by mitotic divisions of the zygote. The other choices represent stages of oocyte development.

25. (E)

Wandering to other body sites from point of production is a characteristic of cancer cells. This traveling seeds other body regions with the rapidly-dividing cells, thus spreading the cancer.

26. (B)

The insertion is the more moveable of the two locations to which a muscle is attached. The origin is the anchored end toward which the movable end moves to produce movement. Prime mover refers to the main muscle contracting to produce a movement. Its opposing antagonistic muscle must yield by stretching or relaxing. The synergist is a muscle that aids the prime mover in its action.

27. (D)

In a cell cycle, the S phase is characterized by DNA synthesis prior to the active stages of mitosis. Chromosomes must duplicate at this stage or there will be an absence of chromosome duplicates to separate into daughter cells during division.

28. (E)

Three DNA nucleotides code for one amino acid in a protein. The ratio is 3 to 1, or 450 to 150.

29. (A)

Rods and cones are retinal cells lining the eye's inner surface. Cones are specialized for color discrimination or visual activity. Rods are utilized in dim light.

30. (C)

Normally, genetic information is transferred from DNA to RNA; RNA then provides the necessary information for protein synthesis. A few oncogenic (cancer-causing) viruses are known to transfer information from RNA to DNA; the DNA then directs protein synthesis.

31. (B)

In the lac operon model, the sugar, lactose, is the inducer. It will bind to the repressor produced by the regulator gene. Unable to bind to and inhibit the operator gene, the repressor is inhibited. This allows the operator gene to activate the cistrons. These structural genes synthesize the enzymes that metabolize the substrate lactose.

32. (C)

During mitotic metaphase, the chromosomes align in the equatorial plane of the cell's mitotic spindle. In anaphase, the centromeres split and chromosome duplicates are pulled apart. The cell divides into two daughter cells in the next stage, telophase. Prophase precedes metaphase.

33. (D)

The number of permutations is 4^3, or 64.

34. (D)

If a uracil-containing nucleotide is added, then the DNA nucleotide on the strand that is being transcribed must contain adenine. Therefore, the base on the strand of DNA that is not being transcribed must contain thymine.

35.   (E)

A and C are farthest apart. B, which is 10 units from A and 5 units from C, is in between. Thus, ABC or CBA are the only choices compatible with the data.

36.   (D)

Pith is the core, with xylem placed around it by the cambial growth layer. Phloem is deposited to the outside, with the cortex external to it.

37.   (A)

Each year the cambium puts down a new xylem layer, pushing older ones progressively toward the center. As xylem ages and matures, it becomes the stem's wood.

38.   (B)

Lactase is the enzyme that catalyzes the hydrolysis of the disaccharide lactose into its monosaccharide components, galactose and glucose. Glucose and glucose are the monosaccharide components of the disaccharide maltose. Glucose and fructose compose the disaccharide sucrose.

39.   (D)

The father could not have had a blood-type with a negative Rh-factor, because the trait for a negative Rh-factor is recessive. The father must also have blood-type O or, heterozygously, have blood type A or B, in order for him to have donated a recessive i (blood-type O) gene. Therefore, the father must have blood-type A⁺, B⁺ or O⁺. Only A-positive is listed as a choice.

40.   (E)

Among mammals, the smaller the organism, the greater its surface-to-volume ratio and coupled metabolism. The shrew is the smallest mammal among the choices and shows these characteristics.

41.   (D)
The probability of independent events occurring in succession is the product of their separate probabilities. Births are independent events. 1/2 to the fifth power equals 1/32.

42.   (B)
Populations have a tremendous capacity to increase their numbers. There is competition for resources in short supply. Genetic variations best equipped to survive will do so at a higher probability. Their chance to reproduce and contribute genes to the next generation is also more likely.

43.   (C)
This is an example of a dihybrid cross. Phenotypically, 9/16 of the offspring express the dominant allele of both genes, 3/16 of the offspring express the dominant allele of the first gene and the recessive allele of the second gene, 3/16 of the offspring express the dominant allele of the second gene and the recessive allele of the first gene, and 1/16 of the offspring will express the recessive alleles of both genes.

44.   (B)
A person of blood-type AB has no anti-A and no anti-B antibodies in his blood plasma. Therefore, there will be no antibodies present that would attack foreign red blood cells that enter the bloodstream during transfusion. A person whose blood has a positive Rh-factor also has no antibodies that attack Rh antigens present on red blood cells. Therefore, a person of blood-type AB+ would have no trouble in receiving any type of blood during a transfusion.

45.   (D)
The formula for this is 2^n, in which n equals the number of heterozygous gene pairs. 2 x 2 x 2 = 8.

46.   (B)
Examples of disease-causing (pathogenic) bacteria in this genus are *C. tetani* (tetanus), *C. botulinum* (botulism) and *C. perfringens* (gas gangrene).

47. (A)
Viruses are incredibly small (nanometers). They lack normal cellular structures and thus need a host organism to grow and reproduce. They consist of protein coats surrounding nucleic acid cores.

48. (D)
All are members of the order of Primates.

49. (C)
The zygote is the fertilized egg, or first cell. Five divisions of mitosis will increase the early embryo cell number to 32.

50. (C)
Crossing-over, the exchange of corresponding regions of homologous chromosomes, could occur between the two medial chromosomes (EFGH and IJKL) to produce many possible new chromosomes, one of which is EFKL.

51. (C)
This is the photosynthetic layer saturated with water vapor found between the upper and lower epidermis. The high water vapor content usually assures 100% humidity.

52. (E)
Pruning a flowering plant, that is, cutting off the bud at the plant shoot apex, removes the source of the plant hormone that would travel vertically down the plant and inhibit lateral bud growth. In its absence, lateral bud growth from the shoot increases. The growth hormone traveling most often through the plant is of a family termed the auxins.

53. (D)
Working backwards from offspring to parents, an Aa x aa cross is the only possible one producing a 50-50 ratio in offspring variability. Consider the genetic grid:

	A	a
a	Aa	aa
a	Aa	aa

54.   (B)

The concentration of follicle-stimulating hormone, FSH, increases following menses. Thus, the menstrual cycle continues. It is secreted by the pituitary gland and serves to increase follicle growth around a selected sex cell in the ovary.

ACTH = adrenocorticotrophic hormone

LH   = luteinizing hormone

GnRH = gonadotrophin-releasing hormone

GH   = growth hormone

55.   (A)

In renal physiology, the volume eliminated equals the volume filtered minus the volume reabsorbed. $X - 1.5l = 188l$ eliminated. The unknown amount equals $189.5l$.

56.   (D)

Morphology refers to body structure and form while phylogeny represents evolutionary history. The two are related. For example, a chimpanzee is the most humanlike animal because of the recent common evolutionary ancestor of chimp and human, revealed by the fossil record.

57.   (B)

The frog is an amphibian, a cold-blooded or poikilothermic animal lacking mechanisms to control internal body temperature independently from changes in the external environment. Reptiles have some ability to do this while birds and mammals, homeotherms, possess the greatest capacity for this control.

58.   (B)

The genetic cross is $I^A i$ dd (blood-type A and Rh-negative, which is a recessive condition) x $I^B i$ Dd. In summary:

	I^B D	I^B d	i D	i d
I^A d	I^AI^B Dd	I^AI^B dd	I^Ai Dd	I^Ai dd
i d	I^Bi Dd	I^Bi dd	ii Dd	ii dd

Note: Only the recombination in the lower right corner yields a completely recessive genotype, indicative of blood-type O, Rh-negative offspring. This is only one of the eight possible recombinations.

59.   (B)
Auxin, the very hormone that stimulates shoot growth, seems to have an inhibitory effect on the specialized cells of the plant root at high concentrations. It inhibits their mitosis rate which is responsible for vertical elongation.

60.   (B)
"Trachea" means tube, the key to the success of this group of land plants. Their vascular tissue is composed of xylem and phloem. Fungi and mosses do not belong to this phylum. Tracheophytes are autotrophs (self-nourishing) due to their photosynthetic ability.

61.   (D)
Insulin, as well as growth hormone and interferon, are now synthesized by bacteria. Actin and myosin, muscle contractile proteins, plus the red blood cell protein for gas transport, hemoglobin, are proteins, but cannot be synthesized as yet by such methods.

62.   (B)
DNA makes RNA during genetic transcription. DNA's base sequence determines RNA nucleotide sequence. RNA polymerase is the enzyme that catalyzes the assemblage of these subunits into a polymer, RNA.

63.   (C)
I is the symbol for iodine, a necessary ingredient of thyroxine. Deficiency of the mineral produces the thyroid gland enlargement symptomatic of simple goiter.

64.   (D)
  Pulse pressure is the difference between an artery's high (systolic) and low (diastolic) pressures. The pressure in all major arteries is pulsatile.

(Systolic pressure) – (Diastolic pressure)   = (Pulse pressure)

        x   – 90 mmHg   = 30 mmHg

                              x   = 120 mmHg

The systolic pressure is 120 mmHg.

65.   (A)
  The invertebrate was removed from a freshwater environment and therefore could not have been an Echinoderm, because echinoderms live only in marine environments. Poriferans and Coelenterates do not have mesodermal germ layers. Annelids do not have an outer cuticle.

66.   (C)
  This is the last of the four Koch postulates. Used in etiology, the study of the causes of contagious diseases, it is a definitive step. If the restudied microbe matches the originally described microbe, it is the pathogen in question.

67.   (A)
  Lymphocytes are white blood cells that attack foreign, invasive agents in the body. T-lymphocytes change to plasma cells that produce humoral antibodies. These molecules react against foreign chemical agents, antigens in the blood. B-lymphocytes attack cellular invaders, such as the foreign cells in a tissue graft.

68.   (E)
  Acetylcholine is a neurotransmitter, which is a class of molecules that traverse synaptic clefts in order to perpetuate a neural impulse. Any drug that inhibited the action of acetylcholine would have to act on the synapse.

69. (A)
This is a snail shell-shaped structure in the inner ear. The semicir-cular canals (dynamic equilibrium) and vestibule (static equilibrium) are there to control body balance. The pinna, outer ear cartilage flap, and tympanum (eardrum) transmit sound waves from the outer ear through the middle ear and on to the inner ear.

70. (A)
Arteries take blood away from the heart, dividing into smaller, more numerous arterioles. They divide to become numerous, microscopic capillaries for exchange with cells. They collect into venules, which merge to form larger veins. Venules and veins return blood to the heart.

71. (E)
Primates, as one order of mammals, have four-chambered hearts. All other listed choices are characteristics that describe primates, a taxonomic order that humans share with the great apes, monkeys, and prosimians.

72. (D)
The population is mating randomly, so the Hardy-Weinberg formula may be used:

$$p^2 + 2pq + q^2$$

in which p is the dominant allele's frequency and q is the recessive allele's frequency. It is given that q^2, the frequency of organisms that express the homozygous recessive alleles, is 36%. Since $q^2 = .36$, then q = .6.

73. (A)
This gas constitutes 79% of the atmosphere. Soil-dwelling bacteria fix it in the soil in a chemical form for plant use — nitrogen fixation.

74.   (E)

Herbivores are plant-eaters and are first to feed in a food chain.

75.   (B)

Carbon monoxide is a colorless, odorless gas that can bind to hemoglobin without the subject's awareness. As hemoglobin in red blood cells transports oxygen, oxygen's unavailability due to CO binding causes internal suffocation, or asphyxiation.

76.   (E)

Translocation is the attachment of a chromosome fragment to a nonhomologous chromosome. Duplication is the attachment of the fragment to the homologous chromosome's counterpart, thus repeating gene types already there. Inversion is the reattachment of the fragment to the original chromosome, but in a reversed orientation, resulting in a reversed gene order. In a deletion, the chromosome fragment does not reattach. Diversion does not refer to chromosomal mutations.

77.   (D)

The seminal vesicle adds fluid in the vas deferens prior to the fluid's entrance into the prostrate gland beneath the bladder. The Cowper's gland adds a small final amount to the urethra prior to the fluid's entrance into the penis.

78.   (B)

In general, there is a decreased amount of energy at each successive trophic level of an energy chain. Producers are found at the base of this chain and thus harbor most of the energy in the chain.

79. (A)   80. (E)   81. (C)   82. (B)

Choice (A), NH_2, is the amino group. Along with an acid group, COOH, not shown, it bonds to the central carbon of an amino acid to derive its name. A nucleotide, the building block subunit of a nucleic acid, has three components: a pentose (five carbon) sugar, a nitrogen base, and a phosphate group as shown in choice (B). Choice (C) is the aldehyde group found on the number one carbon of the hexose chain of the monosaccharide glucose. (E) is the ketone group, denoting the number two carbon in the hexose chain of the monosaccharide fructose.

83.   (C)
This organelle is the mitochondrion, the site of cellular respiration. Pyruvic acid enters the Krebs cycle (citric acid cycle) here, where energy released by oxidation reactions performed on pyruvic acid is stored in the high-energy phosphate bonds of ATP. ATP, adenosine triphosphate is the molecule used by all cells to store energy.

84.   (B)
The inner surface of a mitochondrion's inner membrane has bound to it molecules that take part in oxidative phosphorylation. These molecules are alternately reduced and oxidized, which releases energy that can be used to synthesize ATP.

85.   (D)
The anaerobic metabolism of glucose is known as glycolysis. Unlike the aerobic phase of metabolism in the mitochondrion, the team of enzymes running the glycolytic pathway are found in the cytoplasm near the mitochondrion.

86. (B)   87. (C)   88. (E)   89. (A)

$$6CO_2 + 12H_2O \xrightarrow{\text{light}} 1C_6H_{12}O_2 + 6H_2O + 6O_2$$
$$\text{glucose}$$

In the light reactions of photosynthesis (photosystems I and II), light causes the excitation of electrons in chlorophyll. These excited electrons are passed from one acceptor molecule to another, and finally are accepted by NADP. However, since a chlorophyll molecule must give up its electrons, they must be replaced. According to the equation

$$2H_2O \longrightarrow 4H^+ + 4e\text{-} + O_2,$$

water donates the electrons that are needed by the now electron-poor chlorophyll molecule. Oxygen gas is also a product of this reaction.
CO_2 enters the dark reactions of photosynthesis (Calvin-Benson cycle), and is used to synthesize glucose, the final product of the Calvin-Benson cycle.

Choice (D) could refer to electrons or even to light energy, but these were not listed in the questions. Hence, (D) is an incorrect choice.

90.   (E)
The I-band is the region of the sarcomere that contains actin only. (A) designates the Z-line. (B) specifically points out groups of actin. (C) designates a bundle of myosin molecules. (D) designates the A-band, which contains bundles of myosin, and the region of overlap of actin and myosin.

91.   (C)
The thicker myosin protein filaments have cross-bridges coming off of them at right angles. These crossbridges can bind to actin.

92.   (B)
The crossbridges, which are actually the heads of individual myosin molecules, temporarily bind to nearby actin molecules. The bridges then bend; this causes the actin filament to move along the myosin filament. The movement of many actin filaments causes the region between the Z-lines, the sarcomere, to shorten. The shortening of many sarcomeres results in a muscle contraction.

93.   (A)
(A) represents the nonconducting pith while (D), (C), and (B) respectively represent first-, second-, and third-year xylem growth. (E) designates phloem, another conductive tissue type laid down outside of the other layers.

94.   (D)
Xylem is pushed from the growth layer inward. Thus, the more central the xylem layer, the older it is. (D) thus designates the newest layer (third layer) of xylem.

95. (B)   96. (A)   97. (E)
Each choice is descriptive of its matched invertebrate phylum.

98. (E) 99. (C) 100. (A) 101. (D)

The *Paramecium* is a freshwater protozoan that has numerous hairlike cilia on its cell membrane. They beat in synchrony to propel the organism. Mastigophorans whip a flagellum to produce movement, as does the *Euglena*. Sarcodinians, such as the *Amoeba*, project protoplasmic extensions, pseudopodia, to move along. Sporozoans, immobile as adults, cause harm to their hosts. Generic member *Plasmodium* is the best known; it causes malaria. A bacteriophage is a virus parasitizing a bacterium.

102. (C) 103. (E) 104. (D) 105. (A)

Chordates have supportive dorsal structures, either notochords or vertebral columns. Coelenterates, such as jellyfish, have saclike bodies and radial symmetry. All coelenterates have specialized stinging cells, nematocysts, in their external cell layers. Platyhelminthians, such as members of genus *Dugesia* (commonly called planaria), lack the true body segmentation of annelids, the segmented worms. They are dorsoventrally flattened and have gastrovascular cavities. Poriferans are sponges, which have retained many primitive animal characteristics. They greatly (but not totally) resemble their distant ancestors, who, quite early in animal evolution, diverged from the evolution of the other animal phyla.

106. (B)

This structure is the medulla, at the base of the brainstem. It controls blood pressure by gauging the sizes of blood vessels and their resistance to blood flow. Cardiac and respiratory centers are also found here.

107. (E)

This is the temporal lobe of the cerebral cortex. Mapped regions for taste (gustation), smell, and hearing are found here. Vision is the domain of the occipital lobe (D).

108. (E)

The pons is a hindbrain structure above the medulla. It contains ascending and descending tracts.

109.   (C)
Autotrophs make their own food. As producers such as the oak tree, they synthesize organic nutrients such as sugars. Heterotrophs depend on producers for their food in a ready-made form. They are consumers and represent successive links of the food chain.

Since only about 10% of the energy made available to members of one trophic level can be acquired by members of the next trophic level, it stands to reason that the heterotrophs (of any food chain) that are members of the trophic level closest to the producers' (autotrophs') trophic level will have more energy available for them. Therefore, in this food chain, the insects, of the second trophic level, are the heterotrophs with the most useful energy available to them.

110.   (B)
Omnivores (omni=all) feed on both plant and animal matter. Of the organisms that are listed, only the human qualifies as an omnivore. Both the bird and the pike are carnivores, and the insect is an herbivore.

111.   (B)
The ten chromosomes contain two chromatids each, for a total of twenty chromatids. For each of the five kinds of chromosomes, another one in the group exists that is similar in size, shape, and structure: a homologue.

112.   (B)
The cell that contains these chromosomes is diploid, because there are two copies of each type of chromosome, each of which contains two chromatids. The two copies of each type of chromosome are homologous. If the cell completes mitosis, there will be ten chromosomes in each daughter cell (each daughter cell will receive one chromatid of each chromosome). Meiosis results in haploid gametes. The haploid number in this case is half of ten, or five.

113.   (D)
A karyotype is a categorization of the chromosomes in a cell based on chromosomal similarities and differences. For example, human chromosomes, of which there are 23 pairs, are karyotyped into seven groups.

114.   (B)

The base-pairing rules between DNA and RNA for DNA transcription are:

DNA	RNA
Adenine	Uracil
Cytosine	Guanine
Guanine	Cytosine
Thymine	Adenine

115.   (C)

The ribosome is the site at which a given mRNA molecule is "read." During this decoding, a tRNA molecule with a complementary anti-codon carries an amino acid to the C-terminus of the protein that is being synthesized.

116.   (D)

The anticodons that will bind to the mRNA are determined by the mRNA codons, which in turn are determined by the DNA sequence:

DNA	mRNA	tRNA
Adenine	Uracil	Adenine
Cytosine	Guanine	Cytosine
Thymine	Adenine	Uracil
Guanine	Cytosine	Guanine

117.   (B)

If one chooses to start with the right atrium, (A), blood next moves through the right ventricle, (C). It then flows to the lungs in the pulmonary artery, not shown, and returns to the left atrium (B) in pulmonary veins. This atrium primes the left ventricle, (D), and the ventricle in turn pumps blood out through the ascending aorta, (E).

118.   (C)

According to the previous explanation, this is the right ventricle's role.

119. (C)

Each enzyme works best in a unique range within 0-14. Enzyme C is farthest to the right for the highest number, most alkaline pH.

120. (A)

Amylase, in the saliva, works optimally in the oral cavity's neutral pH.

ESSAY I

The *Amoeba* is a freshwater protozoan. Its entire makeup consists of one cell. All regions of the organism are in close proximity to the external environment. Under these conditions, diffusion can account for all local transport of substances.

Diffusion is the movement of matter's particles from a region of higher concentration to a region of lower concentration. Be they atoms, ions, or molecules, particles of matter have a tendency to spread out by virtue of their randomized, colliding motion. A bottle of perfume molecules, opened in a room, will release molecules that spread out as they diffuse through the room. If the room is sealed off, the vaporized molecules become equally distributed. Diffusion, therefore, tends toward a balanced concentration of matter's particles, or equilibrium.

In *Amoeba*, oxygen will diffuse from the outside pond water into the cell. Relatively less is in the cell because it is consumed by the cell. Extracellular oxygen moves toward this deficit. Carbon dioxide is produced in the cell. Its concentration gradient forces CO_2 outward. Nutrient molecules will also move in, whereas accumulated intracellular waste diffuses outward. The cell membrane must be permeable to these substances to allow free transport along their concentration gradients.

Osmosis is the diffusion of water through a permeable membrane. Another description of osmosis is the movement of water into a hypertonic environment. <u>Hypertonic</u> means a higher solute (dissolved substance) concentration. If solute levels are <u>higher</u>, water concentration is <u>lower</u>. The moving water comes from a hypotonic setting (<u>lower</u> solute concentration, higher water levels). Filtration is the movement of materials by bulk flow. The mere mechanical pressure of a moving substance accounts for particle transport rather than natural, randomized movement. The impact of water coursing through a garden hose will cause its escape if the hose wall is punched full of holes.

Both osmosis and filtration explain substance movement in the circulatory system vessel exchanging contents with body cells, the capillary. Arterial blood flow at the capillaries' input end is under high filtration pressure from heart pumping action. At the venous end, this pressure has been weakened by the frictional resistance to blood flow of the small-diametered capillary. Both end pressures, as with water in the garden hose, force fluid outward. High concentrations of plasma proteins remain in the blood, as they are too large to pass through the capillary membrane pores. They keep the inside blood hypertonic to the outside and thus draw fluid into the blood by osmosis uniformly along the capillaries' length.

At the arterial end, blood fluid (plasma) moves out to cells as the strong filtration pressure overcomes osmosis. At the venous end, the

weakened filtration pressure succumbs to the stronger, uniform osmotic pressure. Therefore, fluid with materials is drawn from the cells back into the blood as it flows from the capillary.

Active transport is the opposite of diffusion. Matter moves from areas of lower concentration to those of higher concentration as the cell expends energy to transport. Along a nerve cell's membrane, extracellular sodium ions diffuse in due to their high outside concentration. Potassium ions diffuse to the outside. An active transport mechanism, however, pumps sodium back to the outside and potassium to the inside. This keeps each poised higher at their respective extracellular and intracellular levels. As charged particles, ions, they maintain regionally a constant electrical-like character along the neuron's membrane.

ESSAY II

Darwin's principle of natural selection accounts for the force of the environment in the evolution of a population. The major points of his premise are as follows:

1. Populations have a tremendous biotic potential to increase their numbers. As an example of the possibility of growth rate, consider a female sunfish in a pond. It can lay 200,000 eggs per reproductive cycle. About one-half of these eggs could develop into female sunfish. Each of these 100,000 females could lay 200,000 eggs as this trend multiplies to astronomical figures after very few generations of sunfish. However, only two or three of the hatched sunfish may survive. Among wild populations, high death rates among the young is the rule. This leads to the next point.

2. Darwin was influenced in his thinking by an economist named Malthus. Taking a page from supply-side economics, Darwin concluded that the amount of resources to support a population is limited. There is just so much oxygen, food, space, etc., in the pond to support a maximum number of fish. With the tremendous biotic potential of a species, competition for resources is inevitable.

3. By studying populations of reptiles and birds of the Galapagos Islands off the coast of South America, Darwin concluded that variation, genetic difference, is a characteristic of any population. During his study in the mid-nineteenth century, however, he did not know the basis for the variation, the actions of genes, yet undiscovered.

4. Those with adaptive variations favorable in the given environment have better probabilities of surviving and leaving offspring. Thus, broader leaves, taller giraffes, faster predators, or better-camouflaged prey all have adaptions to promote survival. This survival power for increased life span yields increased reproductive fitness, the individual's ability to pass its genes on to future generations.

5. This higher probability of survival and reproductive fitness allows affected members to have more of an effect on the gene pool of the next generation. Thus, a superior gene for tallness, among more surviving members possessing it, leads to higher odds that it will be genetically passed on and lead to a higher percentage of tall members in the next generation.

The merger of these principles with genetics came with the under-standing of mutations. Mutations are changes in a gene and the raw material for evolution. Without a source of change, a population cannot change or evolve. For example, if the only form of an eye color gene were dark, "B," all members of the population would be homozy-gous for "B," BB. All mating, would be BBxBB and produce homozy-gous organisms with dark eye color. If the "B" dominant gene mutates to a recessive "b," then there is an eye color alternative. A mating of two heterozygous organisms, BbxBb, can produce blue-eyed off-spring, bb, with a 25% chance. If blue eyes is adaptive against the environment, more blue-eyed individuals will survive longer. Over time they will leave their blue eye genes hereditarily and gradually increase the frequency of blue-eyed individuals in the future gene pool.

ESSAY III

Mitosis is cell reproduction: one parent cell divides to form two daughter cells and each daughter cell is genetically identical to the parent cell. The genetic constancy includes chromosome number and combination as well as gene-by-gene content on each chromosome. For example, an onion cell normally has a full complement of 24 chromosomes. During its cell cycle, each chromosome duplicates and becomes double-stranded. With the onset of mitosis, the duplicates (chromatids) of each chromosome are separated in orderly fashion at the same time. The duplicates of each are finally separated and segregated into separate daughter cells as the cell splits into two.

Mitosis is more than just a mere splitting. Its preservation of the genetic picture is also paramount. Mitosis functions in the growth and development of the organism of a species. In its life cycle, for example, a frog begins as a zygote, the fertilized egg. Mitosis multiplies the original cell, numbering from one to two - four - eight - sixteen, etc. Eventually, cell number reaches the millions, billions, and trillions. The human body is estimated to have 75 trillion cells.

Even when growth of cell number ceases, mitosis continues to function in replacement of worn-out cells. Outer layers of skin cells are constantly dying. A human red blood cell normally has a life span of 120 days. Mitosis balances rates of cell death with rates of cell reproduction to retain the needed number of cell types. Cancer is an affliction in which cell production is too rapid. Specific cells become too numerous and compete with other cells for nutrients and space.

Meiosis, on the other hand, changes the genetic picture by producing cells with one-half the chromosome number of the full species complement. Two successive divisions of a parent cell produce daughter cells that are <u>haploid</u>. In other words, they possess one-half the species' chromosome number. In the division process, chromosomes in a pair are separated. Thus if an organism has 64 chromosomes, or 32 pairs, meiosis separates chromosomes in each pair to cut the number to one-half by a very specific pattern. In mitosis, chromosome pairs are not separated, as the full complement is preserved from parent to daughter cells. For example, human cells produced mitotically receive 23 chromosome pairs and a full complement of 46. Generally, mitosis produces <u>diploid</u> cells ("di" = two).

In most animals, meiosis produces sex cells, or gametes. During fertilization, two gametes from opposite sexes recombine to reestablish a full chromosome complement in the zygote. Meiosis produces great variety during sex-cell production. Mathematically, sorting one chromosome per pair, randomly, from all pairs yields a tremendously large number of sex cell identities. For example, in humans, the number of genetic variations in sex cells is 2^{23}, or over 8 million. The exponent 23 comes from the number of chromosome pairs.

In the life cycle of many sexually-reproducing plants, both processes of cell production figure into the plant's alternation of generations. A diploid sporophyte reproduces spores by meiosis. These haploid products develop into a smaller gametophyte. Haploid gametophytes produce haploid cells by <u>mitosis</u>, thus preserving the haploid chromosome number. Fusion of some of these haploid cells founds a diploid sporophyte after mitosis for its growth and development.

ESSAY IV

Ironically, a marine fish such as a tuna or marlin faces potential body water loss. This is because their body fluids are hypotonic to the extracellular salt water. Apparently, marine fish evolved from freshwater, inland ancestors and retained the dilute body fluids that matched that environment. The hypertonic outside environment thus draws water from the fish by osmosis. To some degree, they can prevent water loss in their scaly, impermeable body coverings. Certain permeable surfaces, such as the gill areas, however, are vulnerable.

Marine fish tend to lose water by osmosis and take in the outside salt by diffusion. They solve this potential problem of imbalance by drinking continuously and excreting salt through specialized cells in their gills. They have not evolved a kidney powerful enough to concentrate entering salts into the urine as another salt elimination route.

Freshwater fish face the opposite problem. Their body fluid solute levels are higher than an extracellular setting relatively devoid of such dissolved solids. Their body fluids, therefore, attract outside water osmotically and lose solutes to the outside by diffusion across vulnerable body surfaces. Their kidneys excrete dilute urine. They seldom drink and actively absorb lost salts through specialized transporting cells.

The main water-balance problem faced by terrestrial organisms, including mammals, is desiccation. Mammals have evolved a remarkable kidney to combat this potential problem. Physiologically, the kidney works by at least two processes: filtration and reabsorption.

The kidney is a good example of a whole organ whose activity is determined by the collective action of its individual parts. The kidney parts are microscopic units termed nephrons. A typical mammalian kidney has at least one million of such nephrons. The nephrons conduct the two processes of renal physiology.

Imagine the nephron as a smoking pipe with a bowl and a stem. The nephron's bowl is a cuplike section termed the Bowman's capsule. A specialized capillary tuft, the glomerulus, is housed inside of it. This region is the site of renal filtration. Filtration is the movement of blood plasma materials from the blood into the Bowman's capsule. Filtration is a very unselective, nondiscriminating process. Most all soluble blood plasma components move into the nephron initially by substantial amounts.

Step two is reabsorption. This involves the tubular portion of nephrons that relate to another capillary network surrounding them, the peritubular capillaries. Here, components in the initial filtration move from capsule to tubule and return from the microscopic nephron tracts, tubules, back into the blood. Substances returned to the blood flow away and are thus retained. Reabsorption rates are usually very high. For example, a pair of human kidneys can filter 190 liters of recycled body water per day. However, 187 to 189 liters are returned to the blood once the kidneys have monitored levels and returned water as needed. In spite of the high reabsorption rates, there is some latitude for control and fine tuning. The antidiuretic hormone from the posterior pituitary gland works in the kidney to oppose diuresis and facilitate water reabsorption. If 187 liters of 190 are reabsorbed, 3 liters is eliminated. If 189 liters are taken back, only 1 liter is eliminated. Aldosterone, a hormone from the adrenal cortex, works on tubular permeability to increase sodium reabsorption as needed.

Thus, final reabsorption rates work to meet body needs for chemical components.

THE ADVANCED PLACEMENT EXAMINATION IN

BIOLOGY

TEST IV

THE ADVANCED PLACEMENT EXAMINATION IN

BIOLOGY

ANSWER SHEET

1. Ⓐ Ⓑ Ⓒ Ⓓ Ⓔ	21. Ⓐ Ⓑ Ⓒ Ⓓ Ⓔ	41. Ⓐ Ⓑ Ⓒ Ⓓ Ⓔ
2. Ⓐ Ⓑ Ⓒ Ⓓ Ⓔ	22. Ⓐ Ⓑ Ⓒ Ⓓ Ⓔ	42. Ⓐ Ⓑ Ⓒ Ⓓ Ⓔ
3. Ⓐ Ⓑ Ⓒ Ⓓ Ⓔ	23. Ⓐ Ⓑ Ⓒ Ⓓ Ⓔ	43. Ⓐ Ⓑ Ⓒ Ⓓ Ⓔ
4. Ⓐ Ⓑ Ⓒ Ⓓ Ⓔ	24. Ⓐ Ⓑ Ⓒ Ⓓ Ⓔ	44. Ⓐ Ⓑ Ⓒ Ⓓ Ⓔ
5. Ⓐ Ⓑ Ⓒ Ⓓ Ⓔ	25. Ⓐ Ⓑ Ⓒ Ⓓ Ⓔ	45. Ⓐ Ⓑ Ⓒ Ⓓ Ⓔ
6. Ⓐ Ⓑ Ⓒ Ⓓ Ⓔ	26. Ⓐ Ⓑ Ⓒ Ⓓ Ⓔ	46. Ⓐ Ⓑ Ⓒ Ⓓ Ⓔ
7. Ⓐ Ⓑ Ⓒ Ⓓ Ⓔ	27. Ⓐ Ⓑ Ⓒ Ⓓ Ⓔ	47. Ⓐ Ⓑ Ⓒ Ⓓ Ⓔ
8. Ⓐ Ⓑ Ⓒ Ⓓ Ⓔ	28. Ⓐ Ⓑ Ⓒ Ⓓ Ⓔ	48. Ⓐ Ⓑ Ⓒ Ⓓ Ⓔ
9. Ⓐ Ⓑ Ⓒ Ⓓ Ⓔ	29. Ⓐ Ⓑ Ⓒ Ⓓ Ⓔ	49. Ⓐ Ⓑ Ⓒ Ⓓ Ⓔ
10. Ⓐ Ⓑ Ⓒ Ⓓ Ⓔ	30. Ⓐ Ⓑ Ⓒ Ⓓ Ⓔ	50. Ⓐ Ⓑ Ⓒ Ⓓ Ⓔ
11. Ⓐ Ⓑ Ⓒ Ⓓ Ⓔ	31. Ⓐ Ⓑ Ⓒ Ⓓ Ⓔ	51. Ⓐ Ⓑ Ⓒ Ⓓ Ⓔ
12. Ⓐ Ⓑ Ⓒ Ⓓ Ⓔ	32. Ⓐ Ⓑ Ⓒ Ⓓ Ⓔ	52. Ⓐ Ⓑ Ⓒ Ⓓ Ⓔ
13. Ⓐ Ⓑ Ⓒ Ⓓ Ⓔ	33. Ⓐ Ⓑ Ⓒ Ⓓ Ⓔ	53. Ⓐ Ⓑ Ⓒ Ⓓ Ⓔ
14. Ⓐ Ⓑ Ⓒ Ⓓ Ⓔ	34. Ⓐ Ⓑ Ⓒ Ⓓ Ⓔ	54. Ⓐ Ⓑ Ⓒ Ⓓ Ⓔ
15. Ⓐ Ⓑ Ⓒ Ⓓ Ⓔ	35. Ⓐ Ⓑ Ⓒ Ⓓ Ⓔ	55. Ⓐ Ⓑ Ⓒ Ⓓ Ⓔ
16. Ⓐ Ⓑ Ⓒ Ⓓ Ⓔ	36. Ⓐ Ⓑ Ⓒ Ⓓ Ⓔ	56. Ⓐ Ⓑ Ⓒ Ⓓ Ⓔ
17. Ⓐ Ⓑ Ⓒ Ⓓ Ⓔ	37. Ⓐ Ⓑ Ⓒ Ⓓ Ⓔ	57. Ⓐ Ⓑ Ⓒ Ⓓ Ⓔ
18. Ⓐ Ⓑ Ⓒ Ⓓ Ⓔ	38. Ⓐ Ⓑ Ⓒ Ⓓ Ⓔ	58. Ⓐ Ⓑ Ⓒ Ⓓ Ⓔ
19. Ⓐ Ⓑ Ⓒ Ⓓ Ⓔ	39. Ⓐ Ⓑ Ⓒ Ⓓ Ⓔ	59. Ⓐ Ⓑ Ⓒ Ⓓ Ⓔ
20. Ⓐ Ⓑ Ⓒ Ⓓ Ⓔ	40. Ⓐ Ⓑ Ⓒ Ⓓ Ⓔ	60. Ⓐ Ⓑ Ⓒ Ⓓ Ⓔ

61. Ⓐ Ⓑ Ⓒ Ⓓ Ⓔ   81. Ⓐ Ⓑ Ⓒ Ⓓ Ⓔ   101. Ⓐ Ⓑ Ⓒ Ⓓ Ⓔ
62. Ⓐ Ⓑ Ⓒ Ⓓ Ⓔ   82. Ⓐ Ⓑ Ⓒ Ⓓ Ⓔ   102. Ⓐ Ⓑ Ⓒ Ⓓ Ⓔ
63. Ⓐ Ⓑ Ⓒ Ⓓ Ⓔ   83. Ⓐ Ⓑ Ⓒ Ⓓ Ⓔ   103. Ⓐ Ⓑ Ⓒ Ⓓ Ⓔ
64. Ⓐ Ⓑ Ⓒ Ⓓ Ⓔ   84. Ⓐ Ⓑ Ⓒ Ⓓ Ⓔ   104. Ⓐ Ⓑ Ⓒ Ⓓ Ⓔ
65. Ⓐ Ⓑ Ⓒ Ⓓ Ⓔ   85. Ⓐ Ⓑ Ⓒ Ⓓ Ⓔ   105. Ⓐ Ⓑ Ⓒ Ⓓ Ⓔ
66. Ⓐ Ⓑ Ⓒ Ⓓ Ⓔ   86. Ⓐ Ⓑ Ⓒ Ⓓ Ⓔ   106. Ⓐ Ⓑ Ⓒ Ⓓ Ⓔ
67. Ⓐ Ⓑ Ⓒ Ⓓ Ⓔ   87. Ⓐ Ⓑ Ⓒ Ⓓ Ⓔ   107. Ⓐ Ⓑ Ⓒ Ⓓ Ⓔ
68. Ⓐ Ⓑ Ⓒ Ⓓ Ⓔ   88. Ⓐ Ⓑ Ⓒ Ⓓ Ⓔ   108. Ⓐ Ⓑ Ⓒ Ⓓ Ⓔ
69. Ⓐ Ⓑ Ⓒ Ⓓ Ⓔ   89. Ⓐ Ⓑ Ⓒ Ⓓ Ⓔ   109. Ⓐ Ⓑ Ⓒ Ⓓ Ⓔ
70. Ⓐ Ⓑ Ⓒ Ⓓ Ⓔ   90. Ⓐ Ⓑ Ⓒ Ⓓ Ⓔ   110. Ⓐ Ⓑ Ⓒ Ⓓ Ⓔ
71. Ⓐ Ⓑ Ⓒ Ⓓ Ⓔ   91. Ⓐ Ⓑ Ⓒ Ⓓ Ⓔ   111. Ⓐ Ⓑ Ⓒ Ⓓ Ⓔ
72. Ⓐ Ⓑ Ⓒ Ⓓ Ⓔ   92. Ⓐ Ⓑ Ⓒ Ⓓ Ⓔ   112. Ⓐ Ⓑ Ⓒ Ⓓ Ⓔ
73. Ⓐ Ⓑ Ⓒ Ⓓ Ⓔ   93. Ⓐ Ⓑ Ⓒ Ⓓ Ⓔ   113. Ⓐ Ⓑ Ⓒ Ⓓ Ⓔ
74. Ⓐ Ⓑ Ⓒ Ⓓ Ⓔ   94. Ⓐ Ⓑ Ⓒ Ⓓ Ⓔ   114. Ⓐ Ⓑ Ⓒ Ⓓ Ⓔ
75. Ⓐ Ⓑ Ⓒ Ⓓ Ⓔ   95. Ⓐ Ⓑ Ⓒ Ⓓ Ⓔ   115. Ⓐ Ⓑ Ⓒ Ⓓ Ⓔ
76. Ⓐ Ⓑ Ⓒ Ⓓ Ⓔ   96. Ⓐ Ⓑ Ⓒ Ⓓ Ⓔ   116. Ⓐ Ⓑ Ⓒ Ⓓ Ⓔ
77. Ⓐ Ⓑ Ⓒ Ⓓ Ⓔ   97. Ⓐ Ⓑ Ⓒ Ⓓ Ⓔ   117. Ⓐ Ⓑ Ⓒ Ⓓ Ⓔ
78. Ⓐ Ⓑ Ⓒ Ⓓ Ⓔ   98. Ⓐ Ⓑ Ⓒ Ⓓ Ⓔ   118. Ⓐ Ⓑ Ⓒ Ⓓ Ⓔ
79. Ⓐ Ⓑ Ⓒ Ⓓ Ⓔ   99. Ⓐ Ⓑ Ⓒ Ⓓ Ⓔ   119. Ⓐ Ⓑ Ⓒ Ⓓ Ⓔ
80. Ⓐ Ⓑ Ⓒ Ⓓ Ⓔ   100. Ⓐ Ⓑ Ⓒ Ⓓ Ⓔ   120. Ⓐ Ⓑ Ⓒ Ⓓ Ⓔ

ADVANCED PLACEMENT BIOLOGY EXAM IV

SECTION I

120 Questions
90 Minutes

DIRECTIONS: For each question, there are five possible choices. Select the best choice for each question. Blacken the correct space on the answer sheet.

1. If a woman carrying the trait for colorblindness marries a man who is not colorblind, and they have a boy, and if he married a girl who was the product of a like marriage, what is the chance that their first daughter will be colorblind?

    (A) 1/16

    (B) 1/8

    (C) 1/32

    (D) 1/2

    (E) 0

2. A solution hypotonic relative to a red blood cell

    (A) will cause an immersed red blood cell to undergo hemolysis.

    (B) will cause an immersed red blood cell to undergo crenation.

    (C) will have no effect on a red blood cell that is immersed in it.

    (D) will cause an immersed red blood cell to shrink.

    (E) may be a 1% NaCl solution.

3.   Which of the following eras in the geological time scale is correctly matched to the type of animal that dominated it?

(A) Cenozoic - Age of reptiles

(B) Mesozoic - Age of amphibians

(C) Paleozoic - Age of fish

(D) Mesozoic - Age of marine invertebrates

(E) Precambrian - Age of mammals

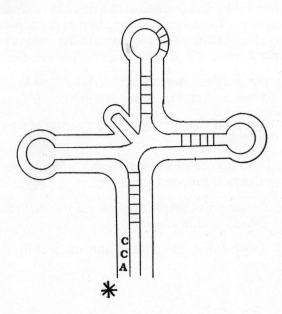

4.   In the tRNA molecule above, the asterisk at the CCA-terminal represents

(A) the anticodon loop.

(B) the site of amino acid attachment.

(C) the binding site for mRNA.

(D) a bond between base pairs.

(E) the codon.

5.	Under anaerobic conditions in animals, the fate of pyruvate is

(A) lactate

(D) glucose

(B) acetyl CoA

(E) fatty acids

(C) ethanol

6.	Maple Syrup Urine disease is a disease in which the metabo-
lism of the branched-chain amino acids is malfunctional.
Knowing that this disease is an inborn error of metabolism, all
of the following statements are logical conclusions EXCEPT

(A) The problem stems from increased levels of metabolites of
branched chain amino acids in the plasma and urine.

(B) A critical enzyme in the degradation of these amino acids
is absent.

(C) Omission of these amino acids from the diet may avert the
effects of this disease.

(D) Omission of maple syrup from the diet may avert the
effects of this disease.

(E) If untreated, the patient may die within the year.

Questions 7 - 8 refer to the oxyhemoglobin dissociation curve.

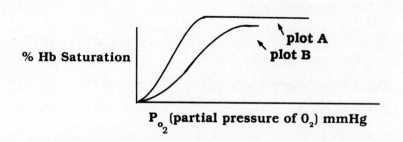

7.	Plot A depicts hemoglobin under normal conditions. Plot B
shows that the curve has shifted to the right and down. This
may occur under conditions of

(A) increased pH.

(B) decreased temperature.

(C) exercise.

(D) hyperventilation.

(E) breathing pure oxygen.

8.   Plot B is a result of

(A) the Bohr effect.   (D) desaturation.

(B) the altitude effect.   (E) saturation.

(C) the chloride shift.

9.   Which biome is correctly matched to its description?

(A) Grasslands are found in the southwestern U.S.

(B) A tropical rainforest has high humidity and much rain-fall. The animal and plant life is sparse.

(C) Coniferous forest is synonymous with deciduous forest.

(D) The taiga is a treeless plain where the winters are severe.

(E) Shrublands have cool, wet winters and hot, dry summers.

10.   A pyramid of biomass

(A) is a measure of the number of individuals in each trophic level.

(B) is a measure of the numbers of trophic levels.

(C) is a measure of the energy consumption at each trophic level.

(D) is a measure of the energy output of each trophic level.

(E) is a measure of the total weight of the organisms in each trophic level.

11. The bacterial cell wall is composed of

(A) chitin

(D) phospholipids

(B) peptidoglycan

(E) starch

(C) cellulose

12. Which of the following statements is true?

(A) In prophase, the sister chromatids separate.

(B) In telophase, the nuclear membrane begins to form.

(C) In metaphase, the sister chromatids begin condensation.

(D) In anaphase, the chromosomes move to the spindle equator.

(E) In prophase, cytokinesis occurs.

13. Which of the following hormones will not cause a rise in plasma glucose concentration?

(A) glucagon

(D) insulin

(B) cortisol

(E) adrenaline

(C) epinephrine

14. Gastrulation results in three primary tissue layers that give rise to all the organs and tissues of the body. Which of the following statements is true?

(A) Endoderm gives rise to muscle.

(B) Epiderm gives rise to skin.

(C) Mesoderm gives rise to bone.

(D) Ectoderm gives rise to the gut lining.

(E) Periderm gives rise to skin.

15. The lens of the eye is formed by a process called

    (A) morphogenesis.

    (D) embryonic induction.

    (B) neurulation.

    (E) cleavage.

    (C) gastrulation.

16. Which of the following statements is not true?

    (A) Sebaceous glands secrete sweat.

    (B) Pancreas secretes hormones, such as glucagon.

    (C) Pancreas secretes digestive enzymes such as trypsin.

    (D) Lacrimal gland secretes tears.

    (E) Mammary glands secrete milk.

17. Which of the following molecules that are present in the nuclear medium would be bound by DNA polymerase if cytosine were on the opposite strand of DNA?

    (A) CMP

    (D) ATP

    (B) GMP

    (E) cAMP

    (C) dGTP

18. A subject with Type A blood

    (A) has A antibodies in his plasma.

    (B) has B antigens on his red blood cells.

    (C) can successfully receive blood from a type O person.

    (D) can successfully receive blood from a type AB person.

    (E) is Rh negative.

19. Industrial melanism refers to the process whereby

(A) light moths became dark moths.

(B) dark-colored moths became favored by the environment.

(C) a mutant gene for dark wings evolved.

(D) dark moths had a survival advantage on both light and dark tree trunks.

(E) skin cancer abounded due to industrial hazards.

20. Regarding the taxonomic classification of man

(A) its phylum is Mammalia

(B) its family is Hominidae

(C) its genus name is *sapiens*

(D) its kingdom is Chordata

(E) its order is Vertebrata

21. The types of mammals, based on their mechanism of embryological development and birth include

(A) the monotremes, which have pouches for development.

(B) the marsupials, which nourish their embryo via placentas.

(C) the placental mammals, such as the opossum

(D) the monotremes, or egg-laying mammals.

(E) the mammaries, including the duck-billed platypus.

22. The only molecule not found in DNA is

(A) deoxyribose   (D) phosphate

(B) uracil   (E) thymine

(C) cytosine

244

23. All of the following statements concerning sponges are true EXCEPT

(A) They are members of the phylum Porifera.

(B) Their adult forms are sessile.

(C) They may have skeletons composed of spicules.

(D) They are predominantly freshwater species.

(E) They may have a skeleton composed of the protein spongin.

 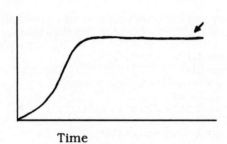

Population Size

Time

24. In the graph above, the arrow indicates specifically

(A) the biotic potential.

(B) the density-dependent effect.

(C) the sigmoid growth curve.

(D) the carrying capacity.

(E) the density-independent effect.

25. Which of the following conditions is due to the effects of a dominant allele?

(A) Albinism

(B) Hemophilia

(C) Sickle-cell anemia

(D) Polydactyly

(E) Color-blindness

26. Facilitated diffusion

(A) requires ATP.

(B) requires a protein carrier.

(C) refers to the osmosis of water.

(D) moves substances against a concentration gradient.

(E) is diffusion that occurs easily.

27. Examples of fungi include all of the following EXCEPT

(A) yeasts

(B) molds

(C) mildews

(D) mushrooms

(E) algae

28. In base pairing within DNA molecules,

(A) adenine is bound to thymine by hydrogen bonds.

(B) adenine forms a peptide bond with thymine.

(C) adenine bonds ionically to thymine.

(D) guanine binds to uracil.

(E) adenine binds to uracil.

29. All of the following statements concerning leaf anatomy are true EXCEPT

(A) The cuticle is composed of waxes and helps in water retention.

(B) The stomata are pores that allow gas diffusion.

(C) The mesophyll contains chloroplasts.

(D) The xylem conducts water.

(E) The palisade layer is adjacent to the lower epidermis.

30. Radial body symmetry is present in

    (A) cnidarians.   (D) annelids.

    (B) flatworms.   (E) mollusks.

    (C) arthropods.

31. An example of a mechanism of post-mating reproductive isolation is

    (A) mechanical isolation   (D) behavioral isolation

    (B) hybrid sterility   (E) geographical isolation

    (C) seasonal isolation

32. What would the probable ratio of the number of brown-eyed children to the number of blue-eyed children be if their parents were both brown-eyed and both heterozygous for the allele for brown eyes?

    (A) 3:1   (D) 1:0

    (B) 1:2:1   (E) 1:1

    (C) 1:3

33. Darwin's important observation pertaining to the birds of the Galapagos Islands was that

    (A) there were many different finches, each with specially adapted wings for flight

    (B) there were many different finches on the islands, each with specially adapted beaks for feeding

    (C) all the finches on the various islands have the same beak and are of the same species

    (D) all the finches on the various islands had the same wing structure and are of the same species

    (E) finches eat various different types of food, such as seeds, insects or fruit, and thus must be of different species

34.   One type of organic molecule that is not found in animal cell membranes is

(A) phospholipids   (D) cellulose

(B) intrinsic proteins   (E) cholesterol

(C) extrinsic proteins

35.   Punctuated equilibrium refers to which of the following?

(A) Short periods of rapid speciation separated by periods of slower evolutionary change

(B) Adaptive radiation

(C) Convergent evolution

(D) Gradualism

(E) Catastrophism

36.   The pelagic province of the open ocean is divided into several different regions. The uppermost epipelagic region has more life than the deeper ones because

(A) it receives the fresh rainfall

(B) light penetration allows for photosynthesis

(C) light penetration allows for respiration

(D) warmer temperatures allow for growth

(E) tidal action keeps the nutrients suspended

37.   The replication fork

(A) is the site at which the single-stranded regions emerge.

(B) refers to the lagging strand.

(C) refers to the leading strand.

(D) refers to the site of base pair formation.

(E) is seen in transcription.

38. Synapsis

(A) occurs during the second meiotic division.

(B) refers to the pairing between homologous chromosomes.

(C) is synonymous with chiasmata.

(D) refers to the tetrad of chromatids.

(E) are the junctions between neurons.

39. The cell types found in phloem are

(A) sieve tube members and vessels.

(B) sieve tube members and tracheids.

(C) tracheids and vessels.

(D) companion cells and vessels.

(E) sieve tube members and companion cells.

40. The cell organelles that are most similar to prokaryotes are

(A) the mitochondria and chloroplasts.

(B) the rough and smooth endoplasmic reticula.

(C) the rough endoplasmic reticula and ribosomes.

(D) the rough endoplasmic reticula and Golgi apparatuses.

(E) the lysosomes and ribosomes.

41. The notochord is the forerunner of which vertebrate structure?

(A) spinal cord

(B) vertebral column

(C) brain

(D) gill

(E) gill slits

42. Because fungi can obtain nutrients from nonliving organic matter, they are referred to as

(A) parasitic

(D) heterotrophic

(B) saprophytic

(E) pathogenic

(C) eukaryotic

43. The founder effect

(A) is a direct effect of mutation.

(B) is an extreme case of gene flow.

(C) is an extreme case of genetic drift.

(D) occurs by natural selection.

(E) is an extreme case of natural selection.

44. The type of community interaction whereby one species benefits and the other is unaffected is specifically called

(A) symbiosis.

(D) predation.

(B) mutualism.

(E) parasitism.

(C) commensalism.

45. Which of the following is the female reproductive organ of a flower?

(A) stamen

(D) receptacle

(B) anther

(E) carpel

(C) petal

46. The anticodon is found on

(A) mRNA

(D) DNA

(B) rRNA

(E) ATP

(C) tRNA

47. Gastrulation refers to

(A) fusion of the sperm and egg nuclei

(B) embryonic cell division with no increase in embryo size

(C) differentiation of body parts as signaled from an adjacent part

(D) the development of the notochord and dorsal hollow nerve cord

(E) the migrations of cells into three primary germ layers

48. All of the following events occur in a flower EXCEPT

(A) pollination

(B) fertilization

(C) megaspore formation

(D) pollen tube growth

(E) photosynthesis

49. Double fertilization

(A) results in fraternal twins.

(B) results in identical twins.

(C) results in Siamese twins.

(D) is unique to angiosperms.

(E) occurs in all seed-producing vascular plants.

50. Decomposers feed on

(A) producers.

(B) herbivores.

(C) primary carnivores.

(D) consumers.

(E) dead organic matter.

51. The dark reactions of photosynthesis require all of the following EXCEPT

(A) glucose

(B) ATP

(C) NADPH

(D) RuBP (ribulose bisphosphate)

(E) carbon dioxide

52. Terms referring to the hydrologic (water) cycle include ALL of the following EXCEPT

(A) transpiration   (D) precipitation

(B) evaporation   (E) runoff

(C) fixation

53. Mating between a blue-eyed woman and a heterozygous brown-eyed man would result in a ratio of brown-eyed to blue-eyed children of

(A) 1:1   (D) 1:0

(B) 1:2   (E) 0:1

(C) 2:1

54. A phenotype refers to

(A) the genetic makeup of an individual.

(B) the expression of dominant traits.

(C) the expression of recessive traits.

(D) the manifest expression of the genotype.

(E) the heterozygous condition.

55. All of the plant groups below are vascular EXCEPT

(A) mosses

(D) gymnosperms

(B) ferns

(E) seed plants

(C) horsetails

56. A bacteriophage

(A) is a bacterium that phagocytoses other organisms.

(B) is a bacterium that becomes phagocytosed by other organism.

(C) is a virus that infects bacteria.

(D) is a fragment of DNA.

(E) lives in a lysogenic cell.

57. The reduced form of the coenzyme in the dehydrogenase enzymes is

(A) NADH

(D) NAD^+

(B) FAD

(E) ACTH

(C) ADH

58. In the first step of glycolysis, the phosphorylation of glucose is coupled to

(A) the production of carbon dioxide

(B) the reduction of NAD^+

(C) the synthesis of ATP

(D) the hydrolysis of ATP

(E) the loss of electrons

59.   The active portion of the cytochrome enzymes is a heme group. It contains a mineral element which can exist in the oxidized or reduced state. This element is

(A) sodium   (D) potassium

(B) iodine   (E) calcium

(C) iron

60.   Chemiosmosis refers to the idea that an ion flowing down its electrochemical gradient drives ATP synthesis. This ion is

(A) Phosphate ion, thus the term oxidative phosphorylation

(B) Sodium ion

(C) Iron ion

(D) Hydrogen ion

(E) Calcium ion

Directions: The following groups of questions have five lettered choices followed by a list of diagrams, numbered phrases, sentences, or words. For each numbered diagram, phrase, sentence, or word choose the heading which most directly applies. Blacken the correct space on the answer sheet. Each heading may be used once, more than once, or not at all.

Questions 61 - 63 describe the basic functional units of organs or systems.

(A) Nephron   (D) Alveolus

(B) Neuron   (E) Villus

(C) Sarcomere

61.   Functional unit of the kidney, it produces urine.

62.   As the functional unit of a muscle, it contracts.

63.   This cell transmits messages.

Questions 64 - 66 refer to events dealing with DNA.

(A) Transcription   (D) Translocation

(B) Translation   (E) Transduction

(C) Transformation

64. The process by which a cell takes up DNA from its immediate environment and incorporates that DNA into its genome.

65. The process of linking amino acids together to form a protein.

66. The transfer of bacterial DNA from one organism to another by way of a viral vector.

Questions 67 - 70 distinguish different terms relating to the alternation of generations in plant life cycles

(A) Sporophyte   (D) Sporangia

(B) Spore   (E) Spore mother cell

(C) Homosporous

67. Plant that produces only one type of spore

68. The haploid product of a sporophyte

69. The diploid generation, it becomes more dominant in the evolution of plants

70. The multicellular structure that produces megaspores and microspores

Questions 71 - 74 refer to animal cell organelles and their functions.

(A) Mitochondrion

(B) Lysosome

(C) Ribosome

(D) Smooth endoplasmic reticulum

(E) Microfilament

71.   Functions in protein synthesis

72.   Functions in the cytoskeleton

73.   Fuses with phagocytic vesicles

74.   Synthesizes lipids

Questions 75 - 78 list enzymes that function in the key metabolic processes involved in complete glucose oxidation.

(A) Krebs Cycle

(B) Electron transport chain

(C) Glycolysis

(D) Formation of acetyl CoA

(E) Chemiosmosis

75.   Hexokinase

76.   Cytochrome oxidase

77.   Pyruvate dehydrogenase

78.   Succinate dehydrogenase

Questions 79 - 81 refer to plant hormones and their functions.

(A) Gibberellins   (D) Auxins

(B) Abscisic acid   (E) Cytokinins

(C) Ethylene

79.   Stimulates fruit ripening

80.   Promotes closure of the stomata

81.   Terminates seed dormancy.

Questions 82 - 85 describe distinguishing characteristics of major animal phyla

(A) Mollusca   (D) Annelida

(B) Chordata   (E) Arthropoda

(C) Echinodermata

82.   Members of this phylum have dorsal body walls called mantles; each organism has a muscular organ called a foot

83.   These segmented worms include the earthworms

84.   Members of this phylum have hardened exoskeletons and jointed appendages

85.   Locomotion in members of this phylum is based on a water-vascular system

Questions 86 - 90 refer to the diagram below which shows the secondary growth of a woody stem.

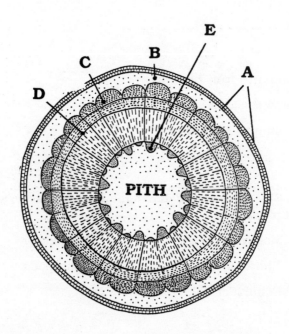

86. Vascular cambium

87. Primary phloem

88. Primary xylem

89. Epidermis

90. Cortex

Directions: The following questions refer to experimental or laboratory situations or data. Read the description of each situation. Then choose the best answer to each question. Blacken the correct space on the answer sheet.

Questions 91 - 93 refer to the pedigree below.

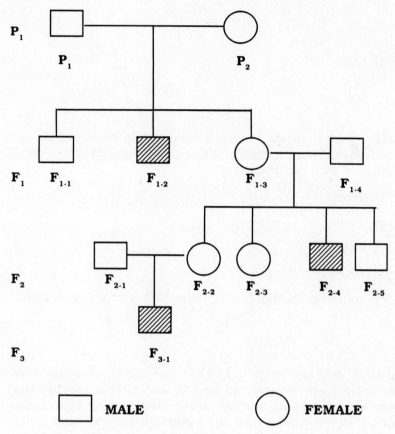

☐ MALE   ◯ FEMALE

Shading indicates the presence of a disease.

91. With reference to the above figure, all of the following statements concerning the disease in question are true EXCEPT

   (A) It is sex-linked.

   (B) It is caused by a recessive gene.

   (C) It may be hemophilia.

   (D) It is passed from mothers to their sons.

   (E) It is expressed only in the homozygous recessive state.

92. If the disease were colorblindness, the genotype of P_1 must be

   (A) X^cX^c

   (B) X^cX^C

   (C) X^cX^c

   (D) X^cY

   (E) X^cY

93. If F_{2-5} were to marry a woman homozygous dominant for the trait in question, the probability that they would have a child afflicted with the disease is

   (A) 0%

   (B) 25%

   (C) 50%

   (D) 100%

   (E) unknown; cannot be determined from information given.

Questions 94 - 97 refer to the graph below, which represents data obtained from a spirometer, an instrument that measures the volume of air moved into and out of the lungs during breathing. The volume of air moved into the lungs during a normal quiet inpiration is called the tidal volume. This same volume will be moved out of the lungs during a normal quiet expiration. One tidal inspiration is labeled on the graph (see bar).

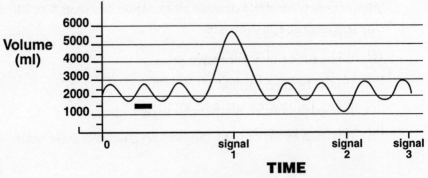

Bar indicates one tidal inspiration.

94. At the time of signal 1, the subject has inspired maximally. What is the subject's inspiratory reserve volume?

(A) 6500 ml   (D) 2500 ml

(B) 3500 ml   (E) 2000 ml

(C) 3000 ml

95. Just before signal two, the subject

(A) inspired maximally

(B) expired forcefully

(C) increased his breathing rate

(D) decreased his breathing rate

(E) stopped breathing

96. What is the subject's vital capacity?

(A) 5500 ml   (D) 2500 ml

(B) 4500 ml   (E) 2000 ml

(C) 3500 ml

97. Assuming that the time span between signals two and three represents 30 seconds, this subject's respiratory rate is

(A) about 6 breaths/minute

(B) about 24 breaths/minute

(C) 12 breaths per 30 second interval

(D) about 12 breaths per minute

(E) None of the above

Questions 98 - 101

Suppose a teacher does a statistical analysis of the eye color in her school of mostly black students. She finds that of the 1000 students, 910 have brown eyes, while only 90 have blue eyes (considering green as blue, too).

Five years later, she does her analysis again, since as an attempt at desegregation, some students are sent to other schools and new students from neighboring towns are brought in. She now finds that of the 1000 students, 840 have brown eyes and 160 have blue eyes.

Her table below summarizes the data.

Year	Brown	Blue	Total
1981	910	90	1000
1986	840	160	1000

98.   In the original sample (1981), the frequency of the allele for brown eyes (B) is

(A) .7   (D) .3

(B) .49   (E) .91

(C) .9

99.   The number of students in the original sample that are heterozygous for brown eyes is

(A) 910

(B) 490

(C) 420

(D) 90

(E) cannot be determined by the data given

100.   The deviation from the Hardy-Weinberg equilibrium, as exemplified by the new data in 1986, is due to

(A) mutation   (D) selection

(B) migration   (E) chance

(C) smaller sample size

101.   In the second sample, the frequency of the allele for blue eyes (b) is

(A) .84   (D) .16

(B) .6   (E) .4

(C) .04

Questions 102 - 105 refer to responses to receptor stimulation and blockade.

The autonomic nervous system is the involuntary nervous system which innervates smooth muscle, cardiac muscle and glands. There are two divisions: the sympathetic division is stimulated by the release of the neurotransmitter chemical norepinephrine. Sympathetic stimulation causes the "fight or flight" reaction: it readies the body for action. It causes pupillary dilation, increases in heart and respiratory rates and vasoconstriction of many blood vessels to increase blood pressure. The parasympathetic division is stimulated by the release of the neurotransmitter acetylcholine. It allows the body to rest and recuperate, and as such, stimulates digestive activities. It has opposing effects from the sympathetic division on the pupil, heart and many other organs.

In the following experiment, a frog was used. A few drops of various chemicals were placed on appropriate organs or tissues in or on the frog and the responses were observed or measured. The responses listed below are all relative to the unstimulated state.

Chemical:	Atropine	Propranolol	Curare
Response:	pupil dilation	decreased heart rate	Paralysis of skeletal (voluntary) muscle

263

102. Given that atropine has no direct effect on the receptors for norepinephrine, the mechanism of action for atropine is

(A) stimulation of the sympathetic nervous system

(B) inhibition of the sympathetic nervous system

(C) stimulation of the parasympathetic nervous system

(D) inhibition of the parasympathetic nervous system

(E) not suggested by the information given

103. Given that propranolol has no direct effect on the receptors for acetylcholine, it may be concluded that it

(A) stimulates norepinephrine releasing cells

(B) blocks the action of norepinephrine

(C) stimulates sympathetic nerves

(D) stimulates parasympathetic nerves

(E) inhibits parasympathetic nerve activity

104. If one were to cut the parasympathetic nerve (vagus) which innervates the heart, one would expect to see

(A) an increase in heart rate, due to increased sympathetic activity

(B) an increase in heart rate due to decreased parasympathetic activity

(C) a decrease in heart rate due to decreased parasympathetic activity

(D) a dead heart, as a heart requires innervation for function

(E) a decrease in heart rate, since that is the effect of parasympathetic stimulation

105. Curare acts via

(A) stimulating sympathetic activity

(B) stimulating parasympathetic activity

(C) inhibiting parasympathetic activity

(D) inhibiting sympathetic activity

(E) none of the above.

Questions 106 - 108 refer to the fates of pyruvate. Pyruvate is placed into three different chambers that contain various substances. Pyruvate is converted to the substances listed.

Chamber:	A	B	C
Product:	lactate	alcohol	CO_2 and water

106. Which tube(s) lack(s) oxygen?

(A) A and B   (D) A, B, and C

(B) B and C   (E) A only

(C) A and C

107. Which tube(s) contain(s) yeast cells?

(A) A   (D) A and B

(B) B   (E) A, B, and C

(C) C

108. Which tube mimics the biochemical events that occur in the cells of a sprinter?

(A) A   (D) A and B

(B) B   (E) B and C

(C) C

Questions 109 - 111 refer to the graphs below. Two species of *Paramecium* (A and B) are grown separately and together. The growth curves of their populations are depicted below.

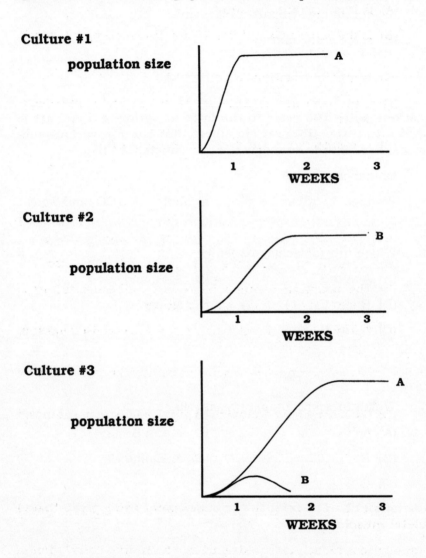

Culture #1
population size

A

1   2   3
WEEKS

Culture #2
population size

B

1   2   3
WEEKS

Culture #3
population size

A

B

1   2   3
WEEKS

109. The type of interaction between species A and B is described as

(A) mutualistic   (D) a survivorship curve

(B) protocooperative   (E) symbiotic

(C) competitive exclusion

110.　In the culture #3, the population size of species A when compared to its size in culture #1

(A) is less than it is in culture #1

(B) is the same as in culture #1, but it takes longer to reach maximum size

(C) is greater than it is in culture #1.

(D) is the same as it is in culture #1, but it reaches maximum faster

(E) is unaffected by the presence of species B

111.　When comparing the maximum population sizes for species B in the mixed culture vs. solo culture, it is evident that

(A) B reaches its maximum faster in the mixed culture

(B) The maximum population size for B in the solo culture is twice that of B in the mixed culture

(C) The maximum population size for B is reached fastest in the solo culture

(D) In the solo cultures, B's population size surpasses that of A

(E) B does not have a maximum population size in the mixed culture

Questions 112 - 113 refer to the experiment using glycerinated skeletal muscle.

A rabbit psoas muscle is teased apart into individual fibers, some of which are maintained in glycerol solution under standard conditions at 25°C and pH 7. The length of the resting fibers are measured as 2 mm. Various solutions are added to the cerol solutions and the fibers are remeasured under the dissecting microscope. The results are listed in Table A. The solution in Experiment A that caused maximum contraction is then used in experiment B and experiment C and the results are depicted in Tables B and C, respectively.

Table A
Standard conditions
(25°C, pH 7)

Solution	Glycerol	Glycerol + ATP	Glycerol + Salts (KC1, MgCL$_2$)	Glycerol + ATP + Salts
Length of fiber (mm)	2	1.75	2	.75

Table B
Optimal solution from experiment A

Conditions	25°C, pH 7	37°C, pH 7
Length of Fiber (mm)	.70	.50

Table C
Optimal solution from experiment A

Conditions	25°C, pH 7	25°C, pH 4
Length of Fiber (mm)	.65	1.00

112. In experiment A, contraction of skeletal muscle occurred most readily in the presence of

(A) glycerol.

(B) glycerol and ATP.

(C) glycerol and salts.

(D) glycerol, ATP, and salts.

(E) nervous stimulation.

113. Considering experiments B and C, skeletal muscle contraction is optimal under the conditions of

(A) standard conditions, by definition.

(B) room temperature and acidic environment.

(C) body temperature and neutral pH.

(D) freezing temperature and neutral pH.

(E) room temperature and neutral pH.

Questions 114 - 115 refer to an experiment in polypeptide hydrolysis.

Polypeptides are placed into warm watery solutions in separate beakers, each containing a different substance that has been isolated from pancreatic juice or intestinal tissue. After two hours, the contents of the individual beakers are analyzed. The results are below.

SUBSTANCE PRESENT IN BEAKER	MUCUS	TRYPSIN	CHYMOTRYPSIN	CARBOXYPEPTIDASE	AMINOPEPTIDASE
RESULTS	POLY-PEPTIDES	DIPEPTIDES TRIPEPTIDES	DIPEPTIDES TRIPEPTIDES	AMINO ACIDS DIPEPTIDES TRIPEPTIDES	AMINO ACIDS DIPEPTIDES

114. It appears that the only substance that has no enzymatic activity is

(A) mucus

(B) typsin

(C) chymotrypsin

(D) carboxypeptidase

(E) aminopeptidase

115. The only substances that act on the terminal residues of the polypeptide are

   (A) carboxypeptidase and chymotrypsin

   (B) mucus and trypsin

   (C) trypsin and chymotrypsin

   (D) carboxypeptidase and aminopeptidase

   (E) polypeptides and aminopeptidase

Questions 116 - 117 refer to an experiment that concerns cell poisons. The cell is viewed with a powerful microscope under normal conditions and again in the presence of a cell poison. The activity of the cell is noted in the chart below.

Poison	None	Cytochalasin B	Colchicine
Cell Activity	normal	No movement of vesicles and other organelles No contraction in skeletal muscle fibers	No chromosome movement

116. The organelles that cytochalasin B must act on are

   (A) actin and myosin   (D) mitochondria

   (B) microtubules   (E) nucleus

   (C) microfilaments

117. The protein that colchicine binds to is

   (A) actin   (D) histone

   (B) myosin   (E) chromatin

   (C) tubulin

Questions 118 - 120 refer to the table below, in which the presence (+) or absence (-) of certain types of organelles in five sample human cells is indicated. A blank does not signify the absence of an organelle, just its relative lack of importance when compared to the importance of other organelles.

	Nucleus	Flagellum	Lysosome	Mitochondria	Golgi Apparatus
Cell A	-	-		-	
Cell B	+	-	+	+ +	
Cell C	+	+	+	+	
Cell D	+	-	+ +		
Cell E	+	-		+	+

118. Which cell is most likely to be one that is secreting proteins?

(A) Cell A

(B) Cell B

(C) Cell C

(D) Cell D

(E) Cell E

119. Which cells would most likely be found in blood?

(A) A and B

(B) B and E

(C) A and D

(D) D and E

(E) C and D

120. Cell C is

    (A) a muscle cell, due to the presence of mitochondria.

    (B) an egg cell.

    (C) a neuron.

    (D) prokaryotic.

    (E) haploid.

SECTION II

1.   Describe the four major groups of organic compounds that compose the human body. Include their functions in your essay, but focus on their chemical constitutions.

2.   Describe the major steps of translation: initiation, elongation, and termination. Describe also the basic structure of a ribosome and the activation step required before translation can occur.

3.   Discuss the levels of organization in an organism and the major systems in the human body, including their organs and functions.

4.   Define and explain the three major plant tropisms: phototropism, gravitropism, and thigmotropism.

ADVANCED PLACEMENT
BIOLOGY EXAM IV

ANSWER KEY

1.	A	31.	B	61.	A	91.	E
2.	A	32.	A	62.	C	92.	D
3.	C	33.	B	63.	B	93.	A
4.	B	34.	D	64.	C	94.	C
5.	A	35.	A	65.	B	95.	B
6.	D	36.	B	66.	E	96.	B
7.	C	37.	A	67.	C	97.	E
8.	A	38.	B	68.	B	98.	A
9.	E	39.	E	69.	A	99.	C
10.	E	40.	A	70.	D	100.	B
11.	B	41.	B	71.	C	101.	E
12.	B	42.	B	72.	E	102.	D
13.	D	43.	C	73.	B	103.	B
14.	C	44.	C	74.	D	104.	B
15.	D	45.	E	75.	C	105.	E
16.	A	46.	C	76.	B	106.	A
17.	C	47.	E	77.	D	107.	B
18.	C	48.	E	78.	A	108.	A
19.	B	49.	D	79.	C	109.	C
20.	B	50.	E	80.	B	110.	B
21.	D	51.	A	81.	A	111.	A
22.	B	52.	C	82.	A	112.	D
23.	D	53.	A	83.	D	113.	C
24.	D	54.	D	84.	E	114.	A
25.	D	55.	A	85.	C	115.	D
26.	B	56.	C	86.	D	116.	C
27.	E	57.	A	87.	C	117.	C
28.	A	58.	D	88.	E	118.	E
29.	E	59.	C	89.	A	119.	C
30.	A	60.	D	90.	B	120.	E

ADVANCED PLACEMENT
BIOLOGY EXAM IV

DETAILED EXPLANATIONS
OF ANSWERS

SECTION I

1.   (A)
The best way to work out this problem is with a Punnett square. Let c represent the recessive trait of colorblindness, and C represent the normal dominant trait. A female only expresses the trait when both her X-chromosomes carry the c gene, since it is recessive; a male will express the trait when his one X chromosome carries the c gene, since his Y chromosome will not have a dominant gene to mask it. When a female is heterozygous for the trait, she is a carrier of the trait.

P_1 X^cX^c x X^cY

	X^c	X^c
X^c	X^cX^c	X^cX^c
Y	X^cY	X^cY

The children of this couple would be, by probability: 25% normal female, 25% carrier female, 25% normal male, and 25% colorblind male.

In order for a boy produced by this couple to marry a girl produced by a like marriage and produce a colorblind daughter, the mating must be between a female carrier (X^cX^c) and a colorblind male (X^cY).

The probability that the boy was colorblind is 50% and that the girl was a carrier is 50%.

275

Now, if the children mate, the possible sex-chromosome genotypes are represented below:

F_1 $X^C X^c$ x $X^c Y$

	X^C	X^c
X^c	$X^C X^c$	$X^c X^c$
Y	$X^C Y$	$X^c Y$

The probability that, in the F_2 generation, the child will be a color-blind girl is 25% ($X^c X^c$). This is dependent on the fact that there was a 50% chance that the child's father is color blind and a 50% chance that the child's mother was a carrier. Therefore, the final probability is:

$$1/2 \ x \ 1/2 \ x \ 1/4 = 1/16$$

2.   (A)
A red blood cell has a .85% to .9% salt (NaCl) concentration. Normally, red blood cells are suspended in a plasma that is isotonic to the red blood cell; i.e., the plasma has the same salt concentration. Since the fluids inside and outside the cell have the same concentrations, there is no tendency for water to enter or leave the cell.

By osmosis, water travels down its concentration gradient. When the red blood cell is suspended in a hypotonic solution, i.e., a solution that has less osmotically active particles than the red blood cell, water will tend to enter the red blood cell. Note that water is more concentrated in the solution, since the solutes are less concentrated. An example of a hypotonic solution is distilled water or anything less than .85% NaCl. In actuality, the water enters and exits the red blood cell, but the net movement of water in this case is into the cell. When water enters the red blood cell, it causes the red blood cell to swell and burst. The bursting of red blood cells is called hemolysis.

If the red blood cell were placed in a hypertonic solution, any solution greater than .9% NaCl, such as a 1% solution, the net movement of water would be out of the cell and into the solution. The cell would shrink; this is called crenation.

3.   (C)

The geological time scale is divided into four eras: The Precambrian era is the period of Earth's history prior to 600 million years ago. Prokaryotic life existed, and eukaryotes began to evolve. The Paleozoic era lasted from 225 to 600 million years ago. Marine invertebrates were dominant in the oldest period within this era. Dominance of fish and amphibians followed. The Mesozoic era extended from 65 to 225 million years ago. Dinosaurs were abundant. This was the age of reptiles. The modern era, the Cenozoic, which dates back to 65 million years ago, is characterized by the dominance of mammals, birds, and insects.

4.   (B)

A tRNA (transfer RNA) molecule is a small RNA molecule (see figure). There are specific tRNAs for specific amino acids. One end of the molecule binds to the amino acid that is to be added to the growing polypeptide chain. This end always has a terminal CCA (cytosine, cytosine, adenine) sequence.

The other end has the anticodon, which can bind to codons on mRNA. Thus the codon ultimately calls for a specific amino acid in the medium. Note that the binding between tRNA and mRNA is via hydrogen bonds of the base pairs.

The shape of the tRNA molecule is maintained by hydrogen bonds between complementary base pairs within the molecule itself.

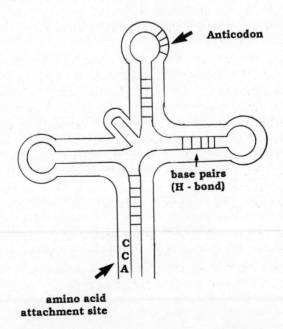

Anticodon

base pairs
(H - bond)

C
C
A

amino acid
attachment site

5.   (A)

Pyruvate is the end-product of glycolysis, the process of breaking down glucose. Glucose is a six-carbon molecule, while pyruvate is a three-carbon molecule, and hence still contains much of the potential energy of glucose.

Pyruvate has various fates in the cell. In the absence of oxygen in animal cells, pyruvate is converted to lactate. (This is the source of lactate in athletes performing anaerobic feats, such as sprinting.) The conversion to lactate is a reduction reaction. It is required in order to regenerate the oxidized form of a coenzyme required for glycolysis. Thus, by producing lactate, glycolysis may continue and produce minimal amounts of energy in the form of ATP.

The major fate of pyruvate is to acetyl CoA. This occurs under aerobic conditions, i.e., when oxygen is available. Pyruvate enters the mitochondrion from its site of production in the cytoplasm; within the mitochondrion, it is converted to acetyl CoA. The latter molecule will then enter the Krebs cycle.

In yeast cells, in the absence of oxygen, pyruvate is converted to ethanol (the alcohol found in alcoholic beverages). This process is called anaerobic fermentation. It is the mechanism whereby the sugar in grapes is converted to the alcohol in wine. For all practical purposes, this step does not occur in animals. However, there are a few documented cases of humans becoming spontaneously drunk, associated with large colonies of a strain of *Candida* yeast in their intestines, which convert the sugar in foods into alcohol.

The oxidation of fatty acids yields acetyl CoA, which is beyond the production step of pyruvate.

6.   (D)

Inborn errors of metabolism refer to inherited diseases caused by the inability to synthesize a particular enzyme, or the synthesis of a defective enzyme. Maple Syrup Urine Disease is characterized by the lack of alpha-keto acid decarboxylase, an enzyme required in the metabolism of the branched-chain amino acids (leucine, isoleucine, and valine). Toxic metabolites of these amino acids build up in the plasma and urine, leading to vomiting and early death. Omission from the diet of these amino acids will prevent the effects of the disease. The disease is named according to the odor of the urine, but has nothing to do with the ingestion of maple syrup.

7. (C)

The curve shows the relationship between the percent saturation of hemoglobin versus the partial pressure of oxygen (P_{O_2}) measured in mmHg. As P_{O_2} increases, there is an increased saturation of hemoglobin by oxygen. Several physiological parameters affect the desaturation of hemoglobin; these factors ease the dissociation of oxygen from hemoglobin so that the oxygen may diffuse into the cells in need. This increased unloading of oxygen from hemoglobin is represented by a curve that has shifted to the right and down (plot B). Note that at any given P_{O_2}, hemoglobin is less saturated with oxygen. This shift is called the Bohr effect. The primary factors responsible for the shift are an increase in P_{CO_2} (partial pressure of carbon dioxide) and an increase in acid (H^+) levels, which is a decrease in pH. Increased temperature will also shift the curve.

During exercise, when the active cells need more oxygen, hemoglobin will unload the oxygen due to three combined factors: increased temperature, increased carbon dioxide pressure, and decreased pH. Hypoventilation can also induce this change since carbon dioxide is retained in the body, whereas in hyperventilation, excess carbon dioxide is expelled, thus decreasing the concentration of this gas.

A decrease in carbon dioxide concentration or an increase in pH would cause the curve to shift to the left. Breathing pure oxygen would certainly not indicate an oxygen shortage and thus the curve would not shift to the right under such circumstances.

8. (A)

The shift of the oxyhemoglobin dissociation curve to the right due to the presence of increased levels of hydrogen ion (H^+) or carbon dioxide (CO_2) is called the Bohr effect. At high altitude, the Bohr effect comes into play due to the build-up of a substance called 2,3-diphosphoglycerate (DPG). While DPG is a normal constituent of red blood cells, its concentration nearly doubles within two days of exposure to the environmental conditions present at high altitudes. Consequently, there is a decrease in the saturation of hemoglobin by oxygen, allowing increased delivery to the tissues since the shift to the right represents a desaturation of hemoglobin. A shift to the left would indicate increased saturation, although under sea level conditions, hemoglobin is already about 97% saturated in arterial blood.

The chloride shift refers to the diffusion of chloride ions from the plasma into red blood cells to compensate for the diffusion of bicarbonate ions in the reverse direction. Thus ionic balance is maintained. Sodium bicarbonate is formed in the extracellular fluid, while potassium is formed within the red blood cell.

9.   (E)
A biome is a terrestrial region that has a characteristic climate, topography, flora, and fauna.

Grasslands are primarily associated with the vast open spaces of Africa and South America. The land is flat; the climate includes a long, hot, dry period; rainfall is moderate. Grazing animals such as zebra and giraffes abound.

Tropical rainforests are found at equatorial latitudes. Rainfall levels and humidity are high; temperature is warm. The animal and plant life is impressive in both number and variety.

Other types of forests include coniferous and deciduous. A coniferous forest (or taiga) is found at high altitudes. Examples in the U.S. include the Sierra Nevadas, Cascades, and Rocky Mountains. As presumed from the name, conifers (firs, pine, and spruce) are the dominant plant life; they live through cold, wet winters and mild, brief summers.

The deciduous forests, by definition, lose their leaves in the autumn. This is exemplified by the forests of the Southern Appalachians. The cold winters contrast with the long summer. Animal life includes foxes, raccoons, and black bears.

Much of California can be classified as shrubland. The rain falls in the cool winter. The summers are long, hot, and dry.

Another major biome is the desert. Many of the states in the American Southwest have climates nearly devoid of rain. The sparse animal and plant life shows specialized adaptions to the heat and aridity.

Finally, a tundra is a treeless plain in the most northern latitudes. The cold harsh winters prohibit much plant and animal life. Mosses and lichens survive here. Animals include caribou, wolves, and reindeer.

10.   (E)

There are several different types of ecological pyramids. All the pyramids have the same basic structure: their trophic levels are organized according to the modes of nutrition of their respective members. The basal level is represented by the producers - photosynthesizing plants that make the food upon which all the others ultimately depend. The second trophic level consists of herbivores. They are plant-eating animals. The primary carnivores (animals that eat animals) feed on the herbivores, while secondary carnivores, at the top, feed on primary carnivores or herbivores. Of course, this is a simplistic view, as there is much interaction between trophic levels. For instance, an omnivore eats both plants and animals. Decomposers (bacteria and fungi) feed on members of all trophic levels - they decompose dead organisms, including other decomposers. Their products are then recycled and used as nutrients for the producers.

A pyramid of biomass is a ratio of the weight of all the organisms at each level. The many plants form a broad base to the pyramid, and the pyramid gradually decreases in size at each level. This pyramid takes on such a shape primarily due to the pyramid of numbers - although certain plants may be small in size, they are high in number. Similarly, the secondary carnivores may be large animals, but their overall small numbers means the pyramidal shape will be observed.

A pyramid of energy refers to the stored energy at each trophic level. As one proceeds up a pyramid, typically only 10% of the energy present in one level is present in the next trophic level. Thus, the secondary carnivore level has only .1% of the total energy of that of the producers.

11.   (B)

All prokaryotes and some eukaryotes (plant and fungi) have cell walls. Bacterial cell walls are composed of peptidoglycan which, as its name suggests, is composed of amino acids and sugars. The peptidoglycan has a very orderly arrangement.

Chitin is a polysaccharide and is the major component of fungal cell walls. Interestingly, it also is the basis of the arthropod exoskeleton.

Cellulose is a glucose polymer forming plant cell walls. It is among the indigestible residues of the plant foods that we eat.

Animal cells do not have cell walls enveloping their plasma membranes. The plasma membrane of animal cells is composed primarily of phospholipid with proteins embedded within and upon this phospholipid framework.

12.   (B)
Mitosis is the phase of the cell cycle in which one cell divides into two. Strictly speaking, mitosis refers to division of the nucleus. Cytokinesis, or cytoplasmic division, follows immediately. There are four consecutive stages of mitosis: prophase, metaphase, anaphase, and telophase.

In prophase, the chromosomes become visible as the sister chromatids condense into rod-like bodies. The nuclear membrane disintegrates and the nucleolus disappears. Microtubule spindles begin their formation.

In metaphase, the spindle attaches to the centromeres (central constricted region of each chromatid). Chromosomes are guided by spindle microtubules to the spindle equator, the central plane of the cell.

In anaphase, the sister chromatids of each pair separate and move from the equator to opposite poles of the cell. This movement is perpetuated by the depolymerization of the microtubule apparatus.

Telophase is simply a reversal of prophase: the nuclear membrane and nucleolus begin formation, and the chromosomes decondense into thread-like forms. Cytokinesis follows or occurs simultaneously and the bilobed cell with two nuclei will split into two individual cells.

13.   (D)
Plasma glucose concentration is regulated and often falls between a fasting level of 80 mg/dl and a post-absorptive level of 130 mg/dl. Many hormones participate in the regulation of plasma glucose levels.

Insulin is a hypoglycemic hormone; in other words it is secreted from the pancreas after a meal and stimulates the uptake of glucose by fat and muscle cells, hence lowering blood glucose concentration. In contrast, glucagon, another pancreatic hormone, is secreted when plasma glucose concentration is low. It increases plasma glucose levels by stimulating the breakdown of glycogen into glucose by hepatocytes (liver cells). This process is called glycogenolysis. The glucose then diffuses into the blood. This hormone is an important regulator of plasma glucose between meals.

Cortisol is one of the group of glucocorticoids that is released from the adrenal cortex. As suspected by its classification, it affects glucose metabolism. Cortisol increases blood glucose concentration mainly by inhibiting its peripheral utilization at the expense of fatty acids. In addition, it stimulates liver gluconeogenesis, the production of glucose from amino acid precursors.

Epinephrine, also called adrenaline, is a hormone that is released from the adrenal medulla. It assists the sympathetic nervous system and prolongs the "fight or flight" syndrome, which prepares an animal to escape or confront a stress. Among other things, plasma glucose levels are increased to supply energy to deal with the stress. Once again, the liver is stimulated to release glucose from its storage supply.

14.   (C)
Gastrulation occurs early in embryonic life. It refers to the process whereby the single-layered blastula is transformed into a three-layered gastrula. The three germ layers are the ectoderm (outer layer), mesoderm (middle layer), and endoderm (inner layer). There is no epiderm or periderm.

The ectoderm will become the epidermis of the skin and all neural tissue. The mesoderm becomes the connective tissue (including blood and bone), muscle, and organs of the circulatory, reproductive, and excretory systems. The endoderm is destined to line the gut and to form accessory glands of the gut. It also forms the lung epithelium.

15.   (D)
Induction is the process whereby one tissue can determine the fate of a neighboring tissue by the release of chemical factors. One classic example is the development of the vertebrate lens which is induced to form when the developing eye contacts the overlying epidermis. Thus lenses can be induced to form from any epidermal cells if a developing eye is transplanted to the appropriate vicinity.

Another example of embryonic induction is neurulation. This refers to the development of the notochord and dorsal hollow nerve cord, two features exclusive to chordates.

Gastrulation must occur before neurulation. It is the process whereby the blastula, which is a hollow single-layered ball of cells is transformed into a gastrula, which has the three primary germ layers, from which all the tissues will form.

Continuing to work backwards, cleavage is the series of cell divisions from the initial zygote (fertilized egg) to the formation of the blastula discussed above. Despite the numerous cell divisions, the blastula is no larger than the original zygote.

Morphogenesis occurs later in embryonic life than the other processes. It refers to the formation of shape. In its second month, the embryo's limbs and digits can be discerned.

16.   (A)

Glands are classified as being endocrine or exocrine. An endocrine gland releases a hormone directly into the blood, where the hormone circulates until it reaches its target cell. An exocrine gland secretes its product via a duct directly to its site of action - into a lumen or onto a body surface.

Examples of exocrine glands are all digestive glands that release digestive enzymes. Other exocrine glands and their secretions are: the lacrimal glands, which secrete tears; mammary glands, which secrete milk; sebaceous glands, which secrete sebum (an oily secretion); and sweat glands, which secrete sweat. The gastric glands of the stomach have specific cells that release enzymes (pepsin), mucus, or acid (HCl)

Examples of endocrine glands are: the thyroid gland, which secretes thyroxin; the testes, which secrete testosterone; and the anterior pituitary, which secretes growth hormone. There are many other examples.

The pancreas is an organ that has both exocrine and endocrine functions. As an exocrine gland, it secretes digestive enzymes, such as trypsin, into the duodenum, the first segment of the small intestine. As an endocrine gland, it releases the hormones, glucagon and insulin, both of which regulate blood glucose levels by opposing actions.

17.   (C)

In DNA replication, there is complementary base pairing between purine and pyrimidine bases. The purine bases, adenine and guanine, have two-ringed structures, while the pyrimidine bases, thymine and cytosine, have one-ringed structures. In RNA, the pyrimidine uracil replaces thymine. The base pairing rule is that adenine pairs with thymine (or uracil, in RNA) and guanine pairs with cytosine.

However, the bases are not free in the medium but exist as triphosphate nucleotides (i.e., attached to a pentose sugar and three phosphate groups). Although the residues in DNA are monophosphate nucleotides, they are initially bound in the triphosphate form. The pyrophosphate (PPi) is cleaved off, leaving the DNA residues, dAMP, dTMP, dGMP, and dCMP (deoxyadenosine monophosphate, deoxythymidine monophosphate, deoxyguanosine monophosphate, and deoxycytidine monophosphate). The residues in RNA are AMP, UMP, GMP, and CMP. UMP is uridine monophosphate.

DNA polymerase catalyzes the attachment of the appropriate nucleotide. Thus a cytosine on one DNA strand would call for the binding of dGTP, so that the cytosine and guanine can form a base pair.

cAMP is a nucleotide used as a second messenger for protein hormones. It is formed intracellularly from ATP. ATP is a triphosphate nucleotide best known as the high-energy compound used to do much of the work of the cell.

18.   (C)

The ABO blood system is based on the presence of antigens on the red blood cells and antibodies in the plasma. A type A person has A antigens on his red blood cells and anti-B antibodies in his plasma. He cannot have anti-A antibodies in his plasma as they would cause his red blood cells to agglutinate (clump). A type B person has B antigens and anti-A antibodies. A type AB person has both A and B antigens, and therefore has no antibodies, while a type O person has no antigens and therefore has both anti-A and anti-B antibodies.

Concerning blood transfusions, a risk exists when a recipient's plasma antibodies agglutinate the donor's cell antigens. There is no risk associated with the recipient's cell antigens being agglutinated by the donor's plasma antibodies, because those antibodies are diluted by the recipient's plasma.

A type A subject can only receive blood from another type A person or from a type O person, since the anti-B antibodies in the type A's subject plasma will not have anything to agglutinate. If the type A person were to receive blood from a type AB subject, his anti-B antibodies would agglutinate the B antigens from the donor's blood.

Another important group of blood antigens is the Rh factor. Someone with the Rh factor is designated Rh⁺. Those without the Rh factor are designated Rh⁻. Unlike the ABO blood groups, no antibodies are normally present, unless the blood has been exposed to the Rh antigen. There can be no prediction of the Rh blood type for the type A subject in question, as the two groups are independent.

19.   (B)

The peppered moth, *Biston betularia*, can be either light- or dark-colored. Dark color is controlled by a dominant allele, but was rare in the European population prior to the Industrial Revolution in the mid-1800s. In the early part of the century, the light moths were predominant; they camouflaged well with the tree trunks, and hence were less likely to be eaten by their predators, birds.

The pollutants of the Industrial Revolution settled on the tree trunks; the soot made the trunks dark. Now the dark moths would have the selective advantage due to camouflage.

It is important to realize that light moths do not "become" dark, or vice versa. Both already existed in the environment, perhaps due to an earlier mutation. Natural selection was at work here; those moths that had the gene for dark color would be more likely to survive and hence pass their genes on. Thus, industrial melanism refers to the evolutionary change whereby the dark moths had a selective advantage in their changing environment. Whether or not skin cancer developed is immaterial to this question.

20.   (B)

Taxonomy refers to the scientific classification of all living things into a systematic scheme. It is based entirely on evolutionary relationships.

The largest category of classification is the kingdom, of which there are five: Monera (bacteria and blue-green algae), Protista, Fungi, Plantae, and Animalia. Within a kingdom, there are phyla (or divisions as in Fungi and Plantae). Phyla are divided into classes, classes into orders and orders into families. Families are divided into genera. Each genus is then divided into species. When we refer to a specific organism, we usually give its binomial name, the genus and species. Hence man is called *Homo sapiens*. Note that by convention, the genus name is capitalized and the specific epithet is not; the binomial name is underlined or italicized.

The specific epithet <u>never</u> stands alone. Thus, man's species name is *Homo sapiens* not *sapiens*. The specific name could be given to organisms of other genera. For example, *multiflora* is a specific name. When used alone, you could be referring to *Rosa* or *Begonia*, clearly two different species. By saying *Rosa multiflora*, you are referring to one and only one species.

Man's complete taxonomic classification is as follows:

kingdom	-	Animalia
phylum	-	Chordata
class	-	Mammalia
order	-	Primates
family	-	Hominidae
genus	-	*Homo*
species	-	*Homo sapiens*

There are also sub-groups. For instance, man belongs in the subphylum Vertebrata. For other species, there may be subclasses, suborders, etc.

21.   (D)
Mammals are hairy or furry creatures, characterized by well-developed cerebral cortices, the ability to thermoregulate, and most notably, the possession of mammary glands, which produce milk for the developing young.

There are three major types of mammals, based on their birthing method. The monotremes lay eggs. The only examples are the duck-billed platypus and the spiny anteater. The marsupials, which include the kangaroo and opossum, have abdominal pouches in which their immature babies continue development after a brief period in the mother's uterus. The largest number of species belong to the placental mammals, of which man is an example. Other placentals are dogs, cats, bats, and rabbits, to name a few. The placenta is a vascular connection between the mother's uterus and the embryo, composed of both maternal and fetal tissues.

22.   (B)
DNA (deoxyribonucleic acid) is a polymer of nucleotides. It is the substance of variety in life, since our genes are made of DNA. The

nucleotides that comprise nucleic acids consist of a nitrogenous base, a pentose sugar, and a phosphate group.

The nitrogenous bases are classified as purines (double-ringed structures), such as adenine and guanine, or pyrimidines (single-ringed structures), such as thymine or cytosine. In RNA, the pyrimidine base uracil replaces thymine.

Pentose sugars are five-carbon sugars, as opposed to the hexoses (six-carbon sugars) in the food we eat. The pentose in DNA is deoxyribose, but it is ribose in RNA.

A base and a sugar chemically linked is called a nucleoside. When a phosphate attaches to it, the resulting structure is a nucleotide. Nucleotides bond together to form long strands of nucleic acid, either DNA or RNA.

23.   (D)
The sponges (phylum Porifera) are lower invertebrates. Only about 1% of the 10,000 or so species live in freshwater; the rest are marine.

As larvae, they are free-swimming, but as adults they are sessile, anchored to the substratum.

Their pores give them their name and characteristic appearance. Their bodies can contain inorganic spicules of calcium carbonate or silica, and/or they can contain the organic protein called spongin. The protein and spicules strengthen the gelatinous matrix.

24.   (D)
There are several terms used to describe aspects of population growth. No population can show infinite growth; they stabilize at a certain size, despite early rapid growth. This stable number is called the carrying capacity. Of course, there are small fluctuations about this value; when the population exceeds its capacity, the death rate will exceed the birth rate and bring the population back down to capacity.

The biotic potential refers to the rate at which a population will grow, when growth is uninhibited, such as in the early growth of a population. The slope indicating biotic potential is exponential - it has a very sharp slope.

The ultimate graph of population growth is sigmoid (S-shaped) because as the carrying capacity is approximated, the slope flattens considerably, indicating a decreased rate of growth.

Limitations on population growth can be density-dependent or density-independent. Density-dependent factors may include limited resources such as food, water, and space, or behavioral factors.

Density-independent factors do not involve population size. For instance, cold spells or other climate changes can affect the population size.

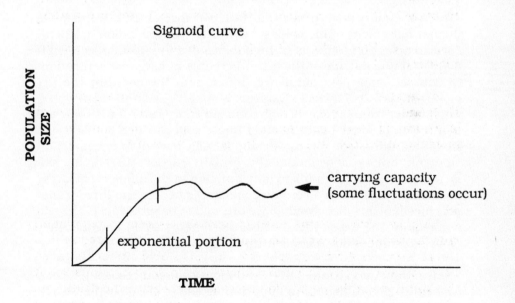

25.   (D)

Polydactyly, a condition in which the afflicted has six fingers, is due to the effects of a dominant gene. Both hemophilia and colorblindness are sex-linked traits. The genes are on the X-chromosome and are recessive. While color-blindness is more a condition than a disease, hemophilia is a dangerous disease, in which the blood clotting mechanism is faulty. The alleles for albinism and sickle-cell anemia are both recessive and are both located on autosomes (chromosomes other than the sex chromosomes). Albinos cannot produce melanin and hence have no color in their skin, hair and eyes. They are very susceptible to the sun's rays and must protect themselves against the sun. Sickle-cell anemia is a type of anemia in which the distorted hemoglobin disrupts the shape of the red blood cell and hence limits its oxygen carrying capacity. Unlike other recessive diseases, in this case, a heterozygote (one with both the normal dominant and abnormal recessive allele) may show symptoms of sickle cell anemia under conditions of low oxygen tension, such as occur during severe exercise or at great altitudes.

26.   (B)

There are many types of cellular transport mechanism. Diffusion is the net movement of molecules down their concentration gradient. For example, potassium is in high concentration inside a cell, so it diffuses to the outside of the cell, where it is in lower concentration. Of course, some potassium can diffuse into the cell, but the net movement is out of the cell. Osmosis is simply the diffusion of water; water moves from a region of high concentration (more dilute solution) to a region of lower concentration (more concentrated solution). In facilitated diffusion, the movement is still down its concentration gradient, but it is facilitated by protein carriers that span cell membranes. This mechanism may function in the diffusion of larger polar molecules, such as amino acids and glucose. The three transport mechanisms discussed above are said to be passive, i.e. they do not require the expenditure of cellular energy (ATP) because they move substances down concentration gradients.

Active transport requires ATP, as it moves substances against their concentration gradients. An important example of this is the Na^+/K^+ ATPase pump. This pump pumps K^+ into the cell in which the concentration of K^+ is already high. It also pumps sodium out of the cell, despite the high concentration of Na^+ outside of the cell already.

27.   (E)

Fungi are eukaryotic cells. Most are multicellular, except the unicellular yeasts. Fungi have cell walls composed of chitin. Their nutritional needs are met primarily by decomposition, although a parasitic mode of life is possible too.

There are four divisions of fungi. The Oomycota, or egg fungi, include water molds and mildews; the Zygomycota include bread molds; the Ascomycota are the sac fungi (these are the molds responsible for most food spoilage). Yeasts fall into this category as well. The Basidiomycota, or club fungi, include mushrooms, both edible and poisonous.

Algae are classified as plants (green, brown and red algae), moneras (blue-green algae), or protists (diatoms, dinoflagellates, euglenoids, and golden-brown algae) depending how closely they resemble the members of a particular kingdom.

28.   (A)
In DNA molecules, there are specific rules that dictate which base will bond with which base. Adenine and guanine are both classified as purines, which are large, double-ringed nitrogenous bases. Thymine and cytosine are both pyrimidines, the smaller, single-ringed nitrogenous bases. A purine must pair with a pyrimidine due to the limited amount of space in the interior of the DNA molecule. The number of adenines in a DNA molecule is equal to the number of thymines; this is because they are always bound to each other. The same story holds for guanine and cytosine.

Bonding between base pairs occurs via hydrogen bonds. There are two hydrogen bonds between every adenine and thymine, and three between every guanine and cytosine. Although hydrogen bonds are not individually very strong, collectively they are quite strong. Yet the hydrogen bonds must easily break, whenever replication or transcription occurs.

Note that uracil is a pyrimidine base that replaces thymine in RNA molecules. Hence in transcription (synthesis of mRNA), adenine will bond with uracil.

29.   (E)
The leaf functions in photosynthesis, so its structure provides a large surface area in order to capture the sun's energy. The layers of a leaf include a mesophyll layer sandwiched between the epidermal layers.

The upper epidermis is covered by a waxy cuticle to afford protection by minimizing water loss. The mesophyll consists of loosely packed tissue with many chloroplasts in each cell and hence is the site of photosynthesis. The palisade mesophyll is adjacent to the upper epidermis, while the spongy mesophyll below, which contains much air space, is adjacent to the lower epidermis. The lower epidermis is also covered by a cuticle. This layer also has stomata, or pores, which regulate the movement of gases (oxygen, carbon dioxide, and water vapor) into and out of the leaf.

The vasculature of the leaf includes the xylem and phloem. Xylem brings water to the leaf while the phloem carries the photosynthetically produced sugars away.

30.   (A)
One major comparison between invertebrate animal phyla that can be observed is the type of body symmetry. The cnidarians, primitive invertebrates, have radial symmetry, as exemplified by the jellyfish. Their parts are arranged about a central axis.

Flatworms, mollusks, annelids, and arthropods are all bilaterally symmetrical: their bodies can be divided into similar right and left halves. They also have dorsal and ventral surfaces, and anterior and posterior ends.

It might seem surprising that echinoderms, rather advanced invertebrates, share the same type of symmetry as the cnidarians, radial symmetry. However, the larvae of echinoderms are bilaterally symmetrical and hence, echinoderms most likely evolved from bilaterally symmetrical ancestors.

31.   (B)
One type of phenomenon that limits evolutionary outcomes consists of reproductive isolating mechanisms, the barriers to successful mating. In this case, success means not simply producing viable offspring, but fertile ones as well.

Isolating mechanisms can be categorized as pre-mating or post-mating; the former indicates that no mating will occur, and the latter indicates that while mating occurs, it is unsuccessful.

Premating mechanisms are the more common at work in nature. Types of pre-mating mechanisms include, but are not limited to, the following. Geographical isolation refers to the separation of species due to their land and climate preferences/needs. Seasonal isolation occurs when the breeding seasons of two species do not overlap. Behavioral isolation indicates that communication is species-specific. A bird of one species may not respond to a mating call of another. Mechanical isolation indicates that the structures used in mating are incompatible due to size or shape.

Post-mating isolating mechanisms include hybrid sterility, in which an offspring is produced but is not fertile. For instance, a male donkey and a mare produce a mule, which is sterile. Other examples of post-mating mechanisms include the failure of the gametes to fuse and the abnormal development and early death of the hybrid after birth.

32.   (A)
Two people heterozygous for the allele for brown eyes carry one dominant and one recessive allele, by definition. Hence, their genotype would each be Bb. The best way to see the probability of the genotypes of their offspring is by using a Punnett square.

	B	b
B	BB	Bb
b	Bb	bb

The results of the Punnett square show that the genotypes of the offspring would be 25% homozygous dominant (BB), 50% heterozygous dominant (Bb), and 25% homozygous recessive (bb). The phenotypic ratio would be 75% brown-eyed (BB or Bb) and 25% blue-eyed (bb). The genotypic ratio is 1:2:1, but the phenotypic ratio, the question of interest, is 3:1. A ratio of 1:0 implies that only brown-eyed babies would be produced; a ratio of 1:1 implies an equal probability of brown-eyed and blue-eyed babies.

33.   (B)

Charles Darwin proposed a theory of evolution, largely based on his observations made during the voyage of *The Beagle*. While traveling about the Galapagos Islands off the west coast of Ecuador, Darwin observed that each of the islands had similar birds, all finches. However, these finches were of different species. The most obvious anatomical differences were found in the beak. Since structure often dictates function, Darwin realized that the differences in beak structure were adaptations to unique food sources (i.e., insects, fruit, and seeds) on the islands. The finches all were of a common ancestor, and through many generations differentiated into distinct species adapted to their own particular environment of that island. While the phrase was not coined at the time, this is a classical example of adaptive radiation.

34.   (D)

The fluid mosaic model describes the animal cell membrane (see figure on following page). The fluid refers to the phospholipids that make up much of the substance of the membrane. Phospholipids have polar phosphate heads and non-polar fatty acid tails. A bilayer of phospholipid forms with the heads oriented toward the aqueous medium of the intracellular and extracellular fluids, and the tails oriented inward toward the tails of the other bilayer. While phospholipids are the major lipid in the membrane, there is cholesterol as well.

Proteins in the cell membrane are classified as intrinsic (integral) or extrinsic (peripheral). The proteins form the "mosaic" aspect of the membrane, as they are said to "float in a sea of lipid." The intrinsic proteins may span the entire membrane, acting as transport channels.

Some intrinsic proteins only span half of the bilayer. The peripheral proteins lie on the external or internal surface adjacent to the aqueous mediums. Some proteins facing the extracellular fluid have sugar residues on them. These glycoproteins may function as recognition sites for antigens, hormones, etc.

Cellulose is a structural component of plant cell walls, which surround plant cell membranes and help to prevent cell bursting. Cellulose is a glucose polymer and the indigestible residue of the plant foods that we eat.

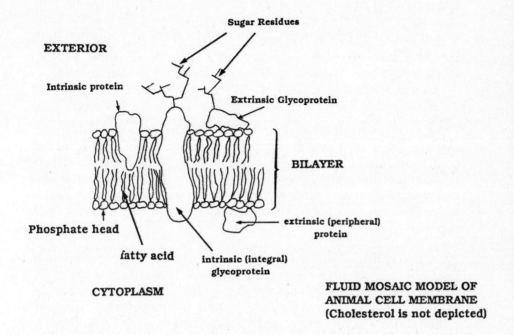

EXTERIOR

Sugar Residues

Intrinsic protein

Extrinsic Glycoprotein

BILAYER

Phosphate head

fatty acid

extrinsic (peripheral) protein

intrinsic (integral) glycoprotein

CYTOPLASM

FLUID MOSAIC MODEL OF ANIMAL CELL MEMBRANE (Cholesterol is not depicted)

35.   (A)

There are two major patterns of evolution that may explain the changes that occur. When Darwin proposed his theory of evolution based on natural selection, he implied a mechanism of gradualism, whereby changes were continual and gradual, the sum of many small changes. More recently, another mechanism has been proposed, though it is still consistent with Darwin's theory of natural selection. In punctuated equilibrium, evolutionary changes occur during short periods of rapid change that are separated by long periods of little change. There is a minimum of transitional states.

Adaptive radiation refers to the emergence of several species from one species, due to the segregation of their habitats. For instance, the finches that Darwin described on his famous voyage on *The Beagle* may have all radiated from one original species of finch.

In a sense, convergent evolution appears to be the opposite of adaptive radiation. It describes the appearance of similar adaptations to the environment amongst different species, despite the lack of a common ancestor.

Catastrophism, proposed by George Cuvier, states that the seeming appearance of new species is due to events of mass destruction that left only a few survivors. The survivors repopulated the environment and appeared as new species. In his beliefs, there was only one time of creation.

36.   (B)
The ocean is itself an ecosystem. It is divided into many different regions. The pelagic province refers to the open ocean away from the continental shelf. There are various regions to this province based on the depth of the water. The epipelagic region is the uppermost one extending 200 meters, and while it does not have as much life as the waters nearest the shores, there is more life here than in the deeper zones.

Life can continue in the epipelagic region because there is enough light allowing for photosynthetic reactions. However, moving away from the surface, one reaches the mesopelagic region, which extends 800 meters. The decreasing light forbids photosynthesis. There is total darkness at the deeper bathypelagic region. Only decomposers, scavengers, and dead organisms from above are found here.

Of course, aside from light penetration, there are gradients in temperature and salinity. However, light is a direct requirement for photosynthesis.

There is no tidal action in the pelagic province since this is not the intertidal region which occurs in the shallow ocean.

37.   (A)
In order for replication to occur, the DNA double helix must separate. The site of separation, where the hydrogen bonds between base pairs break, is called the replication fork; the single-stranded regions emerge here.

Each of the parent strands serves as a template for DNA replication. The DNA polymerase enzyme that catalyzes the addition of nucleotides to the growing daughter strands can only read the parent chain in the 3' - to - 5' direction and hence synthesize the daughter strand in the

5' - to -3' direction. Since the original DNA duplex is antiparallel, i.e. the strands have their 3' and 5' ends oriented in opposite directions, the DNA polymerase will add nucleotides continuously on one strand and in discontinuous bursts on the other. The strand that shows continuous replication is designated the leading strand. Synthesis occurs in the direction of the replication fork. The other strand, the lagging strand, shows short bursts of replication away from the replication fork, but in the proper 5' - to - 3' direction.

There is no replication fork in the transcription of DNA to RNA, because only a central portion of one DNA strand unwinds and serves as a template.

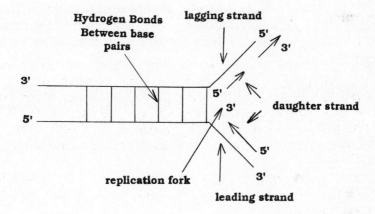

38.   (B)
In the first meiotic prophase, many events occur that provide the basis for variation even between offspring of the same parents.

First, the homologous chromosomes pair up in a process called synapsis. Since each homologous chromosome has already replicated, it exists as two sister chromatids joined together by a centromere. The paired chromosomes hence now exist as a tetrad of chromatids. Now crossing-over, the exchange of segments between homologous non-sister chromatids, can occur. The site of cross-over is called a chiasma (pl. chiasmata). Since the chromosomes and later the chromatids will ultimately segregate randomly and independently, and since they contain recombined chromosome segments and hence, recombined genetic traits, the foundation for variation is laid down.

The junctions between neurons are called synapses; the cells are separated by a synaptic cleft.

39.   (E)
Phloem is a vascular tissue that is continuous from the leaf to the stem to the root. It contains photosynthetically derived sugars and transports them from the site of origin in the leaf to other parts of the plant. The cells of phloem are the sieve tube members, which function in conduction, and the companion cells, which aid in the metabolic needs of the sieve tube members, since only the companion cells have nuclei.

The xylem transports water from the roots up to the top of the plant. In angiosperms, the vessel elements function primarily in conduction. The tracheids, while able to conduct, primarily give strength to the tissue.

40.   (A)
A mitochondrion is a cellular organelle that utilizes oxygen to produce ATP. It has its own DNA, which replicates autonomously from the nuclear DNA. It is suspected, based on the size, structure, and biochemistry of mitochondria, that they were once prokaryotic cells similar to bacteria that formed a symbiotic relationship with a eukaryotic host. Due to evolution, the mitochondrion has lost its independence.

A similar story holds true for chloroplasts, which also have their own DNA, similar to that of bacteria. A chloroplast, with its capacity for photosynthesis, could have originally been an independent prokaryote, now dependent on the cell in which it lives.

The endoplasmic reticula (both rough and smooth), ribosomes, Golgi apparatuses, and lysosomes do not contain their own DNA. Their functions are ultimately dictated by the nucleus. The ribosomes synthesize proteins. If the ribosomes are attached to the endoplasmic reticulum, making it rough endoplasmic reticulum, the proteins will enter the reticular lumen and be transported through the cell and reach the Golgi apparatus for modification and continued distribution. The smooth endoplasmic reticulum functions primarily in lipid synthesis. The lysosomes contain hydrolytic enzymes that can digest cell debris or the contents of endocytotic vesicles.

41.   (B)
There are three distinguishing features of phylum Chordata, of which all vertebrates are members, although these features need not persist throughout life.

The presence of a notochord (hence the name Chordata) is prerequisite. This is a flexible rod that develops into a cartilaginous or bony vertebral column in vertebrates. The dorsal hollow nerve cord differentiates into the brain and spinal cord of vertebrates. Finally, the pharyngeal gill slits become the gills of fish, yet serve other, seemingly unrelated functions in higher vertebrates, due to modifications that occur during embryological development.

42.   (B)
The fungi encompass an entire kingdom in the classification scheme. They function as decomposers of organic matter and hence aid in the carbon, nitrogen and phosphorus cycles. Of interest, fungi decompose both living and nonliving matter. The term saprophytic refers to its ability to decompose dead matter. This is in contrast to parasitic behavior, exhibited by some fungi, which refers to decomposition of living matter.

All fungi are eukaryotic. The eukaryotes, which means literally, "true nucleus" includes all organisms in kingdom Fungi, Animalia, Plantae and Protista. They have a distinct nucleus enclosed in a membrane and many organelles. Only organisms of kingdom Monera (bacteria and cyanobacteria) are prokaryotic.

Prokaryotes do not have a distinct nucleus. Rather their DNA is in a nucleoid region.

The term heterotrophic refers to the inability to manufacture one's own food. Fungi secrete digestive enzymes onto their food substrate, and then absorb it. Animals are also heterotrophic, although animals ingest their food prior to digestion and absorption. In contrast, autotrophic organisms can produce their own food. For instance, plants produce food by photosynthesis. The monerans and protists are autotrophic or heterotrophic.

Some fungi are pathogenic, i.e. cause disease. Fungi can cause disease in animals (ringworm) and plants (potato blight). However, most fungi are not pathogenic and may even serve specific benefits for mankind. For instance the antibiotic penicillin is produced by a fungus.

43.   (C)

There are many factors that participate in evolutionary change. Mutation refers to random, but inheritable changes in DNA. Natural selection refers to the idea that some genotypes will be selected by the environment for survival and propagation. Gene flow implies that allele frequency can change due to migration in or out of the population. Genetic drift refers to random fluctuations in the frequencies of alleles. An extreme case of genetic drift is called the founder effect. It is known that genetic drift is especially important in small populations. When only a few individuals become separated from the main population, they, in essence, are the founders of a new population. The genotype frequency of this new population may differ markedly from the original population from which these founders emerged, because they represent only a small sample of all the genotypes that are present in the main population.

44.   (C)

There are many types of community interactions between species. Symbiosis is a general term that describes the relationship between two coexisting species.

When one species benefits from the interaction and the other is unaffected, the relationship is said to be a commensal one. In mutualism, both species benefit. In parasitism and predation, one species benefits at the expense of the other's misfortune. The distinction between the two is that a predator kills its prey, whereas a parasite does not kill its host, since it lives inside or on the host.

45.   (E)

The sexual reproductive structures of angiosperms are in the flower (see figure). The female reproductive organ is the carpel, which has three parts. The base of the carpel is the ovary. It contains the ovule within. The ovary may develop into a fruit. A stalk called a style extends from the ovary. The sticky cap of the style is the stigma. It is the site that receives pollen.

The male reproductive organs are the stamens, which consist of pollen-bearing, two-lobed anthers atop single stalks called filaments.

The non-reproductive portions of the flower include the receptacle, which functions as a stem. The outermost whorls are the green sepals.

The inner whorls are the colorful petals, which attract insects and other pollinators.

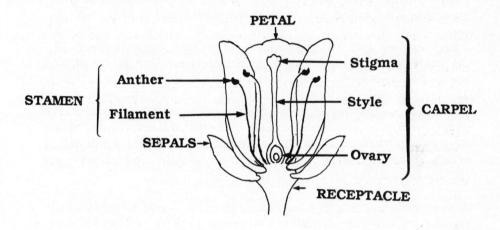

46.   (C)
DNA contains the codes for all the proteins that we need. DNA is transcribed by mRNA (messenger RNA). There are other RNAs also made from DNA: tRNA and rRNA (transfer RNA and ribosomal RNA, respectively). Different types of RNA polymerase molecules aid in the synthesis of different types of RNA.

mRNA aligns with ribosomes and contains the codons (sequences of three nucleotides) that represent specific amino acids and stop signals.

The rRNA and ribosomal proteins are the structural components of ribosomes.

tRNAs are small RNAs. They are specific for amino acids and carry the amino acid to the growing polypeptide chain on the ribosome. tRNAs have anticodons, triplets of nucleotides, that bind to the codons on mRNA. Therefore, mRNA determines which amino acids are added to the polypeptide, because each codon calls for a certain amino acid.

ATP (adenosine triphosphate) is a nucleotide that fuels the energy-requiring reactions of the body. Important functions of ATP include active transport and muscle contraction.

47.   (E)

Embryological development in vertebrates follows an orderly sequence of fertilization, cleavage, gastrulation, and neurulation.

Fertilization refers to the process whereby a haploid sperm cell penetrates a haploid egg cell and their nuclei fuse, forming the diploid zygote (fertilized egg).

Cleavage begins immediately: it refers to the series of cell divisions which produces many smaller cells. It results in a solid ball of about 32 cells called a morula, which continues division to form a hollow ball of cells called a blastula. Despite the many cell divisions, there is no change in overall size.

Next, cells migrate in a particular pattern. These changes in the embryo are called gastrulation and result in three cell layers: ectoderm, mesoderm, and endoderm. All organs are derived from these layers. For example, the ectoderm gives rise to the skin and nervous system, the mesoderm gives rise to the skeleton and muscles, and the endoderm gives rise to the digestive tract and lungs.

The differentiation of the layers into organs follows a specific time sequence. The first organs which develop are the notochord and dorsal hollow nerve cord, in a process called neurulation. These organs need only be present in vertebrate embryos, but they may not necessarily persist into post-gestational life, depending upon the organism in question.

Induction refers to the event whereby one tissue determines the course of development of adjacent tissue. It is best exemplified by the development of the lens of the eye from epidermal tissue upon contact with the optic stalk.

48.   (E)

The flower is the reproductive organ of the angiosperms, or flowering plants. The gametes form in the flower. In the male organ, specifically the anther, microspore mother cells that are produced mitotically undergo meiosis to form microspores, which develop into binucleate pollen grains, the immature male gametophyte. In the female organ, specifically the ovary, megaspore mother cells that are produced mitotically undergo meiosis to form megaspores, one of which develops into the female gametophyte, which will divide mitotically and ultimately produce the egg, and other non-functional cells.

Pollination refers to the transfer of pollen grains from the male organ to the sticky surface of the stigma, the most exposed portion of the female organ. This initiates the growth of the pollen tube from the stigma down through the style to reach the ovary. During pollen growth, one of the nuclei of the pollen grains will mitotically divide and produce two sperm cells. The male gametophyte is now mature. The sperm are delivered directly to the egg. The nuclei fuse and fertilization is complete.

Photosynthesis is a non-reproductive process. It occurs in the organs that have cells with chloroplasts, such as leaves, which have mesophyll cells rich in chloroplasts.

49.   (D)
Double fertilization is unique to the angiosperms, or flowering plants.

A pollen grain, the immature male gametophyte, has two nuclei. During pollen tube growth, one of the nuclei divides mitotically to produce two sperm nuclei. It is now a mature male gametophyte.

The megaspore that develops into the female gametophyte divides mitotically to produce eight haploid nuclei. One will become the egg, two will become polar nuclei, and the other five serve no known function and disintegrate.

When fertilization occurs, not only does a sperm nucleus fuse with the egg nucleus to produce the zygote, but the second sperm nucleus fuses with both polar nuclei to form the endosperm cell. Since the gametes are all haploid, the zygote is diploid as expected, and the endosperm is triploid. The zygote will develop into the embryo; the endosperm functions to store food.

The other group of seed-producing vascular plants are the gymnosperms, which, like angiosperms, rely on pollination for reproductive success. Gymnosperms do not display double fertilization. The sperm cells just fuse with egg cells.

There are different ways in which twins can be produced. Fraternal twins are not identical, because they result from the fusion of two distinct sperm cells and two distinct egg cells. In contrast, identical twins result from post-zygotic events. A single sperm fertilizes a single egg. The resultant zygote splits into two embryos early in development. Their genetic constitution is exactly alike and of course they are always the same sex. Siamese twins, those born attached at some point on their bodies, are always identical twins that have not separated completely. Surgical separation is usually performed shortly after birth.

50.   (E)
The decomposers are important parts of the food chain. They are usually bacteria and fungi. They occupy no particular trophic level, because they feed on organisms of all levels. In general, they feed on dead organic matter, whether it came from a producer or a consumer (herbivores and carnivores) or even another decomposer.

The function of decomposition is to return gases and minerals, which contain vital elements like carbon, nitrogen, oxygen, and phosphorus, to the biosphere to be recycled. The elements are now available to the producers.

51.   (A)
The dark reactions (or light-independent reactions) of photosynthesis function primarily to convert carbon dioxide to glucose. The steps of the dark reaction were elucidated by Melvin Calvin in the 1950s. The Calvin cycle, as the series of reactions is called, requires the input of atmospheric carbon dioxide and of components that are made in the light reactions, such as ATP and NADPH. An important intermediate in the synthesis of glucose is RuBP (ribulose bisphosphate). RuBP is a five-carbon sugar. By the addition of carbon dioxide (carbon dioxide fixation), RuBP is converted into an unstable six-carbon molecule. The complex series of reactions in the Calvin cycle requires six revolutions of the cycle and hence six carbon dioxide molecules to produce one glucose.

52.   (C)
Biogeochemical cycles exist for the elements carbon, hydrogen, oxygen, and nitrogen, which cycle between the earth, living organisms, and the atmosphere.

The water cycle is relatively simple to follow. Water can enter the atmosphere via various sources. Water evaporates as vapor from the land. Respiration of animals and plants produces water vapor. Plants lose water by evaporation through openings in the leaves by a process called transpiration.

Water returns to the earth as precipitation (rain or snow). Most of this water falls on oceans and other bodies of water. A small percentage of the precipitation falls on the land. This water can percolate into the soil and eventually reach the groundwater which will eventually drain into a larger body of water. The surface water is also transported as runoff. And of course some of the precipitate will evaporate.

The term fixation can be applied in both the nitrogen and carbon cycles. Carbon dioxide fixation occurs during photosynthesis, where it is incorporated into carbohydrate. Nitrogen fixation refers to the process whereby atmospheric molecular nitrogen is incorporated into nitrogen-containing compounds. Some microorganisms carry out this process.

53. (A)
A blue-eyed woman is homozygous recessive (bb). A heterozygous man is Bb. The cross is best seen with a Punnett square:

	B	b
b	Bb	bb
b	Bb	bb

The proportions of the offspring are half heterozygous (Bb) and thus brown- eyed, and half homozygous recessive (bb) and thus blue-eyed. An equal proportion of brown- and blue-eyed offspring is a 1:1 ratio.

A 1:2 ratio would indicate that for every brown-eyed child, there are two blue-eyed children. A 2:1 ratio would be the reverse. A 1:0 ratio would mean that all the offspring are brown-eyed and none are blue-eyed. A 0:1 ratio is the reverse.

54. (D)
The genotype is the actual genetic constitution of the individual, but the phenotype is the expression of those genes. For instance, the genotypes that code for eye color are BB (homozygous dominant), Bb (heterozygous) and bb (homozygous recessive). There are thus three genotypes. But there are only two phenotypes: Bb and BB both code for brown eye color, as the allele for brown eyes (B) is dominant to that for blue eyes (b). Blue eyes are only possible with the genotype bb. (Note that green eyes are considered as blue, genotypically and phenotypically). Thus a blue-eyed person knows his genotype immediately, but a brown-eyed person needs to look at his lineage to possibly figure out his genotype.

Phenotype includes all physical characteristics of an organism that are the results of genotype. A characteristic need not be seen by an observer to be included in an organism's phenotype; e.g. one's blood type is part of one's phenotype.

55.   (A)

The ancestors of modern bryophytes were the first land plants. However, lacking vascular tissue, which transports water and food throughout the plant, bryophytes must live in moist areas. They are anchored to the ground by rhizoids, which function as roots. Sexual reproduction requires water: the antheridia release sperm in the water and they must swim to the eggs that are produced by the archegonia. Classes of bryophytes include liverworts, hornworts, and mosses, of which the last is the most well-known.

Vascular plants show a definite evolutionary advancement. The vascular tissue includes xylem, which transports water from the roots to the leaves via the stem, and phloem, which transports food to all parts of the plant.

The lower vascular plants, such as the horsetails and ferns, require water for sexual reproduction. Like the bryophytes, the sperm that are produced by the antheridia must swim to the eggs that are produced by the archegonia. However, unlike the bryophytes, these plants have vascular tissue and true stems. Some lower vascular plants also have true roots + leaves.

The most advanced of the vascular plants are the seed plants: the gymnosperms (cone-bearing plants) and the angiosperms (flowering plants). Their reproduction is independent of water. Seeds can be dispersed by many media (aside from water), including wind and animal vehicles. The reproductive structures of the gymnosperms are cones; the reproductive structures of the angiosperms are flowers.

56.   (C)

Viruses are very small structures. They are not organisms in the true sense because they cannot reproduce. They require hosts for their metabolic and reproductive needs. Virus consists of a single strand of DNA or RNA enclosed in a protein sheath called a capsid, and may or may not have a surrounding envelope. Viruses can specifically attack plants, animals, or bacteria.

A bacteriophage is a virus that attacks bacteria. It is sometimes simply called a phage. Bacteriophages always contain DNA as their nucleic acid. The phage injects its nucleic acid into the host cell, leaving its protein coat outside. The phage now begins to control the host cell activity by directing the synthesis of more viruses. The phage kills the host cell by lysing it, releasing the newly formed viruses. Sometimes, the phage does not cause lysis. Rather, the viral DNA becomes incorporated into the host's single chromosome; there it lies latent. The term prophage refers to the fragment of vital DNA that is inserted into the bacterial chromosome. The virus is referred to as a lysogenic, or temperate, phage (as opposed to a virulent one), and the host cell is called a lysogenic cell, since it has the ability to lyse in the future.

57.   (A)
Dehydrogenases are a class of enzymes which catalyze oxidation/ reduction reactions. Literally, the term means take hydrogens away. It does this by removing electrons and hydrogen ions from a substrate.

The active portion or coenzyme of the dehydrogenases can exist in the oxidized form (FAD, NAD^+). The reduced form has gained the electrons and hydrogen ions and thus forms NADH and $FADH_2$. When NAD^+ is reduced to NADH, it is coupled to the oxidation of a substrate. For instance, in the Krebs cycle, isocitrate is oxidized to alpha-ketoglutarate concomitant with the reduction of NAD^+ to NADH. Many of the steps in the Krebs cycle are catalyzed by dehydrogenases: three of the steps use NAD^+ and one uses FAD.

Conversely, when NADH is oxidized to NAD^+, it is coupled to the reduction of a substrate. For instance, the reduction of pyruvate to form lactate occurs concomitantly with the oxidation of NADH to NAD^+. (This regeneration of NAD^+ allows glycolysis to continue.)

ADH and ACTH are hormones. ADH (antidiuretic hormone) acts on the kidney to increase water reabsorption back into the blood (hence it counteracts a diuresis). ADH is released from the posterior pituitary gland.

ACTH (adrenocorticotrophic hormone) is released from the anterior pituitary gland. Its target is the adrenal cortex and it stimulates the release of glucocorticoids such as cortisol from that gland.

58.   (D)

Glycolysis is the first of a sequence of reactions dealing with the complete oxidation of glucose as described by the following equation:

$$Glucose + 6\ O_2 \rightarrow 6\ CO_2 + 6\ H_2O + 38\ ATP.$$

This complete oxidation requires the enzymatic reactions of glycolysis, the Krebs cycle, and the electron transport chain. Glycolysis functions to convert the six-carbon glucose molecule into two three-carbon pyruvate molecules. Pyruvate is converted to acetyl CoA, which then enters the Krebs cycle.

While much ATP can be produced ultimately from glucose oxidation, some ATP must be initially "invested." This investment occurs in the first and third steps of glycolysis, each of which requires one ATP molecule. In the first step of glycolysis, glucose is phosphorylated to form glucose-6-phosphate. The energy for this step comes from the hydrolysis (breakdown) of ATP to ADP. The phosphate group is transferred from ATP to glucose.

The production of carbon dioxide occurs primarily in the Krebs cycle. The Krebs cycle is also the site of many oxidation reactions, where the intermediates lose their electrons and hydrogen ions, which are transferred to NAD^+ and FAD, reducing them to NADH and $FADH_2$. These reduced forms donate their electrons to the electron transport chain. Electron transport is associated with the production of 36 of the ATP molecules. The other two are produced in glycolysis (actually, four ATPs are produced in glycolysis but two are used, so there is a net production of two ATPs in glycolysis).

59.   (C)

The active portion of the cytochrome enzymes is a heme group which is an iron-containing pigment. The cytochromes function in many of the oxidation-reduction reactions of the electron transport chain. They are pure electron carriers. The iron can exist in the oxidized state (Fe^{3+}) or the reduced state (Fe^{2+}). Note that the reduced form has gained an electron and is thus less positively charged. Heme is also found in the oxygen transporting protein hemoglobin.

Sodium (Na^+) is the major cation of the extracellular fluid, while potassium (K^+) is the major cation of the intracellular fluid. Calcium (Ca^{++}) is stored in bone tissue. These three cations are very important in nerve and muscle activity. Iodine is taken up by the thyroid gland and is essential in the synthesis of thyroid hormones.

60.   (D)
In the mitochondrion, as electrons are being transferred along the electron transport chain of the inner mitochondrial membrane, hydrogen ions are being pumped from the mitochondrial matrix to the intermembrane space (the space between the inner and outer mitochondrial membranes). (See figure.) For each NADH which donates its electron pair, three pairs of hydrogen ions are extruded into the space, while for $FADH_2$, donating its electrons, only two pairs of hydrogen ions are extruded.

The inner mitochondrial membrane also has ATP synthetase enzymes. A channel spans the membrane and the enzymatic head faces the matrix side. Hydrogen ions accumulate in the space and hence flow down their gradient through the channel back into the matrix. The dissipation of the gradient is associated with ATP synthesis. (Of course, the gradient is not dissipated since more hydrogen ions are being pumped into the space.) For every pair of hydrogen ions flowing back into the matrix, one ATP molecule is synthesized. Thus, three ATPs are synthesized for each NADH, while only two ATPs are synthesized for each $FADH_2$.

A chloroplast is a double-membraned organelle which contains many thylakoid (flattened sacs) within the stroma (matrix). A similar principle explains ATP synthesis in chloroplasts. In this case, hydrogen ions flow down their gradient from the interior of the thylakoid back to the stroma.

MITOCHONDRION

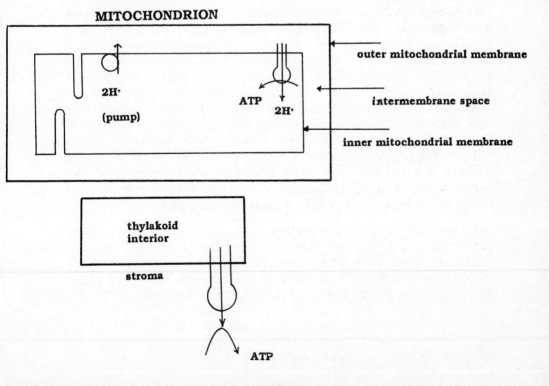

The term oxidative phosphorylation refers to the coupling of oxidation in the respiratory (electron transport) chain to the phosphorylation of ADP to form ATP. However, it is the H^+ not the phosphate group which travels through the channel protein. Sodium ion is the major cation of the extracellular fluid. Iron is found in the cytochrome enzymes of the electron transport chain. Calcium is an important cation in nerve and muscle activity.

61.   (A)   62. (C)   63. (B)

Each kidney contains about one million nephrons (see Figure 1). A nephron is a renal tubule and the associated vascular component. The tubule consists of Bowman's capsule, the proximal convoluted tubule, the loop of Henle, the distal convoluted tubule, and the collecting duct. Collecting ducts from many nephrons join together to carry urine out of the kidney into the ureter. The vascular component includes the afferent arteriole, the glomerular capillaries, the efferent arteriole, and the peritubular capillaries. The plasma in the glomerular capillaries is filtered by Bowman's capsule. The filtrate moves through the tubules, in which reabsorption (movement of solutes and water from tubular lumen to blood) or secretion (movement of solutes and water from the blood into the tubular lumen) may occur. By the time the fluid leaves the collecting duct, no changes in urine composition can occur.

A sarcomere is the functional contractile unit of a muscle (see Figure 2). It consists of thick and thin filaments. The thick filaments contain myosin, while the thin ones are composed primarily of actin. The Z lines divide individual sarcomeres, which line up sequentially along the length of the muscle cell. The dark staining region, called the A band, is due to the presence of the thick filament. The region between adjacent A bands is called the I band.

Fig. 1.

Nephron Structure

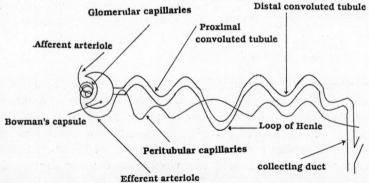

310

Fig. 2

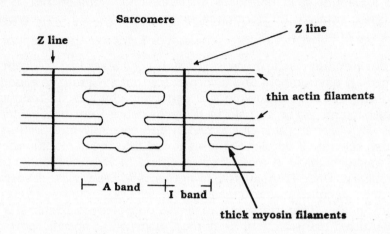

A neuron is a nerve cell. Neurons can electrically depolarize (decrease their membrane potential). When the wave of excitation reaches the end of this long cell, a chemical neurotransmitter is released into a space called a synapse. The neurotransmitter can then bind to the receptors of an adjacent neuron and excite it. In this manner, a message has been transmitted.

An alveolus is one of millions of small sacs in the lungs. Since each alveolus is covered with pulmonary capillaries, there is an immense surface area available for gas exchange between the lungs and the pulmonary blood. Hence, oxygen can diffuse into the blood and carbon dioxide can diffuse into the alveolus, where it will be exhaled.

A villus is a microscopic fold of the innermost lining (mucosa) of the small intestine. It has many microvilli on it, which contain digestive enzymes. Nutrients are absorbed through the villus into either the blood or lymphatic capillaries.

64.   (C)   65. (B)   66. (E)

In transcription, the cell uses one strand of DNA as a template for the assembly of a messenger RNA molecule (mRNA). Like replication, transcription is based on complementary base pairing. RNA differs from DNA in having a different sugar in the backbone (ribose instead of deoxyribose) and having a different nitrogenous base to pair with adenine (uracil instead of thymine).

In translation, the second major step of the overall process called protein synthesis, mRNA aligns with a ribosome in the cytoplasm (either free or bound to rough endoplasmic reticulum). On the ribosome, anticodons of two tRNA molecules form complementary pairs with two codons of the mRNA. Each codon is a set of three bases coding for a single amino acid. Each tRNA has an anticodon at one end

311

and an amino acid bound to its other end. Components of the ribosome catalyze the formation of a covalent bond between two adjacent amino acids.

Transformation is the process by which a cell takes up DNA from its immediate environment and incorporates that DNA into its genome, resulting in both phenotypic and genotypic changes. In this way, living bacteria can be "transformed" by the debris from dead bacterial cells. By observing this phenomenon and correctly interpreting it, British scientist Frederick Griffith showed that the hereditary material was a chemical. It took other scientists 14 years to identify that chemical as DNA. Transduction is a process whereby viruses specific to bacteria, called bacteriophages, transfer DNA from one bacterial cell to another. The virus may transfer random pieces of bacterial DNA (generalized transduction) or it may incorporate specific parts of bacterial DNA into its own genome, and subsequently cause its genome to be incorporated into the host bacterial genome (specialized transduction).

Transduction is a process whereby viruses specific to bacteria, called bacteriophages, transfer DNA from one bacterial cell to another. The virus may transfer random pieces of bacterial DNA (generalized transduction) or it may incorporate specific parts of bacterial DNA into its own genome, and subsequently cause its genome to be incorporated into the host bacterial genome (specialized transduction).

"Translocation" is used in three different ways in molecular biology: (1) To refer to the process in which a ribosome moves a tRNA from the first tRNA site to the second. (2) To refer to an exchange of portions of non-homologous chromosomes, due to breakage of chromosomes. (3) To describe a process in plants, in which food is transported from the leaves to the roots by way of the vascular phloem tissue.

67.　(C)　68. (B)　69. (A)　70. (D)

All plants show an alternation of generations. The diploid (2n) generation is a sporophyte plant, a spore-producing plant. The haploid (1n) generation is the gametophyte plant, the gamete-producing plant. In the evolution of plants, the sporophyte generation becomes increasingly dominant.

The life cycle is illustrated starting with the sporophyte generation. The sporophyte plant has multicellular structures called sporangia. The spore mother cells within undergo meiosis and produce haploid spores. If the plant can produce both types of spores (megaspores and microspores), it is heterosporous. If it produces only one type of spore, it is homosporous.

The spores are the first cells of the haploid generation and as such, they grow into the haploid gametophyte plant. Mitosis in the gametophyte results in production of haploid gametes, eggs, and sperm. Fusion of the gametes results in a diploid zygote that is a member of the sporophyte generation and grows into a sporophyte plant. The cycle repeats.

71.　　(C)　　72. (E)　　　73. (B)　　　74. (D)

A "typical" animal cell, if there is such a thing, has many cell organelles, which carry out specific functions within their compartments.

Ribosomes are small organelles composed of rRNA (ribosomal RNA) and ribosomal proteins. They may occur free in the cytoplasm or bound to the endoplasmic reticulum. In the latter case, the endoplasmic reticulum is then referred to as rough endoplasmic reticulum. Ribosomes are the sites of protein synthesis, as they link up with mRNA. Free ribosomes synthesize cellular proteins; ribosomes on the rough endoplasmic reticulum synthesize proteins destined for export.

Microfilaments and microtubules are the constituents of the cytoskeleton — they are the framework upon which cell shape is maintained. They also function in cell movements. Actin is an example of a protein found in microfilaments; it has a specific function in skeletal muscle contraction. Tubulin is the protein in microtubules. Microtubules are components of cilia and flagella.

Lysomes are small membrane-bound spheres that contain hydrolytic enzymes. One function of lysosomes is the digestion of aged organelles. Lysosomes also fuse with vesicles formed by endocytosis. The phagocytic vesicle may contain "food" particles; alternatively, it may contain bacteria particles. In either case, the engulfed vesicle fuses with the lysosome for digestion.

The endoplasmic reticulum is a series of channels that originate near the nucleus and extend throughout much of the cell, serving as an intracellular circulatory system. Endoplasmic reticulum that has ribosomes attached is called rough endoplasmic reticulum. It functions in the transport of proteins synthesized on its ribosomes. Endoplasmic reticulum devoid of ribosomes is called smooth endoplasmic reticulum. It functions in lipid synthesis, including the synthesis of steroids. In muscle cells, the specialized smooth endoplasmic reticulum is called the sarcoplasmic reticulum: it stores and releases calcium, as dictated by the stimulus for muscle contraction.

Mitochondria are the sites of oxygen utilization and ATP (energy) production. Mitochondria are especially prominent in active cells, i.e., those that undergo contraction, active transport, and other metabolically active processes.

75.   (C)   76. (B)   77. (D)   78. (A)

The enzymes in the questions all participate in the complete oxidation of the glucose. The metabolic process that initiates the total oxidation of glucose is glycolysis, which breaks glucose down into pyruvate. Pyruvate is then converted into acetyl CoA, which takes part in the Krebs cycle. Reducing equivalents formed from these processes transfer their electrons to the electron transport chain. The movement of electrons along the respiratory chain is prerequisite to the chemiosmotic synthesis of ATP.

Hexokinase catalyzes the first step of glycolysis, the entire process of which occurs in the cytoplasm. Although the net yield of glucose oxidation will be 38 ATPs, some energy must be initially invested. This will start the glucose "burning." A kinase is an enzyme that transfers a phosphate group. In this case, it transfers phosphate from ATP to a hexose (six-carbon sugar), glucose. The glucose is then sequentially broken down into two pyruvate molecules.

In the presence of oxygen, pyruvate enters the mitochondria, in which it is decarboxylated (loses a carbon dioxide) and oxidized (loses of pair of electrons and hydrogen ions) to form acetyl CoA. These two sequential steps are catalyzed by a single large enzyme complex (series of polypeptides) called pyruvate dehydrogenase.

Acetyl CoA is shuttled into the Krebs cycle, in which it combines with a four-carbon compound oxaloacetate, to regenerate the six-carbon citrate molecule. In this cycle, sequential oxidation steps (as well as two decarboxylation steps) that regenerate oxaloacetate occur. The oxidation steps are carried out by dehydrogenase enzymes that remove electrons and hydrogen ions from the substrates and transfer them to their own coenzymes, NAD^+ or FAD. For example, succinate dehydrogenase catalyzes the conversion of succinate to fumarate. Concomitantly, $FADH_2$ is formed.

The NADH and $FADH_2$, which are formed primarily in the Krebs cycle, but also in glycolysis and in the formation of acetyl CoA, now move to the electron transport chain, which is located in the inner mitochondrial membrane. Many of the participating enzymes which transfer the electrons along the respiratory chain are called cytochromes. They contain iron in the oxidized (Fe^{3+}) or reduced (Fe^{2+}) state. The final step, catalyzed by cytochrome oxidase, reduces oxygen to water as the reduced iron is oxidized by oxygen.

These oxidation steps are coupled to the phosphorylation of ADP, to form ATP. This complex process, called chemiosmosis, requires an

ATP synthetase enzyme on the inner mitochondrial membrane facing the mitochondrial matrix. The ATP synthesized is now used to do work for the cell.

79. (C)   80. (B)   81. (A)

Ethylene is produced during plant respiration and is associated with fruit ripening. It has other functions, such as the abscission of leaves. Although, by name, it would appear that abscisic acid promotes the latter function just attributed to ethylene, this is of only minor importance. The major role of abscisic acid appears to be in promoting the closure of the stomata. The effects of abscisic acid oppose those of many of the hormones that aid in plant growth and development.

Gibberellins stimulate seed germination by activating enzymes of the seed. They also promote the elongation of stems, particularly in genetically dwarfed plants.

Cytokinins function in cell division; auxins function in stem elongation, particularly in response to light.

82. (A)   83. (D)   84. (E)   85. (C)

Mollusks, which include snails, clams, and scallops, have muscular mantles as a distinguishing feature. This mantle is a pair of folds on the dorsal body wall that envelopes and protects the visceral organs. In those mollusks that have shells, it is the cells of the mantle that secrete the chemical components required to produce a shell. A muscular foot on the ventral surface of a mollusk functions in locomotion.

The best-known species in the phylum Annelida is the earthworm. Annelids are segmented worms: their bodies are partitioned into segments that are lined by longitudinal and circular muscles.

There are nearly one million different species of arthropods. Among the distinguishing features of an arthropod are the hardened exoskeleton composed of protein and chitin and jointed appendages. Like the bodies of annelids, their bodies are segmented. Examples of arthropods are spiders, crabs, and centipedes, to name a few.

The echinoderms are all marine, in contrast to members of other phyla that were discussed, which live in marine, freshwater, and terrestrial environments. A unique feature of echinoderms is their method of locomotion. They have water-vascular systems in which seawater circulates through canals and tube feet. Sea stars (starfish) are echinoderms.

The chordates include fish, amphibians, reptiles, birds, and mammals. The distinguishing characteristic of this seemingly wide classification is the notochord, which functions as a skeletal framework at some time in embryonic or post-natal life. Other distinguishing characteristics include a dorsal hollow nerve cord, pharyngeal gill slits, and a postanal tail at some point during development.

86. (D) 87. (C) 88. (E) 89. (A) 90. (B)

Aerial stems are classified as woody or herbaceous. Since the stems of the latter types are annual, there is no secondary growth. The diagram shows secondary growth in a woody stem. The primary tissues develop in the first year of growth. All growth that occurs after the first year is called secondary growth.

Secondary growth arises from the vascular cambium, a continuous ring between the two types of vascular tissue, xylem and phloem. The cells of the vascular cambium divide, producing secondary tissues. Towards the end of every year, the cells that are produced by the vascular cambium are smaller, and the portion of vascular tissue that contains these smaller cells appears as a dark annual ring.

The phloem contains companion cells and sieve tubes that conduct food substances from the leaves down into the root. The primary phloem lies just external to the vascular cambium. During secondary growth, this primary layer moves outward, and secondary phloem then lies adjacent to the cambium.

The xylem contains tracheids and vessel members. It conducts water from the roots up to the leaves. The primary xylem is just internal to the vascular cambium. As secondary growth occurs, this layer is pushed inward, away from the cambium, by secondary xylem. The epidermis is the outermost single layer of cells. In the root, it functions in water and mineral absorption, but in the stem, it has a heavy cuticle layer that protects the tissues from desiccation.

The cortex is the ground substance consisting of widely spaced cells. It has a role in support and storage.

The tissue in the center of the stem is pith.

91. (E) 92. (D) 93. (A)

The transmission of the disease to the second and third filial (F_2 and F_3) generations are clearly from the mothers to their sons, since the fathers have simply married into the family. This suggests a sex-linked disease, which is in accordance with the transmission to the first filial generation, as well.

Sex-linked diseases are carried on the X chromosome. The Y chromosome is smaller than the X chromosome and contains genes that produce maleness. The X chromosome can carry the gene for color blindness, hemophilia, and baldness, all of which are conditions that can be seen more often in males.

The allele for the disease must be a recessive one, since all of the mothers in the pedigree are carriers, but not afflicted. In order for a female to be afflicted, she must be homozygous recessive. However, a male cannot have a homozygous recessive genotype for the disease, since he only has one X chromosome. In this case, the recessive allele is expressed when he contains only one "bad" X chromosome, since the Y chromosome contains a dominant allele whose effects could mask the effects of the recessive allele.

X^CX^C	normal female
X^CX^c	carrier female
X^cX^c	color-blind female
X^CY	normal male
X^cY	color-blind male

Since P_1 is a normal male, his genotype must be X^CY.

A female can only be color blind if she inherits two recessive alleles, which means her father must be color blind, and her mother must be a carrier (or color blind herself). Of course, marriage between relatives increases this incidence.

F_{2-5} is a normal male, since there is no carrier state for the male. His genotype is X^CY. If he marries a homozygous dominant woman, X^CX^C, she is normal and they cannot have an afflicted child.

94.   (C)   95. (B)   96. (B)   97. (E)

A spirometer measures the volume of air moved into and out of the lungs during breathing. There are four basic lung volumes. The tidal volume is the volume of air moved in a standard breath. The subject's tidal volume is 500 ml. The inspiratory reserve volume (IRV) is the volume of air that one can inspire in addition to the tidal volume. In the figure, it is the volume between 5500 ml and 2500 ml, or 3000 ml. It is important NOT to include tidal volume in the IRV. The expiratory reserve volume (ERV) is the volume of air that one can exhale in addition to tidal volume; this is produced during a forced exhalation. In the figure, it is the difference between 2000 ml and 1000 ml, or 1000 ml. The subject performed a maximal expiration just before the second signal, as indicated by the downward deflection on the spirometer reading. The residual volume cannot be measured on a spirometer. It is the volume of air left in the lungs despite a maximal expiration. It is typically about 1000 ml.

Lung capacities are simply sums of the lung volumes. Inspiratory capacity is the sum of tidal volume and inspiratory reserve volume. Functional residual capacity is the sum of expiratory reserve volume and residual volume. Total lung capacity is the sum of all four volumes.

Vital capacity is the sum of tidal volume and both reserve volumes. This sum can be measured directly by inhaling maximally and then exhaling maximally, thus achieving in one breath the greatest deflections both upward and downward. However, the subject did not do this. The vital capacity is the difference between 5500 ml and 1000 ml, or 4500 ml. Note that this is the sum of tidal volume (500 ml), inspiratory reserve volume (3000 ml) and expiratory reserve volume (1000 ml).

None of the choices (A–D) for Item 97 fits because the breathing rate cannot be figured with the data provided. Spirometry measures the volume of air moved into the lungs in an inspiration and the volume of air moved out of the lungs in an expiration. (A spirometer does not measure the residual volume of the lungs.) Thus, the premise of the question—that one could find the subject's respiratory rate (including inspirations and expirations)—does not hold up, for the spirogram does not indicate a span of time across the full range of measurement. (Only two wave forms are shown between signals 2 and 3. The spirogram would need six to show a rate of 12 breaths per second.)

98.   (A)   99. (C)   100. (B)   101. (E)
The Hardy-Weinberg equilibrium refers to an equilibrium in which the proportion of alleles at a given locus will remain constant. The requirements for such an equilibrium are: no mutation, no migration, no natural selection, and a large population size. In real populations, these factors are not usually all met and a change in the allele frequencies can occur.

Assuming a Hardy-Weinberg equilibrium, there are two mathematical statements that can describe it. If the trait, such as eye color, has only two alleles, p and q, then

$$p + q = 1$$

in which p is the frequency of one allele and q is the frequency of the other allele. (The sum of the frequencies must equal 1). Another mathematical equation is true under this equilibrium:

$$(p + q)^2 = p^2 + 2pq + q^2$$

If p represents the dominant allele, then p^2 is the frequency of homozygous dominants in the population. If q is the frequency of the recessive allele, the q^2 is the frequency of homozygous recessives in the population. Thus 2pq is the frequency of heterozygotes in the population. Note also that:

$$p^2 + 2pq + q^2 = 1$$

In the original sample, 91% of the students have brown eyes and 9% have blue eyes. Since 9% (or .09) is equal to q^2, then q must be .3. Therefore, p must be .7, which is (1-.3).

The 910 brown-eyed students consist of those that are homozygous and those that are heterozygous for the allele for brown eyes. The homozygous dominant population represents 49% ($p^2 = .7^2$) of the total population and the heterozygotes make up 42% of the total. (2pq = 2 x .7 X .3). 42% of 1000 students is 420 students.

The data for 1986 is significantly different. The change in allele frequencies can be accounted for by migration (immigration and emigration) during the attempt at desegregation.

The other factors that could disrupt a Hardy-Weinberg equilibrium are not pertinent here. For instance, the population size is the same.

In the second sample, blue eyes account for 16% of the population (160/1000). Thus, $q^2 = .16$ and q must equal .4. To verify this, note that p must equal .6. Therefore, the homozygous dominant population would account for 36%, or 360 students, and the heterozygotes would account for 48% (2 x .6 x .4), or 480 students. Indeed, there are 840 (which equals 360 + 480) brown-eyed students in the second population.

102. (D)   103. (B)   104. (B)   105. (E)

From the data given, pupillary dilation would seem to be caused by stimulating sympathetic nerves or inhibiting parasympathetic ones. Since atropine has no effect on receptors for norepinephrine, it has no sympathetic effect. Since parasympathetic stimulation causes pupillary constriction, atropine must block this effect, by blocking the receptors for acetylcholine.

Likewise, since propranolol causes a decrease in heart rate that is not due to parasympathetic (acetylcholine) stimulation, it must be due to sympathetic inhibition, by blocking norepinephrine activity.

The heart beats spontaneously, but its rate is modified by the combined effect of sympathetic and parasympathetic discharge. At rest, parasympathetic discharge is greater and causes an inhibitory tone. When the vagus nerve is cut, parasympathetic discharge to the heart is abolished. This will allow the heart rate to increase, because there is no counteraction to sympathetic discharge. However, note that sympathetic discharge is not increased; it is merely not counteracted. Also note that even if the heart were removed from the body and hence totally denervated, it could continue to beat for a while at its intrinsic rate.

Curare causes paralysis of the skeletal muscles.

Skeletal muscle is voluntary, as opposed to smooth and cardiac muscle. Hence it is not innervated by the autonomic nervous system. Instead it is innervated by the somatic motor nervous system.

106. (A)   107. (B)   108. (A)

During glycolysis, glucose is converted to pyruvate. Pyruvate has many fates in the cell, depending upon whether or not oxygen is available, and the type of cell that is involved.

Under aerobic conditions, i.e., when oxygen is present, pyruvate is transferred from the cytoplasm to the mitochondrial matrix. There, it is converted to acetyl CoA, which enters the Krebs cycle. Oxidative phosphorylation then occurs. The result is the production of carbon dioxide, water, and ATP, which occurred in test tube C.

Under anaerobic conditions, pyruvate has other fates. Fermentation occurs when an intermediate of carbohydrate metabolism acts as the final acceptor of electrons from NADH, as opposed to what occurs in the electron transport chain in which oxygen is the final acceptor of electrons and is reduced to water.

In lactate fermentation, pyruvate acts as the final electron acceptor and is converted to lactate. The importance of this step is that it regenerates the NAD⁺ and allows glycolysis to continue. In animal cells, lactate fermentation occurs when oxygen is in short supply, as when sprinting. When lactate is formed and diffuses into the blood, it causes pain (usually in the calves of a sprinter), and causes the sprinter to slow down or stop. Thus the affected muscles will start to get more oxygen. In essence, this mechanism prevents him from accidentally suffocating himself!

Alcoholic fermentation occurs in yeast cells, which use acetaldehyde as the final electron acceptor. Pyruvate is converted to acetaldehyde, which is converted to ethyl alcohol (ethanol). Carbon dioxide gas is also produced. The gas is responsible for causing breads that are baked with yeast to rise. Alcoholic fermentation in yeasts is used to make beer. There are even rare cases of humans whose cells can perform alcoholic fermentation, causing the person to become spontaneously drunk!

109.   (C)   110. (B)   111. (A)

The two species of *Paramecium* each survive in the culture dish alone, but when mixed together, species B succumbs to species A. The interaction between the two species is called competitive exclusion, because they are both competing for the same nutrients in the culture, but, due to limited resources, one must succumb. This interspecific competition is merely due to the overlap of their niches (in this case, a culture dish). Species that are very similar have similar requirements and hence are more likely to succumb to competitive exclusion.

A survivorship curve is a curve that describes the percentage of population at all gradations of age. By looking at a survivorship curve, one can easily see if age is correlated with mortality. For instance, some organisms are susceptible at birth, but if they survive that period, they may live for a long time.

Protocooperation refers to a mutually beneficial interaction between two species living together, but unlike mutualism, in protocooperation, the interaction is not obligatory. Symbiosis is a general term referring to all interactions between two species in a community.

The size of population A is ultimately the same whether it is living alone or with species B, as indicated by the growth curves. However, the initial growth period in the combined culture is greater; as species B dies off, the growth of species A picks up.

At one week in time, it is clear the size of the population of species

A is still climbing toward its maximum, but the size of the population of species B has already peaked and is declining.

The graphs plot population size vs. time. Although the maximum population size of B in the mixed culture is very low and never stabilizes, its peak is reached in about a half a week vs. one and a half weeks in the solo culture. B shows its greatest numbers in the solo culture, but that value is still less than the size of A in either the solo or mixed cultures. Furthermore, it is about three times the size of B's size in the mixed culture.

112. (D)   113. (C)
Glycerinated skeletal muscle is used to examine the striations of skeletal muscle. The lengths of the fibers can be measured with a millimeter ruler. If a fiber becomes shorter upon application of a chemical, it has contracted. It usually appears darker as well, because the cross striations (dark and light bands) get closer together as the microscopic sarcomeres shorten.

In Experiment A, the muscle fiber that is shortest is the one that is in the presence of ATP and salts (KC1 and $MgC1_2$), in addition to the glycerol solution. While ATP is of course required to provide the energy for contraction, the salts are necessary as activators of the ATPase enzyme. The slight contraction that occurred in the fiber that was in contact with glycerol and ATP alone is probably an error due to contamination by salts. For instance, there are many salts on the skin. Touching the fiber can contaminate it.

Standard conditions are pH 7 and room temperature (25°C). However, in vivo, the muscle tissues are bathed in extracellular fluid with a pH of about 7.4 (plasma) and body temperature is warmer than room temperature. Body temperature is about 37°C. Experiment B shows that a rise in temperature increased the amount of contraction, as indicated by the decrease in fiber length. When the length of a fiber in a neutral solution is compared with that of a fiber in an acidic solution (pH 4), it is clear that contraction occurs more readily in the neutral solution. Freezing temperatures (0°C) were not measured, but they would probably be inhibitory to muscle contraction. However, glycerinated muscle can be stored in a frozen or very cold state.

114. (A)   115. (D)
Mucus is produced throughout much of the digestive tract, including the stomach, esophagus, and large intestine. It serves as a lubricant. Mucus had no effect on the polypeptide, as it remained

intact after two hours.

Peptidases can be classified as endopeptidases or exopeptidases. The former type breaks peptide bonds in the middle of the chain, yielding shorter fragmented chains. The latter cleaves off the terminal amino acid, yielding a free amino acid.

Chymotrypsin and trypsin are pancreatic endopeptidases. Their actions result in the formation of shorter peptides called dipeptides and tripeptides (peptides having two and three amino acids, respectively). Carboxypeptidase is an exopeptidase that is produced in the pancreas. It cleaves the peptide bond at the C-terminus of a polypeptide. It thus yields free amino acids as well as the short peptides. Aminopeptidase is an exopeptidase that is attached to the small intestinal brushborder, and usually is active as the short peptides are being absorbed by the intestinal epithelium. Aminopeptidase cleaves the peptide bond at the N-terminus, also yielding free amino acids and very short peptides.

116.　(C)　　117. (C)
There are two organelle components of the cell's cytoskeleton that control cell movements and cell shape. Microfilaments are composed primarily of the protein actin. (Myosin, which functions with actin in muscle contraction, is larger than true microfilaments). Aside from muscular contraction, microfilaments are responsible for cytoplasmic streaming, in which organelles are in continual motion.

Microtubules have larger diameters than microfilaments do. They are composed of the protein tubulin, in a polymerized form. Unlike microfilaments, these are hollow cylinders. Microtubules are very active in mitosis, in which they are responsible for chromosome movement.

There are many chemicals that are poisonous to certain organelles in the cell. For instance, ouabain inhibits the Na^+-K^+ ATPase pump. Cyanide binds to a cytochrome in the electron transport chain and prohibits oxygen utilization.

Cytochalasin B binds to actin and inhibits its interaction with myosin to produce muscular contractions. It also inhibits vesicular and organelle movements. Note that the organelle is a microfilament whereas actin is a protein.

Colchicine inhibits microtubular activity by binding to tubulin. Thus chromosome movement is prevented.

Histones are the proteins that, with DNA, comprise chromosomes. Chromatin is basically synonymous with chromosomes; however, the term is reserved for the long thread-like appearance of the chromosomes just before mitosis begins.

118.   (E)   119. (C)   120. (E)

Secretory cells are specialized for the production and secretion of proteins or other substances that are produced within. If a cell is to produce protein, it must be equipped with a rich supply of rough endoplasmic reticulum and Golgi apparatuses. Protein synthesis occurs on the ribosomes. The proteins are inserted into the reticular lumen as they are being synthesized. Protein is then circulated through the reticulum and reaches the Golgi apparatus, in which chemical modifications, such as glycosylation (addition of sugars to form glycoproteins) may occur. Vesicles pinch off, move to the plasma membrane, and the contents of the vesicles are extruded by exocytosis. Secretory cells must also have many mitochondria in order to supply the energy for exocytosis. Cell E is secretory.

Cell A must be a red blood cell, since it is anucleate and has no mitochondria. It is believed that the lack of a nucleus provides more space for hemoglobin. Certainly, red blood cells cannot have mitochondria, since mitochondria utilize oxygen, and the purpose of the cell is to transport oxygen.

Cell D is probably a white blood cell, as indicated by the significant amount of lysosomes. There are many types of white blood cells; all have distinct nuclei. Some white blood cells are phagocytic, especially the neutrophils and the monocytes (after they become wandering macrophages). As phagocytes, they need high numbers of lysosomes to digest whatever they ingest, such as bacteria.

Cell C must be a sperm cell, because it is the only cell in the human body that has a flagellum. The flagellum is the tail of the cell; its beating movements cause the cell to swim. This activity requires high numbers of mitochondria. The head of the sperm contains acrosomal enzymes, which are basically like extracellular lysosomes: they digest the outer layers of the egg in order to allow the sperm nucleus to penetrate it. The gametes (eggs and sperm) are haploid. All animal cells are eukaryotic. Of course, in prokaryotes, flagella are much more common, as many bacteria have flagella.

Cell B may be a skeletal muscle cell, which has many mitochondria and many peripheral nuclei.

SECTION II

ESSAY I

Organic compounds contain carbon. Since carbon is tetravalent (able to make four bonds), it tends to form large compounds (when compared to inorganic compounds). Organic compounds form the primary structural and functional components of living cells, and hence of the entire organism. Large organic molecules are usually synthesized from smaller monomers, or building blocks. Aside from carbon, the elements found most often in organic compounds are hydrogen, oxygen, nitrogen, phosphorus, and sulfur. The four major classes of organic compounds are the carbohydrates, lipids, proteins, and nucleic acids.

Carbohydrates function as the most readily available energy source. The empirical formula of all carbohydrates is $C_nH_{2n}O_n$. They provide 4 calories per gram when consumed in the diet.

The building blocks for the larger carbohydrates are the monosaccharides, or simple sugars. Trioses are sugars that have three carbons. The pentoses, or five-carbon sugars, include ribose and deoxyribose, which are found in nucleic acids. The hexoses are the sugars in the foods we eat. These six-carbon sugars include glucose, fructose, and galactose.

Two monosaccharides chemically combined by dehydration reactions produce a disaccharide. Sucrose, table sugar, is a combination of glucose and fructose. Lactose, or milk sugar, is a combination of glucose and galactose. Maltose, a product of the degradation of starch, is the combination of two glucose molecules.

The polysaccharides are primarily polymers of glucose only. They are long chains containing hundreds of glucose molecules in them. Glycogen is the storage form of glucose in animal liver and muscle cells. The glycogen can be hydrolyzed when glucose is needed, such as during exercise. Starch is the storage form of glucose in plant cells. Plants produce glucose by photosynthesis and then store it as starch. We eat the starch. Cellulose is the structural component of plant cell walls. It is indigestible by humans, although ruminants such as cows and sheep have bacteria in their rumens that can digest it. Although cellulose and starch both occur in plants, the linkages between the glucose subunits are different and thus the molecules are very different and we can only digest starch.

Lipids are nonpolar compounds that can dissolve in substances like benzene or ether, but not in water. There are many types of lipids, including the fatty acids, which are long hydrocarbon chains with acid (carboxyl) groups at one end. Fatty acids may be saturated or unsaturated, where the former indicates that the acid is carrying all the hydrogens possible. A triacylglycerol (triglyceride) molecule is the fat that we eat and the fat that we store subcutaneously for insulation and protection. As an energy source, triglycerides provide 9 calories/gram. A monoglyceride is one glycerol molecule, attached to one fatty acid. A diglyceride has two fatty acids attached and a triglyceride has three fatty acids attached to the glycerol. Glycerol is a triol, i.e. it has three alcohol groups. In a triglyceride, the three acid groups from the fatty acids bind to the three alcohol groups due to dehydration reactions. Since the acids can no longer ionize and become charged, triglycerides are also called neutral fats.

Another type of lipid is the phospholipid, which is a diglyceride with a phosphate group attached to the third carbon of the glycerol. The phosphate group is charged, but the fatty acid chain components of the diglyceride are not. Hence, the molecule is amphipathic: it has both polar and non-polar portions. For this reason, it functions in the cell membrane, forming the lipid bilayer, in which the polar phosphate heads are oriented toward the aqueous fluids (intracellular and extracellular fluids), and the fatty acid tails are oriented inward to the center of the lipid bilayer. Phospholipids also form micelles, which are used in the digestive tract to transport fats to the intestinal epithelium. They also form surfactants, used in the respiratory system to decrease the surface tension of the alveoli and prevent their collapse.

Steroids are another class of lipids. They have four-ringed structures. Examples of steroids are cholesterol, vitamin D, bile salts, and the hormones of the adrenal cortex and the gonads (ovaries and testes).

The basic building blocks of proteins are the amino acids. There are twenty amino acids used in protein synthesis. Amino acids consist of a central carbon attached to a hydrogen, amino group (NH_2), carboxylic acid group (COOH), and an R group. The R group is a collection of atoms that differs from one amino acid to the next.

A dipeptide is the union of two amino acids by a peptide bond. The amino terminus of one amino acid links up with the carboxyl terminus of the previous one in a condensation reaction.

A polypeptide is a polymer of amino acids. It may contain 10 amino acids or more than 100. Proteins are arbitrarily defined as polypeptides with over 50-100 amino acids. Proteins may consist of 1 or more polypeptide chains.

Proteins function as enzymes, antibodies, hormones (insulin) muscle proteins (actin and myosin), structural proteins (collagen and keratin), clotting proteins (fibrinogen), etc.

Ingested protein supplies 4 calories/gram, but is not a major energy source for the body. Proteins are major sources of energy only during periods of starvation, during which no fats or carbohydrates are ingested.

Nucleic acids are polymers of nucleotides. Each nucleotide contains a nitrogenous base, which may be a purine or a pyrimidine. The purine bases are double-ringed structures, such as adenine or guanine. The pyrimidine bases are single-ringed structures such as cytosine, thymine, and uracil. Nucleotides also must contain pentose sugars, such as deoxyribose or ribose. Finally, a nucleotide has one or more phosphate groups attached.

The nucleic acid DNA (deoxyribonucleic acid) contains the genetic codes for proteins that are necessary for life. The sugar in DNA is deoxyribose and the bases may be adenine, thymine, cytosine, or guanine. The nucleic acid RNA (ribonucleic acid) transfers the code in DNA to make protein. It contains the sugar ribose and may have the bases adenine, uracil, cytosine, and guanine.

While the other three organic compounds provide calories in the diet, nucleic acids do not.

ESSAY II

Translation is the process of protein synthesis. The codon on mRNA recognizes and binds to the anticodon on tRNA, which recognizes and binds to a specific amino acid. Note that there is no direct recognition between the codon on mRNA and its amino acid. For instance, a codon UUC on mRNA will bind to the anticodon AAG on a tRNA, which specifically binds to the amino acid phenylalanine; hence UUC is a codon for phenylalanine.

Before translation can occur, the tRNAs must become attached to their appropriate amino acids in an activation step, which requires the expenditure of ATP.

A ribosome, which consists of protein and rRNA (ribosomal RNA), has two sites on it (see figure). The A site is closer to the 3' end of the mRNA and the P site is closer to the 5' end of the mRNA. During translation, the ribosome moves toward the 3' end.

There are three stages of translation: initiation, elongation, and termination.

Translation is initiated by an initiating codon AUG. The small subunit of the ribosome attaches to the mRNA at that site (see figure). The initial tRNA, which always carries methionine, binds to the codon at the P site. A large subunit of the ribosome now attaches, and the ribosome is fully functional.

There are several steps in the elongation phase. The tRNA that is capable of bonding to the next mRNA codon now arrives at the A site. This tRNA molecule is also attached to an appropriate amino acid. The preceding amino acid at the P site is linked to the new one by a peptide bond. The peptide chain on the P site (which at this point is just a single amino acid) is transferred to the A site, and the tRNA at the P site is released to be used again. Translocation follows, whereby the new tRNA with its attached chain (or in this case just a dipeptide) moves to the P site, since the ribosome has moved one codon toward the 3' end of the mRNA. The A site is now available and binds the third tRNA which is carrying the next amino acid with it. Elongation continues, as hundreds of amino acids can be added.

Termination is marked by the presence of "stop" codons of mRNA: UAA, UAG, and UGA. There are no tRNAs for these codons. Instead, a protein is bound at the A site, causing the polypeptide and mRNA to be released from the ribosome. In addition, the polypeptide is released from the last tRNA. The ribosomal subunits then dissociate.

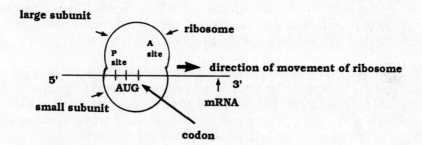

ESSAY III

There are many levels of organization of the human body, and thus there are many approaches to the study of human anatomy and physiology. The chemical level is a reminder that the human body is ultimately a collection of chemicals.

Even at this level, there are many sublevels: The subatomic level, the atomic level, the molecular level, and the macromolecular level.

Above the macromolecular level is the cellular level. The cell is the basic structural and functional unit of life. Within cells are organelles, each of which has a special function.

A tissue is a group of similar cells. There are four basic tissues that compose the human body: muscle tissue is specialized for generating force and movement; nerve tissue is used for communication; connective tissue plays a supportive role (blood and bone are both example of connective tissue); epithelial tissue is used in protection, secretion, and absorption.

An organ is composed of two or more tissues performing a single function. Many organs have all four types of tissue within them.

An organ system is a group of organs whose functions are closely linked in some way.

Finally, an organism, such as man, is a group of organ systems interacting together in life.

Systemic anatomy, or physiology, is a study at the level of the organ system.

The integumentary skin, which includes the skin and its derivatives (hair, nails) functions in protection (from dehydration, bacteria), thermoregulation, and vitamin D production. In addition, wastes can be eliminated through the skin. The skin contains many sensory receptors.

The skeletal system consists of bone and cartilage. It functions in support and protection, blood cell production, and mineral storage. Bones also provide sites of attachment for muscles.

The muscular system includes skeletal muscle attached to the bones, cardiac muscle in the heart, and smooth muscle in the vessels and visceral organs. All muscle types can contract, and thus pump blood, perform peristalsis, or cause the body to move. In addition, skeletal muscle functions in posture and heat production (shivering produces heat).

The nervous system consists of the brain, spinal cord, nerves, and sensory organs. It is a major controller and integrator of activities. It is responsible for thought and consciousness. It can detect changes in the environment and initiate appropriate somatic responses.

The endocrine system consists of many endocrine glands throughout the body. With the nervous system, it is a major controller and integrator of bodily activities.

The circulatory system includes the blood, heart, and blood vessels. It functions to transport nutrients and wastes, and also aids in thermoregulation.

The lymphatic system consists of the white blood cells, lymph, lymphatic vessels, and lymphatic organs (tonsils, spleen, etc.). It functions in the transport of fats from the gastrointestinal tract to the blood and in the return of proteins and fluids to the blood. It also functions in immune reactions.

The respiratory system includes the nose, trachea, and lungs (and the subdivisions within, such as the bronchioles and alveoli). It functions in gas exchange and acid-base balance.

The digestive system includes the mouth, esophagus, stomach, intestines, and accessory structures such as the pancreas, liver, gallbladder, and salivary glands. It functions in mechanical and chemical digestion, absorption, and elimination.

The urinary system includes the kidneys, ureters, bladder, and urethra. It regulates the volume and composition of blood by regulating the fluid and electrolyte balance. It aids in acid-base balance and in the elimination of wastes.

The reproductive system includes the testes, ovaries, and many other organs. However, the gonads (testes and ovaries) complete certain functions, which include the production of gametes (eggs and sperm) and the secretion of hormones. However, unlike the other systems, a functional reproductive system is not necessary for life, though, it is necessary for the continuity of the generations.

ESSAY IV

Tropisms are growth responses to external stimuli. A positive tropism is a response toward a stimulus; a negative tropism is a response away from a stimulus.

The most understood tropism is phototropism, in which stems bend towards a light source. In addition, the leaves of a plant turn their surfaces toward the light source. This positive phototropic response is mediated by the plant hormone auxin. Auxin migrates and accumulates in the shaded side of the stem and promotes cell elongation oriented horizontally. The asymmetric growth causes the stem to bend away from the shaded side and hence toward the light.

Gravitropism is a response to gravity. It was previously called geotropism. Stems display a negative gravitropic response and roots show a positive response. In other words, if a shoot is placed horizontally, the stems will grow upward, against gravity, and the roots will grow downward, in the direction of gravity's force. The response in the stem is probably due to auxin. If a stem is oriented horizontally, auxin will accumulate on the lower side of the stem and stimulate cell elongation. The rapid growth on the underside will cause the stem to grow upward. In the root, the response may not be mediated by a hormone. The root cap has an inhibitory substance. If a root is oriented horizontally, the substance will accumulate on the lower side and inhibit or slow down cell elongation. Hence the more rapid growth on the upper surface will cause the root to bend down.

Thigmotropism is a response to touch. It is seen, for instance, in vines that curl around contacted objects. It appears that cell elongation stops on the contacted side; however, the mechanism by which this occurs has not be elucidated. This type of growth is due to a mechanical stimulus and is aided by the hormone auxin.

Besides these types of tropisms, there are some that are caused by various stimuli in the environment such as water, temperature, chemicals, and oxygen. These factors, in combination with plant hormones (like auxin and gibberellins), control the growth response of a plant at various stages. The cellular response during tropism is now being studied.

THE ADVANCED PLACEMENT EXAMINATION IN

BIOLOGY

TEST V

THE ADVANCED PLACEMENT EXAMINATION IN

BIOLOGY

ANSWER SHEET

1. Ⓐ Ⓑ Ⓒ Ⓓ Ⓔ	21. Ⓐ Ⓑ Ⓒ Ⓓ Ⓔ	41. Ⓐ Ⓑ Ⓒ Ⓓ Ⓔ
2. Ⓐ Ⓑ Ⓒ Ⓓ Ⓔ	22. Ⓐ Ⓑ Ⓒ Ⓓ Ⓔ	42. Ⓐ Ⓑ Ⓒ Ⓓ Ⓔ
3. Ⓐ Ⓑ Ⓒ Ⓓ Ⓔ	23. Ⓐ Ⓑ Ⓒ Ⓓ Ⓔ	43. Ⓐ Ⓑ Ⓒ Ⓓ Ⓔ
4. Ⓐ Ⓑ Ⓒ Ⓓ Ⓔ	24. Ⓐ Ⓑ Ⓒ Ⓓ Ⓔ	44. Ⓐ Ⓑ Ⓒ Ⓓ Ⓔ
5. Ⓐ Ⓑ Ⓒ Ⓓ Ⓔ	25. Ⓐ Ⓑ Ⓒ Ⓓ Ⓔ	45. Ⓐ Ⓑ Ⓒ Ⓓ Ⓔ
6. Ⓐ Ⓑ Ⓒ Ⓓ Ⓔ	26. Ⓐ Ⓑ Ⓒ Ⓓ Ⓔ	46. Ⓐ Ⓑ Ⓒ Ⓓ Ⓔ
7. Ⓐ Ⓑ Ⓒ Ⓓ Ⓔ	27. Ⓐ Ⓑ Ⓒ Ⓓ Ⓔ	47. Ⓐ Ⓑ Ⓒ Ⓓ Ⓔ
8. Ⓐ Ⓑ Ⓒ Ⓓ Ⓔ	28. Ⓐ Ⓑ Ⓒ Ⓓ Ⓔ	48. Ⓐ Ⓑ Ⓒ Ⓓ Ⓔ
9. Ⓐ Ⓑ Ⓒ Ⓓ Ⓔ	29. Ⓐ Ⓑ Ⓒ Ⓓ Ⓔ	49. Ⓐ Ⓑ Ⓒ Ⓓ Ⓔ
10. Ⓐ Ⓑ Ⓒ Ⓓ Ⓔ	30. Ⓐ Ⓑ Ⓒ Ⓓ Ⓔ	50. Ⓐ Ⓑ Ⓒ Ⓓ Ⓔ
11. Ⓐ Ⓑ Ⓒ Ⓓ Ⓔ	31. Ⓐ Ⓑ Ⓒ Ⓓ Ⓔ	51. Ⓐ Ⓑ Ⓒ Ⓓ Ⓔ
12. Ⓐ Ⓑ Ⓒ Ⓓ Ⓔ	32. Ⓐ Ⓑ Ⓒ Ⓓ Ⓔ	52. Ⓐ Ⓑ Ⓒ Ⓓ Ⓔ
13. Ⓐ Ⓑ Ⓒ Ⓓ Ⓔ	33. Ⓐ Ⓑ Ⓒ Ⓓ Ⓔ	53. Ⓐ Ⓑ Ⓒ Ⓓ Ⓔ
14. Ⓐ Ⓑ Ⓒ Ⓓ Ⓔ	34. Ⓐ Ⓑ Ⓒ Ⓓ Ⓔ	54. Ⓐ Ⓑ Ⓒ Ⓓ Ⓔ
15. Ⓐ Ⓑ Ⓒ Ⓓ Ⓔ	35. Ⓐ Ⓑ Ⓒ Ⓓ Ⓔ	55. Ⓐ Ⓑ Ⓒ Ⓓ Ⓔ
16. Ⓐ Ⓑ Ⓒ Ⓓ Ⓔ	36. Ⓐ Ⓑ Ⓒ Ⓓ Ⓔ	56. Ⓐ Ⓑ Ⓒ Ⓓ Ⓔ
17. Ⓐ Ⓑ Ⓒ Ⓓ Ⓔ	37. Ⓐ Ⓑ Ⓒ Ⓓ Ⓔ	57. Ⓐ Ⓑ Ⓒ Ⓓ Ⓔ
18. Ⓐ Ⓑ Ⓒ Ⓓ Ⓔ	38. Ⓐ Ⓑ Ⓒ Ⓓ Ⓔ	58. Ⓐ Ⓑ Ⓒ Ⓓ Ⓔ
19. Ⓐ Ⓑ Ⓒ Ⓓ Ⓔ	39. Ⓐ Ⓑ Ⓒ Ⓓ Ⓔ	59. Ⓐ Ⓑ Ⓒ Ⓓ Ⓔ
20. Ⓐ Ⓑ Ⓒ Ⓓ Ⓔ	40. Ⓐ Ⓑ Ⓒ Ⓓ Ⓔ	60. Ⓐ Ⓑ Ⓒ Ⓓ Ⓔ

61. Ⓐ Ⓑ Ⓒ Ⓓ Ⓔ   81. Ⓐ Ⓑ Ⓒ Ⓓ Ⓔ   101. Ⓐ Ⓑ Ⓒ Ⓓ Ⓔ
62. Ⓐ Ⓑ Ⓒ Ⓓ Ⓔ   82. Ⓐ Ⓑ Ⓒ Ⓓ Ⓔ   102. Ⓐ Ⓑ Ⓒ Ⓓ Ⓔ
63. Ⓐ Ⓑ Ⓒ Ⓓ Ⓔ   83. Ⓐ Ⓑ Ⓒ Ⓓ Ⓔ   103. Ⓐ Ⓑ Ⓒ Ⓓ Ⓔ
64. Ⓐ Ⓑ Ⓒ Ⓓ Ⓔ   84. Ⓐ Ⓑ Ⓒ Ⓓ Ⓔ   104. Ⓐ Ⓑ Ⓒ Ⓓ Ⓔ
65. Ⓐ Ⓑ Ⓒ Ⓓ Ⓔ   85. Ⓐ Ⓑ Ⓒ Ⓓ Ⓔ   105. Ⓐ Ⓑ Ⓒ Ⓓ Ⓔ
66. Ⓐ Ⓑ Ⓒ Ⓓ Ⓔ   86. Ⓐ Ⓑ Ⓒ Ⓓ Ⓔ   106. Ⓐ Ⓑ Ⓒ Ⓓ Ⓔ
67. Ⓐ Ⓑ Ⓒ Ⓓ Ⓔ   87. Ⓐ Ⓑ Ⓒ Ⓓ Ⓔ   107. Ⓐ Ⓑ Ⓒ Ⓓ Ⓔ
68. Ⓐ Ⓑ Ⓒ Ⓓ Ⓔ   88. Ⓐ Ⓑ Ⓒ Ⓓ Ⓔ   108. Ⓐ Ⓑ Ⓒ Ⓓ Ⓔ
69. Ⓐ Ⓑ Ⓒ Ⓓ Ⓔ   89. Ⓐ Ⓑ Ⓒ Ⓓ Ⓔ   109. Ⓐ Ⓑ Ⓒ Ⓓ Ⓔ
70. Ⓐ Ⓑ Ⓒ Ⓓ Ⓔ   90. Ⓐ Ⓑ Ⓒ Ⓓ Ⓔ   110. Ⓐ Ⓑ Ⓒ Ⓓ Ⓔ
71. Ⓐ Ⓑ Ⓒ Ⓓ Ⓔ   91. Ⓐ Ⓑ Ⓒ Ⓓ Ⓔ   111. Ⓐ Ⓑ Ⓒ Ⓓ Ⓔ
72. Ⓐ Ⓑ Ⓒ Ⓓ Ⓔ   92. Ⓐ Ⓑ Ⓒ Ⓓ Ⓔ   112. Ⓐ Ⓑ Ⓒ Ⓓ Ⓔ
73. Ⓐ Ⓑ Ⓒ Ⓓ Ⓔ   93. Ⓐ Ⓑ Ⓒ Ⓓ Ⓔ   113. Ⓐ Ⓑ Ⓒ Ⓓ Ⓔ
74. Ⓐ Ⓑ Ⓒ Ⓓ Ⓔ   94. Ⓐ Ⓑ Ⓒ Ⓓ Ⓔ   114. Ⓐ Ⓑ Ⓒ Ⓓ Ⓔ
75. Ⓐ Ⓑ Ⓒ Ⓓ Ⓔ   95. Ⓐ Ⓑ Ⓒ Ⓓ Ⓔ   115. Ⓐ Ⓑ Ⓒ Ⓓ Ⓔ
76. Ⓐ Ⓑ Ⓒ Ⓓ Ⓔ   96. Ⓐ Ⓑ Ⓒ Ⓓ Ⓔ   116. Ⓐ Ⓑ Ⓒ Ⓓ Ⓔ
77. Ⓐ Ⓑ Ⓒ Ⓓ Ⓔ   97. Ⓐ Ⓑ Ⓒ Ⓓ Ⓔ   117. Ⓐ Ⓑ Ⓒ Ⓓ Ⓔ
78. Ⓐ Ⓑ Ⓒ Ⓓ Ⓔ   98. Ⓐ Ⓑ Ⓒ Ⓓ Ⓔ   118. Ⓐ Ⓑ Ⓒ Ⓓ Ⓔ
79. Ⓐ Ⓑ Ⓒ Ⓓ Ⓔ   99. Ⓐ Ⓑ Ⓒ Ⓓ Ⓔ   119. Ⓐ Ⓑ Ⓒ Ⓓ Ⓔ
80. Ⓐ Ⓑ Ⓒ Ⓓ Ⓔ   100. Ⓐ Ⓑ Ⓒ Ⓓ Ⓔ   120. Ⓐ Ⓑ Ⓒ Ⓓ Ⓔ

ADVANCED PLACEMENT BIOLOGY EXAM V

SECTION I

120 Questions
90 Minutes

DIRECTIONS: For each question, there are five possible choices. Select the best choice for each question. Blacken the correct space on the answer sheet.

1.   Which statement about blood circulation in the heart of a mammal is correct?

    (A) the right ventricle pumps blood to the systemic circulation

    (B) blood enters the right atrium from the inferior vena cava

    (C) blood enters the left atrium from the superior vena cava

    (D) blood enters the right atrium from the pulmonary arteries

    (E) blood enters the left atrium directly from the right ventricle

2.   Which trophic level in a community fixes light energy into chemical energy?

    (A) producers   (D) detritus

    (B) consumers   (E) secondary consumers

    (C) decomposers

3. Heating a test tube culture full of bacteria and killing them all is:

    (A) a density-dependent factor

    (B) an intrinsic factor

    (C) a result of exponential growth

    (D) a result of predation

    (E) None of the above

4. According to the chemiosmotic theory, ATP production is most <u>directly</u> driven by:

    (A) the lower concentration of hydrogen ions in the inner membrane of mitochondria, compared to that of the inter-membranal compartment

    (B) enzymes that catalyze synthesis of ATP

    (C) a proton gradient

    (D) the joining of phosphate groups to ADP molecules

    (E) the rejoining of hydrogen ions to hydroxide ions

5. The phosphorylation of ADP is an example of a(n):

    (A) hydrolysis reaction   (D) condensation-reaction

    (B) endergonic reaction   (E) decomposition

    (C) exergonic reaction

Question 6 refers to the following diagram.

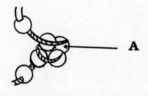

338

6. The structure labelled A is

   (A) DNA

   (B) RNA

   (C) a histone protein

   (D) a prokaryote

   (E) a flagellum

7. Meiosis takes place in which of the following organs?

   (A) ovary

   (B) skeletal muscle of embryo

   (C) spleen

   (D) liver

   (E) pancreas

8. If you cross a homozygous dominant organism with a homozygous recessive organism, what percent of the offspring will be homozygous recessive?

   (A) 0%

   (B) 25%

   (C) 50%

   (D) 75%

   (E) 100%

9. Binomial nomenclature consists of:

   (A) genus and species names

   (B) genus and phylum names

   (C) family and genus names

   (D) family and class names

   (E) family and species names

10. Protists are divided into two major subgroups by their:

(A) methods of locomotion

(B) methods of reproduction

(C) habitats

(D) chromosome numbers

(E) methods of nutrition

11. Laboratory procedures used to form hybrid nucleic acids depend on the fact that:

(A) hydrogen bonds will not be broken when gentle heat is applied

(B) denaturation of nucleic acids is irreversible

(C) strands that are complementary will find each other

(D) nitrogenous bases will repel each other

(E) DNA will associate only with DNA, not RNA

12. What does not occur immediately after fertilization?

(A) A sperm enters the outer membrane of the egg

(B) Cytoplasmic substances in the fertilized egg become re-arranged

(C) The genetic material of the sperm and egg combine

(D) Cleavage occurs

(E) The acrosome enters the egg cytoplasm

The following events occur within a single menstrual cycle. Pick the correct sequence of events in a single menstrual cycle starting with the earliest:

1. fertilization

2. implantation

3. shedding of endometrium

4. increase in luteinizing hormone (LH)

5. ovulation

13. (A) 4,3,5,1,2   (D) 5,4,1,3,2

     (B) 3,4,5,1,2   (E) 5,1,2,4,3

     (C) 1,4,3,2,5

14. In DNA, hydrogen bonding joins:

     (A) two strands of the helix

     (B) nearby pyrimidine bases

     (C) five-carbon sugars

     (D) phosphate groups

     (E) purine bases

15. In the cytoskeleton, the _____ provide(s) the majority of the network that connects all structures in the cytoplasm.

     (A) microtrabeculae   (D) microtubules

     (B) cytoplasmic lattice   (E) microfilaments

     (C) intermediate fibers

16. Chlorophytes (green algae) are thought to be the ancestors of land plants because:

(A) the green algae contain "a" and "b" chlorophylls

(B) the green algae store excess carbohydrates as starch

(C) the green algae have walls containing cellulose

(D) the biochemistry of green algae resembles that of plants

(E) all of the above

17. An example of a plant in which vascular tissue is absent is:

(A) a gymnosperm   (D) a fern

(B) a tracheophyte   (E) a bryophyte

(C) an angiosperm

18. In bryophytes, the zygote divides mitotically to become a structure called the:

(A) haploid sporophyte   (D) antheridium

(B) archegonium   (E) protonema

(C) diploid sporophyte

19. In conifers, microsporangia:

(A) produce cones   (D) contain pollen tubes

(B) contain the ovule   (E) contain endosperm

(C) release spores

20. What is the relationship, if any, between color blindness and hemophilia?

(A) Both are restricted to chromosomes coming from the father

(B) Both are restricted to chromosomes coming from the mother

(C) Both are sex-linked conditions

(D) Both are caused by dominant genes

(E) Both are present on the Y-chromosome

21. A mother with Rh-negative blood had a baby with Rh-positive blood. The father of the baby must have been:

(A) Rh-negative

(B) homozygous Rh-positive or heterozygous Rh-positive

(C) homozygous Rh-positive

(D) heterozygous Rh-negative

(E) heterozygous Rh-positive or homozygous Rh-negative

22. What type of isolation is always involved in speciation?

(A) geographic
(D) allopatric

(B) genetic
(E) physical

(C) sympatric

23. Which of these statements about DNA in eukaryotic cells is not true?

(A) Most of the DNA of the cell is used for gene expression

(B) The amount of DNA per diploid cell is the same for every diploid cell of a species

(C) There is much variation in the amounts of DNA among different species

(D) DNA is frequently redundant and repetitive

(E) About half of the weight of a chromosome is contributed by DNA

24.   Restriction enzymes are used in genetic research to:

(A) cleave DNA molecules at certain sites

(B) produce individual nucleotides from DNA

(C) slow down the reproductive rate of bacteria

(D) remove DNA strands from the nucleus

(E) prevent histones from reassociating with DNA

25.   The nitrogenous base that is complementary to uracil is:

(A) thymine   (D) adenine

(B) guanine   (E) uracil

(C) cytosine

26.   Mutations are caused by:

(A) base changes in DNA

(B) base changes in RNA

(C) changes in the sugars of DNA

(D) changes in the phosphates of RNA

(E) deletions of a codon on RNA

27.   One factor that is considered when classifying species is the amount of similarities in:

(A) the sizes of chromosomes

(B) presence or absence of true coelom and pattern of development from the embryo

(C) mode of nutrition and similarities of adaption to various environments

(D) types of mobility

(E) phenotype

28. What change in the normal structure of hemoglobin results in sickle-cell anemia?

(A) There is a change in the sequence of the amino acids

(B) There is a substitution of one amino acid for another

(C) There is a change in the number of nucleotides

(D) The peptide bonds are broken in the sickle-cell hemoglobin molecule

(E) The hydrogen bonds are weaker in the sickle-cell hemoglobin molecule than in the normal hemoglobin molecule

29. Which group would have the greatest difference phylogenetically?

(A) Humans and cellular slime molds

(B) Chlorophyta (green algae) and mastigophora (*Euglena*)

(C) *E. coli* and cyanobacteria

(D) Whale and moss

(E) Bacteria and *Paramecium.*

30. All of the following are components in animal membranes EXCEPT:

(A) nucleotides   (D) microtubules

(B) carbohydrates   (E) polysaccharides

(C) proteins

31. Assume that clover has a (hypothetical) critical photoperiod of 14 hours. Clover should then flower if it is exposed to uninterrupted _____ for a period of _____.

(A) light; 10 or less hours

(B) light; 14 or less hours

(C) darkness: 14 or more hours

(D) darkness; 10 or more hours

(E) darkness; 10 or less hours

32. What characteristic is found in all echinoderms?

(A) They are bilaterally symmetrical as adults

(B) They are protostomes

(C) They are deuterostomes

(D) They are pseudostomes

(E) They possess nematocysts

33. What would happen if acetylcholine were released by the vesicles in the axon terminals of a nerve cell and no cholinesterase were present?

(A) The acetylcholine would not diffuse across the synapse

(B) The impulse would shut down

(C) The postsynaptic membrane would become refractory (hyperpolarized)

(D) The resting potential would increase

(E) Constant uncontrolled firing at the post-synaptic membrane would occur

34. Two one-celled organisms oxidize glucose as their principal source of nutrition. One organism oxidizes glucose under anaerobic conditions and the other organism oxidizes glucose under aerobic conditions. After one day, the amount of ATP produced by both organisms was the same. Which organism probably consumed the most glucose?

(A) Aerobic organism

(B) Anaerobic organism

(C) Both organisms used approximately the same amount

(D) It depends on whether the two organisms oxidize glucose in the light or dark

(E) It depends on how much ATP is used to initiate the reaction

35. Fermentation:

(A) results in the formation of lactic acid

(B) does not require oxygen

(C) does require oxygen

(D) produces large amounts of energy

(E) occurs only in bacteria

36. In the Krebs cycle, the electron acceptor (of those listed) in the electron transport chain with the highest potential energy is:

(A) water

(B) oxygen

(C) cytochromes

(D) ADP

(E) NAD^+

37. Which of the following processes produces the maximum energy yield when one molecule of glucose is oxidized?

(A) Oxidative phosphorylation

(B) Oxidation of pyruvic acid

(C) Glycolysis in the mitochondria

(D) Glycolysis in the cytoplasm

(E) Formation of lactic acid

38. In humans, lymph is moved by:

(A) diffusion

(B) pressure from the heart

(C) a special lymph pump

(D) differing osmotic pressure in the capillaries

(E) active transport

39.

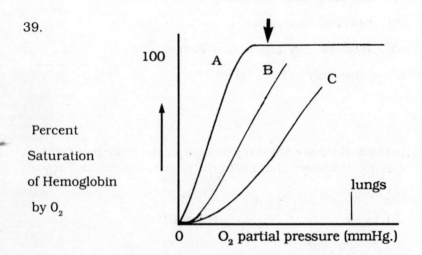

The three lines in the above graph represent the dissociation curves for the hemoglobin of a cat, mouse, and elephant. Which animal is represented by the line farthest to the left (see the arrow) and <u>WHY?</u>

(A) A cat because it has the highest metabolic rate of the three and needs much oxygen

(B) An elephant because its metabolic rate is low and it takes up and gives up oxygen most readily

(C) An elephant because its metabolic rate is low and it takes up and gives off oxygen less readily

(D) A mouse because it has a high metabolic rate and low oxygen requirements; therefore, hemoglobin unloads less readily

(E) A mouse because it has a high metabolic rate and high oxygen requirements; therefore, hemoglobin hangs on to the oxygen at higher levels

40. Which of the following tissues is <u>not</u> related to connective tissue?

    (A) blood   (D) lymph

    (B) bone   (E) collagen

    (C) cartilage

41. Which functions as an enzyme by breaking down ATP to ADP during the contraction of a muscle?

    (A) calcium   (D) myosin

    (B) actin   (E) sarcomere

    (C) myofibril

42. Leaves develop from the:

    (A) apical meristem of the shoot

    (B) lateral meristem of the shoot

    (C) radicle

    (D) area of cell elongation

    (E) internodes

43. In angiosperms, photosynthesis takes place primarily in the:

    (A) guard cells   (D) palisade parenchyma

    (B) stomata   (E) spongy parenchyma

    (C) epidermal layer

44. A respiratory system does not necessarily need:

    (A) an exchange surface with an adequate area

    (B) a means to transport gases to internal areas

(C) a means of protecting exchange surfaces

(D) moist gas exchange surfaces

(E) a location deep inside an organism

45. The pollen grains of angiosperms:

(A) contain two sperm cells

(B) contain pollen tubes

(C) contain one polar body each

(D) have a fragile outer coating

(E) contain fusion nuclei

46. In angiosperms, what structure develops from the outer layer of the ovule after double fertilization has taken place and the seed is being produced?

(A) endosperm   (D) embryo

(B) fruit   (E) carpel

(C) seed coat

47. T-cells are generally not involved in fighting:

(A) cancer cells

(B) transplanted foreign tissue

(C) viral infections

(D) bacterial infections

(E) parasitic infections

48.

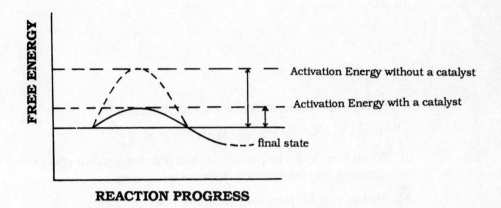

What conclusions can be drawn after studying the above energy diagram for an exergonic reaction?

(A) An uncatalyzed reaction requires less activation energy than a catalyzed reaction does

(B) The initial state contains less free energy

(C) Overall free energy is unchanged by the presence of a catalyst

(D) Activation energy is unchanged with or without a catalyst

(E) A catalyst decreases ΔG

49. The number of codons that can be translated into amino acids is:

    (A) 4   (D) 61

    (B) 16   (E) 64

    (C) 24

50. Which statement about respiration is _incorrect_?

    (A) Humans use positive-pressure breathing

    (B) When exhaling, the position of the diaphragm and ribs in humans is: ribs lowered, diaphragm raised

    (C) Abdominal breathing in humans does not depend on active transport of air

    (D) Frogs use positive-pressure breathing

    (E) Air flow in the lungs of birds is unidirectional

51. Which of the following does _not_ have an open circulatory system?

    (A) clam   (D) earthworm

    (B) grasshopper   (E) crayfish

    (C) snail

52. In animals that have three-chambered hearts, there is mixing of blood with oxygen and blood without oxygen in the ventricle. These animals are referred to as being:

    (A) warm-blooded   (D) isotherms

    (B) homotherms   (E) heterotherms

    (C) poikilotherms

53. Countercurrent exchange as applied to gills means that:

(A) blood with oxygen and blood without oxygen flow in opposite directions

(B) oxygen and carbon dioxide flow in opposite directions

(C) oxygen and carbon dioxide enter the gills from opposite directions

(D) water and blood move in opposite directions

(E) water moves into the gills in a direction opposite to that of the current of the stream

54. In the two-chambered heart of a fish,

(A) the heart has two atria

(B) the heart pumps blood into two ventricles

(C) there is a clear division between pulmonary and systemic circulations

(D) blood goes directly from the heart to the gills and systemic capillary beds

(E) there is a mixing of oxygenated blood and deoxygenated blood

55. In capillary exchange:

(A) proteins in the blood and tissue help to determine osmotic pressure

(B) osmotic pressure moves water outside the capillaries only

(C) blood pressure is greater than osmotic pressure at the venous end of the capillaries

(D) blood pressure is less than osmotic pressure at the arterial end of the capillaries

(E) the pressure of tissue fluid is greatest at the arterial end of the capillaries

56. Which of the following is not a cofactor?

(A) Mn $^{2+}$

(D) FAD

(B) NAD$^+$

(E) ascorbic acid

(C) ATP

57. Glycolysis is most closely linked with what part of the cell?

(A) nucleolus

(D) cytoplasm

(B) mitochondria

(E) endoplasmic reticulum

(C) plasma membrane

58. Some scientists hypothesize that certain organelles in eukaryotic cells might have evolved from prokaryotic cells because:

(A) both types of cells display the same type of construction in their cell walls

(B) both types of cells display the same type of construction of the nuclear membrane

(C) the genetic material of some of the organelles is different from the nuclear genetic material of the cell

(D) the organelles in both types of cells have the same functions

(E) both types of cells have proteins that are extremely similar in amino acid sequence

59. Photosynthetic guard cells can increase their turgor pressure by taking in water which causes them to swell , and, therefore, causes the opening of the stoma that they regulate. This intake of water is caused by:

(A) converting sugar to starch during the night

(B) the hydrolysis of ATP

(C) the active transport of potassium ions into the guard cells

(D) converting starch to sugar in darkness

(E) synthesis of intracellular proteins

For Question 60, refer to the following list of terms that may be in the basic body plan of acoelomates and/or pseudocoelomates:

1. ectoderm
2. endoderm
3. mesoderm

4. digestive cavity
5. coelom
6. pseudocoelom

60. Starting with the <u>outside</u> layer, include all structures that would be seen in a cross section of a round worm (phylum Nematoda).

(A) 1, 6, 3, 4, 2

(B) 2, 3, 4, 1, 5

(C) 1, 3, 6, 2, 4

(D) 6, 4, 1, 2, 3

(E) 3, 2, 1, 4, 6

Directions: The following groups of questions have five lettered choices followed by a list of diagrams, numbered phrases, sentences, or words. For each numbered diagram, phrase, sentence, or word choose the heading which most directly applies. Blacken the correct space on the answer sheet. Each heading may be used once, more than once, or not at all.

Questions 61 - 62

(A) exons

(B) nucleosome

(C) heterochromatin

(D) euchromatic

(E) nucleolus

61. Region that is associated with active transcription of RNA.

62. The place where ribosomes are produced.

Questions 63 - 66

(A) Meiosis

(D) Replication

(B) Mitosis

(E) Both Meiosis and Mitosis

(C) Metaphase II

63. At the end of this process, each nucleus has the haploid number of chromosomes.

64. Cytokinesis occurs in this process.

65. The products that result at the end of this process are diploid.

66. This process requires two complete cellular divisions.

Questions 67 - 70

(A) glucagon

(D) lipase

(B) pepsin

(E) bile salts

(C) salivary amylase

67. Which hormone or enzyme starts the breakdown of starch?

68. Which enzyme digests proteins?

69. This enzyme, secreted by the pancreas, breaks down fats.

70. This hormone, produced by the pancreas, helps to break down glycogen.

Questions 71 - 74

(A) density dependent factor

(B) density independent factor

(C) J-shaped curve

(D) S-shaped curve

(E) Zero population growth

71. Within the context of population growth, climate is an example.

72. Within the context of population growth, competition, parasites, and predators are examples.

73. Exponential growth of a population is best represented by this.

74. This is the most likely representation of growth in a population. The rate of increase is exponential then changes as the carrying capacity is reached.

Questions 75 - 77 refer to the diagrams below that demonstrate three general types evolutionary trends on a population.

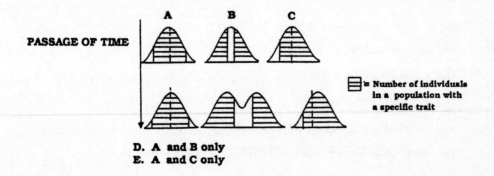

PASSAGE OF TIME

= Number of individuals in a population with a specific trait

D. A and B only

E. A and C only

357

75. Which is an example of disruptive selection?

76. An insect population that develops resistance to insecticides, can be represented by which graph?

77. If having brown hair becomes a definite advantage for survival as opposed to having blonde or red hair, which set of curves would represent that condition?

Questions 78 - 79

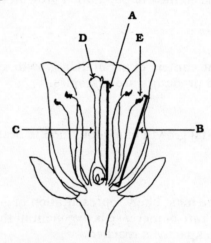

78. The male reproductive structure.

79. Location at which pollen grains attach, germinate, and produce pollen tubes.

Questions 80 - 82 refer to characteristics of the following organisms.

(A) acoelomates

(B) pseudocoelomates

(C) coelomates

(D) acoelomates and pseudocoelomates

(E) acoelomates, pseudocoelomates, and coelomates

80. Triloblasty is characteristic of members of this group.

81. A digestive cavity is characteristic of members of this group.

82. Members of this group have fluid-filled cavities within their mesodermal cavities that contain the digestive tract and other internal organs.

Questions 83 - 85 refer to the action potential curve of a neuron as it carries an impulse.

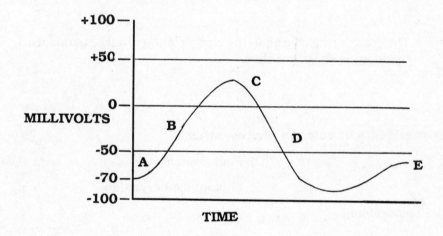

83. Which letter represents the time at which the potassium ions rush out of the neuron?

84. Which letter represents the place at which the nerve fiber is stimulated?

85. Which letter represents the place at which Na ions are actively pumped out of the membrane to re-establish the balance of Na^+ and K^+ ions?

Questions 86 to 89

(A) secondary xylem   (D) bark

(B) phloem   (E) vascular cambium

(C) cork

86. Which structure transports water and solutes upward in angiosperms?

87. Sugar can move upward or downward in this structure.

88. In dicots this tissue is actively growing and dividing.

89. The growth rings found in the cross section of a tree trunk are made up primarily of this.

Questions 90 - 91 refer to photosynthesis.

(A) C_3 (D) Photosystem II

(B) C_4 (E) Photophosphorylation

(C) Photosystem I

90. The first step in the Calvin Cycle is the fixation of carbon by the binding of CO_2 to ribulose bisphosphate (RuBP).

91. Adaptation to hot, dry environments.

Questions 92 and 93

(A) Carbohydrates   (D) Steroids

(B) Fats   (E) Nucleic acids

(C) Proteins

92.   Which organic molecule contains the most useable energy per mole?

93.   Which type of molecule can contain sulfur?

Directions: The following questions refer to experimental or laboratory situations or data. Read the description of each situation. Then choose the best answer to each question. Blacken the correct space on the answer sheet.

Questions 94 - 96 refer to a set of experiments.

  This set of experiments shows the influence of environmental factors on enzyme action. The materials used are an enzyme naturally found in potatoes and a clear substrate which forms an end product with a yellowish-brown color which indicates that a reaction between the substrate and the enzyme has occurred.

        Experiment 1 - Effect of heat on enzyme action. Three tubes are filled 3/4 of the way with distilled water. Tube 1 is placed in crushed ice (0°c). Tube 2 is placed in a beaker of boiling water (100 °c), and tube 3 is kept at room temperature. After ten minutes, ten drops of potato extract (which contains the enzyme) and ten drops of substrate are added to each tube. The tubes are shaken and returned to their original temperature condition. The color of each tube is recorded after ten minutes using the grading scale of : 0, 1, 2, 3 (where 0 indicates no color change and 3 indicates intense color change).

        Experiment 2 - The effect of substrate specific activity. Ten drops of the substrate for the potato enzyme are added to tube 1. Ten drops of a compound, similar in shape to the substrate are added to tube 2. Ten drops of yet another isomer of the substrate are added to tube 3. Then, ten drops of the potato enzyme are added to each of the 3 tubes. The three tubes are shaken, placed in a warm bath (35° c), and color formation is observed after ten minutes, as shown by this table:

Tube	Color after 10 minutes (Enzyme Activity)
1	3
2	1
3	0

94. What would be the most likely representation of the effect of temperature on the turnover rate of the enzyme based on Experiment 1?

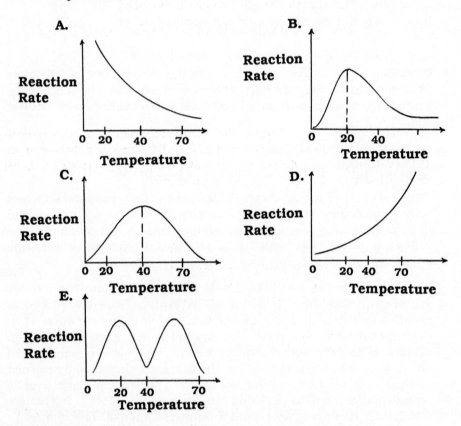

A.

Reaction Rate

0   20   40   70
Temperature

B.

Reaction Rate

0   20   40
Temperature

C.

Reaction Rate

0   20   40   70
Temperature

D.

Reaction Rate

0   20 40   70
Temperature

E.

Reaction Rate

0   20   40   70
Temperature

95. The results in experiment 2 indicate that:

(A) enzymes are highly specific for their substrate

(B) enzymes are only moderately specific for their substrate

(C) enzymes can distinguish among closely related compon-ents

(D) enzymes cannot distinguish among isomers

(E) A and C are true

96. Which of the following might occur if more of the substrate in tube 2 from experiment 2 were added and the temperature was increased to 40°C?

(A) The results would not change

(B) The enzyme activity would decrease

(C) The enzyme would become denatured

(D) The enzyme activity would increase slightly

(E) Such a prediction cannot be made

Questions 97 - 99 refer to the Punnett Square. It represents possible patterns of inheritance in dihybrid crosses. Brown eyes (B) are dominant and Blue eyes (b) are recessive. Straight hair (S) is dominant and curly hair (s) is recessive.

	BS	Bs	bS	bs
BS	A	B	C	D
Bs	E	F	G	H
bS	I	J	K	L
bs	M	N	O	P

97. If organisms of type "J" and type "O" are crossed, what fraction of the offspring would be expected to be <u>homozygous</u> for <u>both traits</u>?

(A) 1/16   (D) 7/16

(B) 3/16   (E) 9/16

(C) 4/16

98. If organisms of type "J" and type "O" are crossed, what fraction of the offspring would be <u>heterozygous</u> for <u>both traits</u>?

(A) 1/16   (D) 7/16

(B) 3/16   (E) 9/16

(C) 4/16

99.   What is the probability that a cross between a homozygous
        recessive male and a heterozygous female will produce off-
        spring with brown eyes and curly hair?

        (A) three-sixteenths   (D) three-fourths

        (B) one-fourth   (E) one

        (C) one-half

Questions 100 - 102 refer to these generalized survivorship curves:

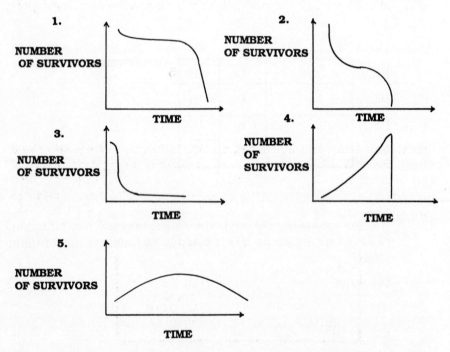

100.   What is probably true about a species represented by graph
        #3?

        (A) They have a slower development

        (B) They have many offspring during a period of time

        (C) They have a low mortality early in life

        (D) They are as likely to die early in life as late in life

        (E) They are not well adapted for dispersal in the environ-
            ment

101. How is curve #1 different from curve #2?

(A) Curve #1 shows exponential growth, while curve #2 does not.

(B) Curve #1 is more representative of small offspring than large offspring.

(C) Curve #1 favors offspring which have much parental care.

(D) For curve #1, the mortality rate at all ages is more or less constant compared to curve #2.

(E) Only (A) and (D).

102. Which graph indicates an overshooting of the carrying capacity?

(A) 1   (D) 4

(B) 2   (E) 5

(C) 3

Questions 103 - 105 are based on the following information and chart. Both the Smith family and the Jones family had babies at the same time. There was a mix-up in the hospital nursery; but luckily, the hospital had the blood groups of the Jones' and the two babies. The chart below gives that information.

Mr. Jones	-	group AB
Mrs. Jones	-	group B
Baby 1	-	group A
Baby 2	-	group O

103. Which of the babies belongs to the Joneses?

(A) Baby 1 (Blood group A)

(B) Baby 2 (Blood group O)

(C) Either Baby 1 or Baby 2

(D) Neither Baby 1 nor Baby 2

(E) None of the answers are correct.

104. If Mr. Jones had group O blood, which child would have belonged to the Joneses?

(A) Baby 1 (Blood group A)

(B) Baby 2 (Blood group O)

(C) Either Baby 1 or Baby 2

(D) Neither Baby 1 nor Baby 2

(E) None of the answers are correct.

105. If Mrs. Jones had group A blood, which child would have belonged to the Joneses?

(A) Baby 1

(B) Baby 2

(C) Either Baby 1 or Baby 2

(D) Neither Baby 1 nor Baby 2

(E) None of the answers are correct.

Questions 106-108. The following time plot represents a predator and prey population. The predator is the lynx and the prey is the snowshoe hare. The letters refer to growth rates for the populations.

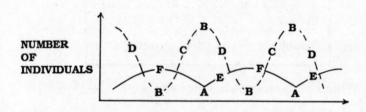

106. Which letter represents the prey increasing?

(A) A   (D) D

(B) B   (E) E

(C) C

107. Which letter(s) represent zero population growth?

(A) A and B

(D) D and E

(B) B and C

(E) A and E

(C) C and D

108. What is the most common explanation for why prey populations are limited below their carrying capacity?

(A) The age of the prey makes them susceptible to predators

(B) Prey in poor physical condition are taken by predators

(C) The existence of only one species that is prey to a predator
means that the species will be preyed on to extinction

(D) The availability of food is the limiting factor

(E) None of the above

Questions 109 - 111 are based on the simplified diagram which
represents a nephron. The numbers represent the different
structures of the nephron.

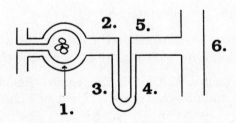

109. Chloride ions are actively pumped out of this structure.

(A) 1

(D) 4

(B) 2

(E) 5

(C) 3

110. At which structures are the walls of the tubules freely permeable to water?

(A) 1, 2, 3, 4   (D) 1, 2, 4, 6

(B) 2, 3, 5, 6   (E) 1, 3, 5, 6

(C) 1, 3, 4, 5

111. Where does filtration take place in the nephron?

(A) 1   (D) 4

(B) 2   (E) None of the above.

(C) 3

Questions 112-113 refer to a female *Drosophila* heterozygous for ebony body color (recessive) and curly wings (recessive) that was mated to an ebony-bodied, curly-winged male resulting in the following offspring:

200 ebony body, normal wing

10 normal body, normal wing

5 ebony body, curly wing

150 normal body, curly wing

112. Choose the correct statement(s).

1. The genes are linked. In the female, the alleles for normal body and normal wing are on the same chromosome.

2. The genes are linked. In the female, the alleles for normal body and curly wing are on the same chromosome.

3. The genes are unlinked.

4. The genes are linked. In the female, alleles for ebony body and curly wing are on the same chromosome.

5. The genes are linked. In the female, the alleles for ebony body and normal wing are on the same chromosome.

(A) 1, 2   (D) 2, 5

(B) 2, 4   (E) 1, 5

(C) 3

113. Which process must occur in order to obtain ebony body, curly wing phenotype in the offspring?

(A) Nondisjunction during meiosis in the female

(B) Crossing over during gamete formation in the female

(C) Crossing over during gamete formation in the male

(D) A mutation in the female

(E) A mutation in the male

Questions 114 - 117. The following shows the change, over time, in the vegetation of a farmed area which was left alone for many years.

abandoned field	dense shrubs	pine	fir, birch,whitespruce
	herb, shrubs		
yr. 0	(weeds)		yr. 150

114. The above diagram illustrates

(A) Primary Succession   (D) Pioneer Plants

(B) Secondary Succession   (E) Evolution of pine trees

(C) Tertiary Succession

115. The stage which is most likely to be the most stable is

(A) the herb community

369

(B) the shrub community

(C) the dense shrub community

(D) the pine community

(E) the fir, birch, white spruce community

116. The most diversity would be expected to be in the

(A) herb community

(B) shrub community

(C) dense shrub community

(D) pine community

(E) fir, birch, white spruce community

117. Which of the following tends to produce more organic material than it uses?

(A) The herb community

(B) The shrub community

(C) The dense shrub community

(D) The pine community

(E) The fir, birch, white spruce community

Questions 118 - 120 refer to the following:

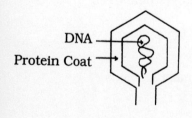

DNA

Protein Coat

The protein coat of a sample *E. coli* bacterio-phage (virus that attacks bacteria) is labelled with radioactive sulfur. In addition, the DNA core of another sample *E. coli* bacteriophage is labelled with radioactive phosphorus. The two phages are then added to two different colonies of bacteria grown in nonradioactive media.

118. Analysis showed that:

   (A) the host cell had both radioactive sulfur and phosphorus

   (B) the host cell had only radioactive sulfur

   (C) the host cell had only radioactive phosphorus

   (D) the host cell had neither radioactive sulfur nor radioactive phosphorus

   (E) viral infection cannot take place in a nonradioactive medium

119. In the bacteriophage infection cycle described in the experiment,

   (A) the entire virus particle enters the host cells

   (B) the protein coat of the bacteriophage enters the host cell

   (C) the viral DNA is left outside the host cell

   (D) the viral DNA replicates within the host cell

   (E) the viral protein coat replicates within the host cell

120. The Hershey and Chase experiment was used to prove that:

   (A) proteins contain the genetic material

   (B) DNA is the genetic material

   (C) viruses cannot reproduce in a radioactive medium

   (D) viruses cannot infect a host which is radioactive

   (E) bacteriophages reproduce only the DNA inside the host cell

ADVANCED PLACEMENT BIOLOGY EXAM V

SECTION II

DIRECTIONS: Answer each of the following four questions in essay format. Each answer should be clear, organized, and well-balanced. Diagrams may be used in addition to the discussion, but a diagram alone will not suffice. Suggested writing time per essay is 22 minutes.

1. Nitrogen, phosphorus, and carbon are recovered through biogeochemical cycles in which matter flows in a cyclical pathway. Describe each cycle, including the effects humans have had on each.

2. Chromosomes pass genetic information through generations. Explain how this occurs in three phases:

   A) Explain the structure, function, and location of DNA.
   B) Detail DNA replication. State specifically where this process occurs.
   C) Describe protein synthesis at the ribosomal level.

3. Development in flowering plants begins from a seed.

   A) Describe the germination process. How can you tell if starch is present?
   B) Some seeds contain substances similar to antibiotics. Design an experiment to detect such substances.

4. The ATP/ADP energy system is used both by biological systems that require energy and ones that release it. Relate each of the following, giving examples where appropriate, to the transformation of energy and/or ATP formation:

   A) First and Second Law of Thermodynamics
   B) Anabolism and catabolism
   C) Phosphorylation of ADP to ATP
   D) Coupled reactions

ADVANCED PLACEMENT BIOLOGY EXAM V

ANSWER KEY

1.	B	31.	E	61.	D	91.	B
2.	A	32.	C	62.	E	92.	B
3.	E	33.	E	63.	A	93.	C
4.	C	34.	B	64.	E	94.	C
5.	B	35.	B	65.	B	95.	E
6.	C	36.	E	66.	A	96.	D
7.	A	37.	A	67.	C	97.	C
8.	A	38.	D	68.	B	98.	C
9.	A	39.	C	69.	D	99.	B
10.	E	40.	D	70.	A	100.	B
11.	C	41.	D	71.	B	101.	C
12.	E	42.	A	72.	A	102.	D
13.	B	43.	D	73.	C	103.	A
14.	A	44.	E	74.	D	104.	B
15.	A	45.	A	75.	B	105.	A
16.	E	46.	C	76.	B	106.	C
17.	E	47.	D	77.	C	107.	A
18.	C	48.	C	78.	B	108.	D
19.	C	49.	D	79.	D	109.	D
20.	C	50.	A	80.	E	110.	B
21.	B	51.	D	81.	E	111.	A
22.	B	52.	C	82.	C	112.	D
23.	A	53.	D	83.	C	113.	B
24.	A	54.	D	84.	B	114.	B
25.	D	55.	A	85.	D	115.	E
26.	A	56.	C	86.	A	116.	E
27.	B	57.	D	87.	B	117.	A
28.	B	58.	C	88.	E	118.	C
29.	E	59.	E	89.	A	119.	D
30.	A	60.	C	90.	A	120.	B

ADVANCED PLACEMENT
BIOLOGY EXAM V

DETAILED EXPLANATIONS
OF ANSWERS

SECTION I

1.   (B)

The blood returning to the heart from the legs and arms enters the upper right chamber of the heart (right atrium). (The superior vena cava drains the arms, neck and head and the inferior vena cava drains the remainder of the body). The right atrium contracts forcing the blood into the right ventricle which then contracts sending the blood into the pulmonary artery. The blood is then sent to the lungs where gas exchange takes place - the hemoglobin of the blood picks up oxygen from the alveoli of the lungs, exchanging carbon dioxide for oxygen. The oxygen-rich blood now returns to the left atrium of the heart via the pulmonary veins. Contraction of the left atrium forces the blood into the left ventricle which then contracts forcing the blood into the aorta, a large artery which branches into smaller arteries. Blood is carried in these arteries to all parts of the body.

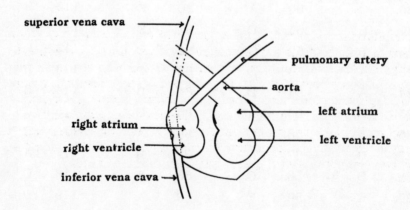

2.   (A)

The producers in the energy pyramid are the plants. These autotrophs make their own organic "food" from simple organic materials and energy.

3.   (E)

In itself, the heat killing of the bacterial culture is unrelated to factors that regulate populations such as choices (A), (B), (C), and (D). The decimation of this microbial population occurred when a critical environmental variable, in this case temperature, exceeded the range of tolerance that is characteristic of the species.

4.   (C)

In a mitochondrian, hydrogen ions (protons) are located in the inner compartment. Energy released from the passage of electrons in the inner membrane of a mitochondrian is used to pump the hydrogen ions from the inner to the outer compartment (between the outer and inner mitochondrial membranes) of the mitochondrian. This results in the establishment of a concentration gradient between the outer and inner compartments. The protons then move inward back to the inner compartment to re-establish an equilibrium and to rejoin OH^- ions formed by the splitting of water. To reach the inner compartment the protons pass through channels in the inner membrane. Energy released during that passage is used by the ATPase enzymes to form ATP from ADP and inorganic phosphate. The best answer is (C). Options (B), (D), and (E) are part of the system but need the chemiosmotic movement of protons. Option (A) is incorrect.

5.   (B)

Phosphorylation, which is the addition of one or more phosphate groups to a molecule, is a type of endergonic reaction — a reaction in which complex molecules are synthesized and energy is temporarily stored. ADP becomes ATP with the addition of the phosphate group, and ATP provides the energy for cell metabolism. Option A is incorrect because hydrolysis (splitting of a molecule into two by the addition of a water molecule) is an example of an exergonic reaction — one in which energy is released. A condensation reaction (option D) is a reaction that joins two compounds with the resulting production of water.

6.   (C)

A nucleosome consists of a DNA double helix wound around a cluster of histone proteins.

7.   (A)

Of all the organs mentioned, only the ovary is part of the reproductive system. Thus, meiosis would be expected to occur in some of the structures associated with the reproductive system. Choice (B) is part of the skeletal system while choices (C), (D), and (E) are part of other body systems.

8.   (A)

When a homozygous dominant organism is crossed with a homozygous recessive organism, the resulting generation is heterozygous with the dominant trait being apparent and the recessive trait being masked.

	B	B
b	Bb	Bb
b	Bb	Bb

B = brown eyes
b = blue eyes
Bb = brown eyes (heterozygous)

9.   (A)

The genus and species names identify the species. A species is a population that can or does interbreed and is separate (in a reproductive sense) from other groups. Genus is the next largest grouping of similar yet distinct species.

10.   (E)

Members of the kingdom Protista may be either autotrophic, heterotrophic or a combination of both depending on the presence or absence of chloroplasts. This criterion can be applied to the three types of protists: algae, slime molds, and protozoa.

11.   (C)

Hybridization is used to detect the presence of specific nucleic acids and to determine the similarity of two nucleic acid sequences. The double helix of DNA is broken into its complementary strands by gentle heating (denaturation). When the solution is cooled, the hydrogen bonds reform. If denatured DNA molecules from a variety of sources are mixed together and denatured, two strands with nearly complementary sequences will combine to form a double helix.

12.   (E)

The acrosome is an organelle in the tip of the sperm which ruptures to release enzymes. The acrosome ruptures before reaching the egg cytoplasm. The acrosomal enzymes are necessary for effecting penetration of the egg by one sperm. This occurs before fertilization.

13.   (B)

As estrogen levels pick up after the shedding of the endometrium, production of LH (luteinizing hormone) is increased, thus, stimulating the release of the egg cell (ovulation). The endometrium has been growing during this time and is ready for implantation.

14.   (A)

Hydrogen bonds form between the complementary base pairs which are combinations of a pyrimidine and a purine. The complementary base pairs of nucleotides are 1) adenine and thymine, and 2) guanine and cytosine. Deoxyribose, the five-carbon sugar present in DNA and the phosphate groups are both bound covalently in DNA.

15.   (A)

The microtrabeculae connect the other cytoskeletal substances (options C, D, E) and the organelles. The cytoplasmic lattice or cytoskeleton (option B) contains the structures (options A, C, D, E) that give the cell its shape, give it internal movement, and hold the organelles in the correct region of the cell for their most efficient functioning.

16.   (E)

Green algae share all the above features with plants.

17.   (E)

The two major groups of terrestrial plants are characterized by the presence or absence of vascular (connecting) tissue that sends water, food, minerals, and other nutritious substances throughout the plant. Bryophytes include liverworts and mosses. Gymnosperms, angiosperms, and ferns are all types of tracheophytes or vascular plants.

18.   (C)
Bryophytes such as mosses and liverworts have life cycles which display an alternation of generations. In this life cycle, the zygote is produced by the fusion of gametes from the antheridium (contains sperm cells) and the archegonium (contains the egg cell). The zygote, which stays inside the archegonium, becomes the mature diploid sporophyte.

19.   (C)
In the microsporangium located inside the pollen-bearing male cone, the microspores develop into pollen grains which are the young male gametophytes. The pollen grains containing the microspores are released and find their way to the female cone (gametophyte) containing the archegonium which, in turn, contains the egg cell. The sperm cells are carried to the archegonium by the pollen tube (produced by the male gametophyte) which grows through the tissue of the ovule. Fertilization then occurs.

20.   (C)
Both are examples of sex-linked conditions carried by recessive genes on X-chromosomes.

21.   (B)
An Rh-negative woman can bear an Rh-positive baby if the father is Rh positive (having the "antigen D" allele). If the father is homozygous for the Rh factor, all the children will be Rh positive. If the father is a heterozygote, about half the children will be Rh positive. The question asked for what the father's Rh factor must have been: he could not have been just homozygous Rh positive or heterozygous Rh negative. Option (B) contains both possibilities.

22.   (B)
 Two or more populations do not have to speciate simply because they may be geographically isolated (Choices A and D). A population is considered a species only if its members can produce viable offspring only with other members of the population, but not with members of another population. Genetic isolation results in such a situation. Sympatry, the use of the same region by two or more populations, is not always involved in speciation.

23.   (A)
 DNA and histones, proteins that are closely associated with DNA, are present in approximately equal amounts in chromosomes. DNA contains both repetitive sequences that code for protein (gene expression), and sequences that are not translated into proteins. The amount of DNA used for gene expression is small compared to the total DNA of the cell. In fact, in humans only about one percent of the DNA codes for protein.

24.   (A)
 Restriction enzymes cleave strands of DNA segments at certain sites, thus yielding uniform fragments to be studied in the laboratory. The DNA molecule is not cleaved straight across by restriction enzymes; rather, these enzymes leave "sticky ends" that are complementary to another molecule cleaved by the same enzyme.

25.   (D)
 Uracil is the base in RNA that is substituted for the thymine in DNA. The uracil links up with adenine, forming a complementary base pair.

26.   (A)
 Mutations can be caused by changes in the nitrogenous bases of DNA. One change can be the replacement of one base pair for another within a segment of DNA. Other changes can be the insertion of extra base pairs, or the deletion of one or more base pairs. A change in RNA nucleotide sequence may result in faulty translation of a gene, but RNA is constantly degraded and synthesized; as long as DNA remains unchanged, RNA will almost always be correctly synthesized.

27.   (B)

Species are classified according to their perceived evolutionary relationships. Only choice (B) provides criteria that pertain to the evolutionary relationships of different species.

28.   (B)

The entire structural difference between a normal and a sickle-cell hemoglobin molecule consists of the substitution of the amino acid valine for the amino acid glutamic acid. This substitution occurs in the sixth position of each of the two B-chains in hemoglobin.

29.   (E)

Bacteria are prokaryotic cells, while *Paramecium*, although one-celled like bacteria, is eukaryotic with more advanced structures. All the other options are examples of organisms that have eukaryotic cells; these organisms are, therefore, more closely related phylogenetically than bacteria and *Paramecium*.

30.   (A)

A nucleotide is composed of a sugar, a phosphate group, and a base containing nitrogen. Nucleotides are found in chromosomes (DNA), which are in turn found in the nucleus of the cell. All other options represent substances that can be found in the cell membrane.

31.   (E)

Plants measure darkness rather than light, and may not flower if the darkness is interrupted. For example, assume that clover will flower only if the light periods are longer than the critical length that is specific for that species. If we use 14 hours as the critical period, then the dark period should be an uninterrupted 10 hours or less.

32.   (C)

Echinoderms are named for their spiny internal skeletons. They show radial symmetry (but larvae show bilateral symmetry) and water vascular systems of canals that provide locomotion. Water creates a liquid skeleton that makes the tube feet rigid enough to walk on.

Echinoderms also exhibit deuterostome development. A major characteristic of this development is the formation of the anus at or near the blastopore while the mouth forms elsewhere. They do not possess nematocysts (stinging cells); only coelenterates do. Protostomes include all animals except chordates, hemichordates, and echinoderms. There are no such things as pseudostomes.

33.   (E)

Cholinesterase is an enzyme that decomposes acetylcholine so that it cannot bind permanently to the postsynaptic membrane. The impulse then ends. If cholinesterase is absent, the acetylcholine continues to be released into the synaptic cleft and binds to the receptors of the postsynaptic membrane. However, molecules of acetylcholine remain bound to the receptors indefinitely, because cholinesterase is not present. This leads to rapid uncontrolled firing of the impulses, which could lead to uncontrollable tremors, spasms, or even death for the organism.

34.   (B)

Anaerobic organisms can produce a yield of two ATPs per glucose molecule as a result of glycolysis, which takes place in the cytoplasm of the cell and does not require oxygen. If oxygen is not present, the organism produces two ATPs by fermentation (either lactate or alcoholic process). This process reduces the pyruvate to lactic acid and releases NAD for further use by the cell. The carbohydrates that enter the anaerobic pathway are not completely broken down and much energy is still contained in lactate and ethanol. Only two ATPs are formed compared to the 36 ATPs that are formed as a result of the aerobic pathway (Krebs cycle and electron transport chain). Therefore, the more inefficient anaerobic organism would have to use more glucose to produce the same amount of ATP as the aerobic organism.

35.   (B)

Fermentation is the production of ethanol from glucose as done by yeast cells. In glycolysis, one glucose molecule is converted to two molecules of pyruvic acid and also provides enough energy for the synthesis of two molecules of ATP and two molecules of NADPH. The pyruvic acid, still containing much potential energy, can next enter either the anaerobic pathway or the aerobic pathway. In one type of anaerobic pathway, pyruvic acid is converted to ethanol by the action of yeast cells on sugar.

36. (E)

Both NAD⁺ and FAD (not listed) are the electron acceptors with the highest potential energies. After completion of the Krebs cycle, the carbon atoms of glucose have been oxidized, and some of the energy of the glucose has been utilized to generate ATP and ADP. The remaining energy is in the electrons removed from carbon-carbon bonds and the carbon-hydrogen bonds. These electrons pass to the electron carriers NAD⁺ and FAD. These electron carriers pass the electrons along to electron carriers with lower energy levels (cytochromes) and finally to the electron carrier, oxygen, which has the lowest energy level. The oxygen then combines with hydrogen ions (protons) to produce water.

37. (A)

The energy yield from oxidative phosphorylation is 32 ATP molecules. The oxidation of pyruvic acid produces 6 ATP molecules. Glycolysis in the mitochondria yields 6 ATP molecules in most cells. Glycolysis in the cytoplasm yields 2 ATP molecules. The formation of lactic acid which takes place during anaerobic respiration yields 2 ATP molecules.

38. (D)

In humans, the lymph is moved along by contractions of the skeletal muscles, through which the lymph vessels are located. Also, lymph vessels return 99% of the fluid that leaves the capillaries at their arterial end. The fluid is reabsorbed at the venous ends of capillaries, where the osmotic pressure in the capillaries is less than the osmotic pressure outside the capillaries.

39. (C)

Smaller, active animals, such as mice, have high metabolic rates. The rate of demand for oxygen by their tissues is greater because they are releasing much carbon dioxide (ex. in the lungs). The pH then becomes lower, and the hemoglobin unloads its oxygen, since increased acidity causes hemoglobin to release oxygen. The opposite is true for larger warm-blooded animals such as elephants which have lower metabolic rates. The hemoglobin hangs on to its oxygen longer allowing the tissues deep inside the body to be exposed to oxygen as the blood carries it to these tissues.

40.   (D)

Connective tissue provides support for body parts and binds structures together. Options A, B, and C are examples of connective tissue. Collagen (Option E) is a protein found in skin and bone and is secreted by the cells of connective tissue. Collagen provides a rigid matrix in which connective tissue cells exist.

41.   (D)

According to the sliding filament theory, myosin filaments have extensions called cross-bridges, which bind to thin actin filaments at receptor sites and pull actin filaments toward the center of the sarcomere, the contractile unit. The cross-bridges can only bind to actin when calcium ions are present. These ions set up "receptor" sites on the actin, to which the cross-bridge can attach. Energy is used by the myosin cross-bridge to attach to the actin, pull it toward the center of the unit, release it, and reattach to the actin at a new "receptor" site. This energy is provided by the hydrolysis of ATP by myosin.

42.   (A)

The apical meristem of the shoot contains the tissues that produce new leaves, branches, and flowers. The lateral meristem (Option B) produces tissues that increase the thickness of woody plants. Areas of cell elongation (Option D) are those in which cell division is slowed and the length of each cell increases, hence the shoot length increases. Internodes are portions of the stem between locations at which leaf primordia and leaves arise from the stem. The radicle is the embryonic root of a plant.

43.   (D)

The palisade parenchyma consists of many-sided thin-walled cells that are long and narrow. Chloroplasts are located in these cells and most photosynthesis takes place therein. Below the palisade parenchyma is the spongy parenchyma (Option E), which has cells with irregular shapes that have some chloroplasts. Guard cells regulate transpiration and the entry of air into leaves by regulating the size of small openings (stomata) on the undersides of leaves.

44.   (E)

Options (A) through (D) refer to characteristics of the respiratory system of both unicellular and multicellular organisms. In single-celled or simple organisms such as algae and flatworms, oxygen diffuses directly through cell membranes. Thus, location deep inside an organism is not a requirement of a respiratory system.

45.   (A)
In angiosperms, the pollen grain is the male gametophyte stage in the alternation of generations life cycle. When a pollen grain lands on the stigma of a pistil, it germinates and generates a pollen tube. Two sperm cells, which were part of the pollen grain as a bi-nucleate structure, move into the pollen tube as it grows toward the ovary. Each pollen grain contains a tube nucleus, not a pollen tube. Polar bodies are associated with angiosperm ova, and fusion nuclei are cells that give rise to the endosperm.

46.   (C)
In the process of double fertilization, one sperm nucleus from the pollen grain fertilizes the egg nucleus in the ovule forming a zygote that divides mitotically to produce the embryo (Option D). The other sperm nucleus from the pollen grain fertilizes the two central nuclei in the ovule resulting in a triploid nucleus. This triploid nucleus divides producing the endosperm, a tissue that surrounds the embryo and contains stored food (Option A).

As the embryo grows, the wall of the ovule grows and becomes the seed coat (Option C), while the ovary enlarges around the ovule and becomes the fruit (Option B). Both the seed coat and the fruit protect the embryo from both predators and moisture loss. The carpel (Option E) is the female reproductive structure which includes the stigma, style, and ovary.

47.   (D)
The primary targets of B-cells are bacterial infections. All other options are the targets of T-cells.

48.   (C)
Normal body temperature does not provide enough energy of activation to start reactions. A catalyst (such as an enzyme) lowers the energy of activation needed to bring reacting molecules together to cause a chemical reaction. The amount of free energy change, ΔG, from the initial state to the final state is not changed by the presence of the catalyst. Only the activation energy necessary to start the reactions is different for catalyzed and uncatalyzed reactions.

49. (D)

A codon consists of three nucleotides, and this triplet forms the code for a specific amino acid. The twenty amino acids are coded for by 61 (not 64) triplet codons. The three additional codons are the "stop" signals to terminate protein synthesis.

50. (A)

Humans (mammals) and birds use negative pressure breathing whereby air is drawn into the lungs. This process involves raising of the rib cage and the downward movement of the diaphragm during inhalation. The volume of the chest cavity is increased, reducing the internal air pressure, resulting in air being drawn into the lungs to equalize the pressure. In contrast, positive-pressure breathing occurs when air is <u>forced</u> into the lungs. For example, a frog closes its nostrils and raises the floor of its mouth, thus reducing the volume of the mouth cavity and forcing air into the lungs. Active transport of air does not occur in humans. The presence of air sacs in birds allows for a unidirectional flow of air through a bird's lungs.

51. (D)

In an open circulatory system, blood goes through sinuses (open spaces) and has contact with the organs and cells. In insects, the blood does not carry oxygen; the tracheae do. An open circulatory system is characteristic of most molluscs and arthropods. In contrast, in a closed circulatory system the blood flows in well-defined vessels. Closed circulatory systems are characteristic of earthworms and all vertebrates.

52. (C)

Three-chambered hearts are characteristic of all amphibians and most reptiles (except crocodilians). These animals are all poikilotherms. Poikilotherms are cold-blooded animals. They do not maintain constant body temperatures and do not need to break down as much glucose to heat their bodies. Thus, they do not need as much oxygen for the respiratory process.

53. (D)

Oxygen diffuses from the water into the blood. The circulatory structures are arranged in the gills so that the blood is pumped into them in a direction that is opposite to that of the direction of the water with its load of oxygen. The blood vessels can then extract a larger percentage of oxygen from the water, because the water will always contain a higher concentration of oxygen than the blood will.

54. (D)
In fish, the blood returns to the atrium after passing through both the respiratory (gill) and systemic (body tissues) circulatory system. A fish's heart consists of one atrium and one ventricle. No mixing of oxygenated and deoxygenated blood occurs.

55. (A)
At the arterial end of the capillaries the blood pressure is higher than the pressure of the tissue fluid outside the capillaries. This differential causes fluid to leave the capillaries and go into the tissue. At the same time the concentration of proteins in the tissue fluid is less than the concentration of proteins in the blood because the large protein molecules cannot easily diffuse through the capillary walls. Thus, water tends to move into the capillaries by osmosis to equalize the osmotic pressure; this occurs at the venous end. Approximately 99% of the water that exits capillaries at the arterial ends due to the net force of blood pressure, re-enters the capillaries at their venous ends due to the net force of osmotic pressure.

56. (C)
A cofactor is a nonprotein substance that helps an enzyme to catalyze a reaction. Ions, such as manganese ions (Mn^{2+}) (Option A), can be cofactors for certain enzymes. A coenzyme is a type of cofactor, and more specifically, is a nonprotein organic molecule that can function as an electron acceptor. NAD^+ (Option B), FAD (Option D), and ascorbic acid (vitamin C, Option E), are electron acceptors and are bound to enzymes.

57. (D)
Glycolysis occurs in the cytoplasm of the cells. The Krebs cycle, which occurs after glycolysis in aerobic respiration, takes place in the mitochondria of animal cells.

58.　(C)

According to the Endosymbiont Theory, mitochondria and chloroplasts, which are found only in eukaryotes, evolved when some prokaryotic cells were ingested by other prokaryotic cells and evolved into specialized organelles. The relationship was symbiotic, because the ingested promitochondria and prochloroplasts derived nourishment from the surrounding cytoplasm, while at the same time providing energy for the cells in which they existed. Evidence for this theory is shown by the fact that the components of the mitochondria, such as the presence of DNA in a single-strand is similar to that of the DNA of prokaryotic bacteria. Furthermore, mitochondria (and chloroplasts) contain their own DNA (distinct from the nuclear genetic materials). Both mitochondria and chloroplasts divide in the cell independently of the nucleus.

59.　(E)

Guard cells can generate turgor pressure by converting starch to sugar in daylight resulting in an increased concentration of solute within the cells. Water then moves into the guard cell by diffusion. Recent research indicates that the level of potassium in the guard cells is related to their opening and closing (Option C). The potassium ions are actively transported between the guard cells and the surrounding fluid using ATP as the energy source. Potassium levels are shown to rise when the stomata open and fall when the stomata close. The changing solute-solvent concentration causes water to diffuse into the stomata by osmosis when the level of potassium ions is up, causing the stomata to open.

60.　(C)

The Phylum Nematoda are pseudocoelomates, which have a more complex body plan than the acoelomates.

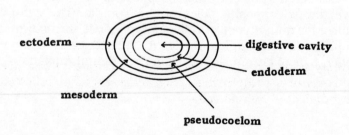

The pseudocoelom, located between the endoderm and the mesoderm, does not have the epithelial lining of a true coelom — a cavity
located within the mesoderm in which are suspended the digestive
tract and other internal organs.

61.　(D)
Euchromatin is a loosely packed or unfolded form of DNA. Euchromatic regions in the chromosomes contain active genes, and are the
sites of active RNA transcription.

62.　(E)
The nucleolus, located in the nucleus of eukaryotic cells contains
DNA, RNA, and protein. The RNA that is part of the structure of the
newly-forming ribosomes is directly transcribed from the DNA in the
nucleolus.

63.　(A)
At the end of Telophase II of meiosis, there are four cells that are
derived from the original diploid cell. Each nucleus has a haploid (n)
number of chromosomes.

64.　(E)
Cytokinesis, the dividing of cytoplasm, occurs in both mitosis and
meiosis. In mitosis during Telophase, chromosomes, identical to
those in the original nucleus, group around poles of the spindle. The
nuclear envelope reforms and the cell cytoplasm divides.

In meiosis, there are two cytoplasmic divisions. During Telophase I,
homologous chromosomes (chromosomes that have replicated and
are connected by a centromere) separate from their homologues and
move toward opposite poles of the spindle. The cytoplasm then
divides; for many types of cells, Metaphase II starts immediately. In
Telophase II, four haploid nuclei are formed, each with one member
of each pair of chromosomes from the original nucleus.

65.   (B)
The products that result from mitosis are diploid; they have two sets of chromosomes. The number of chromosomes is thus maintained from one generation to the next. Mitosis is the basis for cell division (for physical growth) in many-celled organisms.

66.   (A)
Meiosis involves two division sequences which result in the formation of four haploid cells. At the end of the first division, the chromosome number is reduced. At the end of the second division the chromatids are separated forming four haploid cells with single stranded chromosomes. Option (C) is part of the second division sequence of meiosis.

67.   (C)
Salivary amylase, a digestive enzyme found in the mouth, begins the breakdown of starch to form maltose.

68.   (B)
Pepsin, a secretion of the stomach, breaks apart the peptide bonds that link the amino acids which form the proteins. The protein is not completely digested; the peptide bonds linking a few amino acids are broken.

69.   (D)
Lipase is an enzyme that is secreted by the pancreas. It hydrolyzes a small portion of the fats to glycerol and fatty acids.

70.   (A)
The hormone glucagon helps in the breakdown of glycogen which, in turn, increases the blood sugar.

71.   (B)
Climate is not influenced by population density. However, climate can have an impact on the growth and even survival of populations, regardless of size.

72.   (A)
Population growth is regulated by these three mechanisms (competition, parasites, and predators). For example, when the population size increases, mechanisms such as competition, parasites, and predators can act on the population to decrease the birth rate, increase the death rate or foster emigration of the population from the affected area.

73.   (C)
A J-shaped curve occurs when the birth rate remains above the death rate. The addition of new individuals to the potential reproductive base (without an accompanying increase in the death rate) can lead to unrestrained growth.

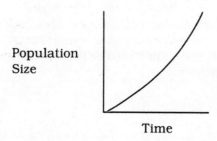

Population
Size

Time

74.   (D)
The "S" shaped curve shows logistic growth. At first exponential growth occurs, then the curve reaches a plateau and flattens out as the carrying capacity of the environment is reached.

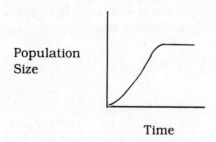

Population
Size

Time

75.   (B)
In disruptive selection, the more extreme types in a population increase in number while the intermediate forms are selected against. Two divergent populations can eventually be produced.

76.   (B)
After the first application of insecticide, many insects are killed but some in the population may survive because of differences in their bodies that allow them to resist the effects of the insecticide. If the differences are genetic in origin, then these traits can be passed on to subsequent generations, resulting in a strain of individuals resistant to the insecticide because of traits that are not shared by the population as a whole. This type of selection, called disruptive selection, favors extremes in a population rather than those at the midpoint of a distribution. A bimodal (two-humped) distribution results.

77.   (C)
This curve is called directional selection. Directional selection acts against individuals who do not have brown hair, a trait that is necessary for survival. The population shifts toward those individuals with brown hair as it attempts to establish a new optimum situation. Choice (A) represents a stabilizing selection that favors those with brown hair but does not completely eliminate the extremes.

78.   (B)
The male reproductive structure is called the stamen. The pollen grains (young male gametophytes) develop in the stamen. It is composed of the anther (Option E) and the long filament.

79.   (D)
The stigma is the place on which the pollen grains land after being released by the anther of another (or the same) flower. The pollen grains then develop pollen tubes that grow down through the style and reach the ovule in which the egg cells are located.

80.   (E)
Acoelomates, pseudocoelomates, and coelomates are triloblastic which means they have three tissue layers: ectoderm, endoderm, and mesoderm (which is located between the ectoderm and the endoderm). The basic arrangement of three-tissue layers that is characteristic of the acoelomates becomes more complex as we shift from acoelomates to pseudocoelomates to coelomates.

81.   (E)

The digestive cavity is a food-containing cavity that is surrounded by the endoderm in acoelomates, pseudocoelomates, and coelomates.

82.   (C)

Mollusks, annelids, and more complex animals each have a coelom (fluid-filled cavity) which distinguishes these animals from acoelomates and pseudocoelomates.

83.   (C)

Option A represents the polarization of the membrane. Negatively charged large organic ions and the leakage of K^+ to the outside set up a negative charge inside the membrane. Option B represents the stimulation of the membrane by an impulse. The membrane becomes permeable to Na^+ which diffuses inward, causing a momentary reversal of membrane poplarity; thus, the inside of the membrane becomes slightly positively charged. This change in the distribution of charges along the neuron is called the action potential. At Option C, THE CORRECT ANSWER, the membrane becomes permeable to K^+, which rush out of the cell faster than the sodium can move in. This shift starts to repolarize the membrane, setting up a positive charge outside the membrane. At Option D the membrane becomes less permeable to Na^+ as the sodium-potassium pump begins to pump three Na^+ out of the cell for every two K^+ it pumps in. The source of energy for the sodium-potassium pump is ATP. At Option E, the resting potential is again achieved because of the difference of electrical charges inside and outside the membrane. This potential is maintained by the sodium-potassium pump, which pumps Na^+ outside the membrane and pumps K^+ inside the neuron.

84.   (B)

Option B represents the stimulation of the membrane by an impulse. The membrane then becomes permeable to Na^+ which diffuses inward causing a momentary reversal of polarity; the inside of the membrane has a positive charge relative to the outside. This change in the distribution of charges along the neuron is called the action potential.

85.   (D)

The sodium-potassium pump uses ATP as the energy source. It pumps three Na^+ ions out of the membrane of the motor neuron for every two K^+ ions pumped in, resulting in a charge difference inside and outside the membrane.

86.   (A)

In xylem, (both primary and secondary) the movement of water and solutes is upward. This movement results from a combination of the force produced by water that enters the roots (root pressure) and the force produced by water that escapes from the leaves (transpirational force). Water enters by osmosis because the concentration of solutes is greater in the root cells than in the soil water. The primary reason for the movement of water up into the leaves of trees is transpiration. The loss of water due to transpiration at the leaves pulls more water into the leaf cells from the stem and roots. This is due to the cohesiveness of water.

87.   (B)

Sugars (carbohydrates) move upward and downward through the plant via the cells of the phloem's sieve tubes due to an osmotic gradient caused by the differing concentrations of sugars in the cells. The turgor pressure in areas of the plant such as the leaves is then greater than that of the regions of lower pressure, such as actively growing areas. This differential results in a mass flow of materials from the area of high turgor pressure (e.g. the leaves) to an area of lower turgor pressure. In this way, translocation of the photosynthetic products to other tissues occurs.

88.   (E)

The vascular cambium layer is growing and dividing mitotically. It gives rise to the secondary vascular tissues which include secondary xylem (toward the inside of the cambium) and secondary phloem (toward the outside of the cambium).

89.   (A)

Growth rings are made up primarily of secondary xylem. The secondary xylem is produced by the vascular cambium. Secondary growth is the process by which woody plants increase the diameters of their stems, trunks, and roots. This type of growth is different from primary growth which is growth in length.

90.   (A)
In C_3 photosynthesis, the fixation of CO_2 by the plant is linked to the reactions of the Calvin Cycle. The enzyme that catalyzes this reaction is RuBP carboxylase (Ribulose Bisphosphate Carboxylase Oxygenase or RuBisCO) which is found in the chloroplast.

91.   (B)
In C_4 photosynthesis, CO_2 reacts with phosphoenolpyruvate (PEP) and a catalyzing enzyme, forming oxaloacetic acid in the mesophyll cells. Oxaloacetic acid is transported to bundle sheath cells where it is broken down to release CO_2. CO_2 reacts with RuBP through the catalytic properties of RuBisCO. It is in the bundle sheath cells that CO_2 enters into the Calvin Cycle. The speed at which the PEP enzyme works keeps the level of CO_2 lower in the leaf than in the outside environment. Then, when the stomata open, carbon dioxide enters the leaf more rapidly to equalize the concentration gradient of CO_2. Thus, the stomata do not have to be open a long time in the dry climate; this prevents the loss of water.

92.   (B)
Carbohydrates are stored for future energy use in the form of glycogen in vertebrates. Lipids in the form of fats and oils function in energy storage in both plants and animals. In fact, fats contain more chemical energy than carbohydrates. Both lipids and carbohydrates are the major reservoirs for energy storage although all organic molecules (including proteins) release energy when they are oxidized.

93.   (C)
Steroids and fats, both of which are lipids, contain carbon, hydrogen, and oxygen only. Carbohydrates also contain only those elements. Nucleic acids contain, in addition to the aforementioned elements, phosphorus and nitrogen. Only proteins that contain the amino acids methionine or cysteine contain sulfur.

94.   (C)
The rate of most reactions that are catalyzed by enzymes drops off quickly at around 40°C because the enzyme becomes denatured. The enzymes begin to lose their three-dimensional structure because of the breakdown of hydrogen bonds and other weak bonds by heat.

95. (E)

Enzymes are specific for certain substrate molecules because the shape of the enzyme's active site allows only certain molecules to fit.

96. (D)

Even though the compound added is not the usual substrate for the enzyme, the results for experiment 2 indicate that there is, nonetheless, some activity. If more of this compound were to be added, slightly more color change would occur, indicating greater enzyme activity. With an increase in temperature, the change would be even more pronounced since enzyme activity increases with increased temperature, up until the point of denaturation (usually greater than 40° C). Thus, activity would increase, but it still would not be as pronounced as the activity found with the usual substrate.

97. (C)

"Homozygous" describes a situation in which the alleles for a particular trait are identical. Organism "J" has a BbSs genotype. Organism "O" has a bbSs genotype. The genes assort independently producing the following gametes:

		BS	Bs	bS	bs	- organism "J"
	bS	BbSS	BbSs	(bbSS)	bbSs	
organism	bs	BbSs	Bbss	bbSs	(bbss)	
"O"	bS	BbSS	BbSs	(bbSS)	bbSs	
	bs	BbSs	Bbss	bbSs	(bbss)	

98. **(C)**

"Heterozygous" describes a situation in which the genes for a trait are different.

		BS	Bs	bS	bs	- organism "J"
	bS	BbSS	(BbSs)	bbSS	bbSs	
organism	bs	(BbSs)	Bbss	bbSs	bbss	
"O"	bS	BbSS	(BbSs)	bbSS	bbSs	
	bs	(BbSs)	Bbss	bbSs	bbss	

99. **(B)**

A homozygous recessive male would have the genotype bbss and the heterozygous female would have the genotype BbSs. The following Punnett square shows the offspring.

		bs	bs	bs	bs	(male)
	Bs	Bbss	Bbss	Bbss	Bbss	
	BS	BbSs	BbSs	BbSs	BbSs	
	bS	bbSs	bbSs	bbSs	bbSs	
(female)	bs	bbss	bbss	bbss	bbss	

The progeny with brown eyes and curly hair would have the genotype Bbss.

100. **(B)**

In this type of survivorship curve there is a high productivity early in life coupled with a high mortality rate. Those that survive early on have a good chance of surviving later in life. Examples of this survivorship are insects, fish, and plants.

101. (C)

Curve #1 is representative of long-lived organisms such as mammals that produce a few well-endowed young over their entire life span. Parental investment is high and there is exponential growth, then steep mortality which may be correlated with population density factors (ex. competition, predation, etc.). Curve #2 shows populations with a fairly constant rate of change in mortality at all ages. Examples of curve #2 organisms are birds and small animals.

102. (D)

The carrying capacity of a population is the maximum density of organisms that an area can handle under a certain environmental situation. A population can reach the carrying capacity after undergoing exponential growth. When the carrying capacity is reached, the curve tends to plateau. Unless the environmental conditions change, the population usually hovers around its carrying capacity. When a population exceeds its carrying capacity, it declines; when the population goes below its carrying capacity, it tends to increase. Sometimes a poulation overshoots its carrying capacity, then crashes because the resources of that area (e.g. food) cannot support a population that size. Usually, the population crashes due to starvation and disease until the population size is well below the carrying capacity.

103. (A)

The I^A, I^B, and i are alleles (or forms of a gene) responsible for human ABO blood groups. Mr. Jones' blood group, AB, has the $I^A I^B$ genotype, and Mrs. Jones' blood group B has the $I^B I^B$ or $I^B i$ genotype. The "i" allele is recessive to I^A and I^B. The only phenotypes possible based on these genotype combinations are $I^A I^B$, $I^B I^B$, $I^A i$, or $I^B i$. Baby 1 with blood group A could have the Jones' as parents since the genotype combination making blood group A could be $I^A i$.

104. (B)

If Mr. Jones' blood group was O his genotype would be "ii" which is recessive to I^A and I^B. For baby 2 to have blood group O, both parents must pass on an "i" allele to the Baby. The blood group B of Mrs. Jones can be represented as the $I^B I^B$ genotype or $I^B i$ genotype. Thus, if Mrs. Jones was $I^B i$, both parents could pass the "i" allele to baby 2.

105.   (A)
Baby 1 with blood group A could belong to Mr. and Mrs. Jones. The blood group genotype for Mr. Jones is $I^A I^B$ and the blood group genotype for Mrs. Jones is either $I^A I^A$ or $I^A i$. The "i" allele is recessive and would not show up in the blood type unless both parents had that recessive allele, and Mr. Jones does not. Therefore, Baby 2 (Option B) could not belong to the Jones.

106.   (C)
The solid line is the lynx population. It follows the oscillations of the hare population. As the density of the hare population increases, the density of the lynx population increases correspondingly but slightly later in time.

107.   (A)
Zero population growth occurs where births and deaths are in balance. The curve becomes more flattened (growth slows down and finally stabilizes). The decline or rise in the curve after the zero population growth has been reached may be due to environmental factors such as food (prey) supply.

108.   (D)
Predators are more likely to take the old (Option A), and those in poor physical condition (Option B). Predators usually are not dependent on only one species (Option C) but may adjust their hunting pattern to prey upon the most commonly available species. However, the most limiting factor to the prey (and predator) population is food resources.

109.   (D)
As the urine passes through the ascending loop of Henle (#4), the chloride ions are actively pumped out into the tissue surrounding the loop. The sodium ions follow by diffusion. The ascending loop is impermeable to water which then passes to the collecting duct. The chloride ions (and Na$^+$ ions) then flow to the descending loop (#2) where they are pumped out again. This constant circulation means that the ascending and descending loop of Henle is always surrounded by a salt solution.

110. (B)
The walls of the proximal tubule (#2) are freely permeable to water which is transported out (by osmosis) along with sodium and chloride ions. As the fluid descends the loop of Henle (#3), it becomes concentrated (hypertonic) as water moves out by osmosis into the surrounding fluid which has a higher salt concentration than in the tubule. When the fluid ascends the loop of Henle (#4), sodium and chloride ions move out but water does not because the loop is impermeable to water. When the fluid reaches the distal tubule (#5) it is hypotonic, but the walls of the tubule are again permeable to water as are the walls of the collecting duct (#6). Water then flows out as a result of the zone of high salt concentration into the surrounding tissue. The urine left in the collecting duct is hypertonic.

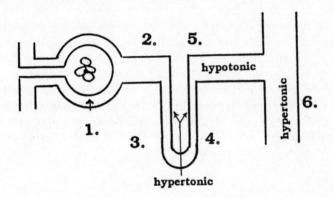

111. (A)
Filtration involves the movement of blood plasma from the glomerulus into the Bowman's capsule (#1). The glomerulus is the convoluted group of thin-walled capillaries surrounded by the Bowman's capsule. Blood plasma and solutes (except protein or blood cells) filter through the walls of the glomerulus into the Bowman's capsule tubule.

112. (D)
That the genes are linked is indicated by the large numbers of progeny with the combination of ebony body and normal wing AND the combination of normal body and curly wings.

113. (B)

Since the genes are linked, the only phenotypes which can be obtained without crossing over are normal body, curly wing and ebony body, normal wing. Crossing over in the female allows for the production of normal body, normal wing and ebony body, curly wing progeny. In this way the genotypes of the female gametes is Ec, eC, EC, ec. In the male, the gametes will always have the genotype ec regardless of whether or not crossing over occurs. Thus crossing over need not occur between the genes for body color and wing type in the male.

114. (B)

In primary succession, the initial substrate is bare rock. The first organisms to grow on bare rock are known as pioneer organisms. Such organisms include lichens and mosses. As time progresses, weathering of the rock and the death of organisms builds up the soil. In secondary succession, soil is already present for the plants to grow on. Abandoned farms and burned areas are sites where secondary succession often occurs. The invading plants are often weeds (herbaceous). In each case of succession, the herbaceous plants give way to larger woody plants.

115. (E)

The most stable stages are often termed climax communities. These communities are still dynamic (changing) rather than static, but in comparison to other communities, the climax community is most stable.

116. (E)

The later stages of succession have communities with greater diversity, not only of plants, but of animals. In the earliest stage, only a few weedy species are present. However, in a climax forest, there are many tree species, shrubs, and herbaceous species.

117. (A)

Communities in succession usually produce more organic matter than they use. In climax communities, an equilibrium is reached between net production and utilization.

118.   (C)
Since the virus inserted only radioactive phosphorus (with its DNA) into the host cell, then DNA (not the protein coat) was shown to be the genetic material from which the entire bacteriophage could be replicated. This experiment was done by Hershey and Chase to determine which substance, DNA or protein, was the genetic material.

119.   (D)
The *E. coli* bacteriophages reproduce by injecting the DNA core into a bacterial host cell leaving the protein coat outside the host cell. Once inside the host cell, the bacteriophages (with their protein coats) are synthesized by the DNA in the host cell; thus, demonstrating that DNA is the genetic material.

120.   (B)
Since only DNA invaded the host cell (as shown by the presence of radioactive phosphorus in the host cell), and since the bacteriophages replicated themselves completely within the host cell, DNA is the material that transmits genetic information.

SECTION II

ESSAY I

Biogeochemical cycles are the movement of inorganic or abiotic matter through an ecosystem, such as nitrogen, phosphorus, and carbon cycles. Broken down, "geo" refers to the environment — the atmosphere, land, and waters of the earth. "Bio" refers to living organisms, both producers and consumers. An important component of this cycle is the detritus feeders, which consume dead organic matter and degrade it to inorganic material, which then is released to cycle through the land and water.

The earth's major supply of nitrogen comes from the atmosphere, where it exists as a gas. However, the movement of nitrogen in the cycle occurs as inorganic matter. Nitrogen is used in physiological systems to form amino acids and other nitrogen-containing compounds that are needed for growth and maintenance.

Nitrogen fixation by bacteria provides much of the usable nitrogen for organisms. In this process, free nitrogen is reduced to organic compounds, such as ammonia (NH_3) or ammonium ions (NH_4+). An example of nitrogen-fixers is the bacteria found at the roots of legumes. The elemental nitrogen they convert is used by the legumes, and any excess is released to the soil.

Detritus feeders decompose the complex nitrogen compounds found in dead organic tissues into simpler amino acid compounds, which they incorporate into their body, releasing any excess into the soil as ammonium compounds. The remainder of the compounds is oxidized by nitrite bacteria to nitrites (NO_2), which is toxic to plants. The nitrites are then oxidized to nitrates by nitrate bacteria, which is the form of nitrogen most commonly utilized by plants. They absorb the nitrogen through the roots and assimilate it into organic nitrogen compounds. This process is called nitrification.

After nitrogen compounds are leached from the soil by plants, animals absorb the nitrogen through digestion of a plant or herbivore. In the process of denitrification, the nitrogen containing compounds are returned to the solid through detritus feeders, which decompose the tissues and dissolve the nitrogen in soil water, which perpetuates the cycle.

Nitrogen can also return to the atmosphere as a gas because denitrifying bacteria can convert ammonia, nitrites, and nitrates to gaseous nitrogen.

Carbon dioxide is the source of inorganic molecules from which organic carbon is derived and is found most frequently in the atmosphere as dissolved carbon dioxide. Plants incorporate extracted carbon which is broken down into respiration. For each turn of the carbon cycle, carbon is released to the atmosphere as CO_2.

Carbon is the building block of life, as it can form covalent bonds with hydrogen, oxygen, nitrogen, or other carbons. It can bond to hydrogen through single, double, or triple bonds, forming hydrocarbons, which produce a great variety of organic molecules forming the basis of compounds essential to living systems. It is found in sugars, carbohydrates, nucleic acids, and lipids, to name a few.

Animal acquire carbon by eating plants and other animals, forming complex molecules from the simple ones absorbed in digestion. Carbon can be excreted as part of animal wastes. Other carbon remains in plant and animal bodies, so upon death, decomposers release the carbon through respiration. If the tissue remains are not completely decomposed, they may be converted into inorganic carbon substances such as coal, or oil, which can be burned as a fuel, releasing carbon dioxide into the atmosphere.

The major supply of phosphorus comes from sedimentary rocks, when they are dissolved by rain. Phosphorus is also released from dead organic matter and animal excretions by detritus feeders, re-entering the soil as phosphates, which the plants absorb and use for growth and maintenance. Animals eat plants and aquatic organisms eat algae, and it becomes incorporated in animal shells, skeletons, and tissues. Excretions of the organisms return phosphates in solution to water. Detritus feeders in the water consume dead organic matter, releasing phosphates in solution. Phosphates in the soil are mixed with water where some phosphates become part of the sedimentary rock at the bottom of the body of water, renewing the phosphorus cycle. Too much phosphate in the water causes excessive algae growth, death of aquatic life as the oxygen is diminished, and eutrophication (aging) of the water body. One important use of phosphorus is that it forms part of the energy molecule adenosine tripsophate (ATP), the energy source for many living organisms.

Humans can cause havoc in these cycles. For example, farmers can leach all of the nitrogen from the soil by continuously growing the same type of crop on their land and not rotating the product with legumes. This leaves the soil nitrogen-starved and makes it very difficult for plants to grow. To combat this, many fertilizers contain nitrogen and phosphorus. This fertilizer often dissolves in rainwater and runs off to a water body, promoting the growth of algae. This kills much of the aquatic life, leaving them limited oxygen and can lead to early eutrophication.

Any machine that burns a carbon-based fuel emits carbon dioxide into the air. As the levels increase beyond the ocean's ability to absorb it for the carbon cycle, CO_2 remains in the air and prevents other gases from escaping to the atmosphere. This increases the basal temperature of the Earth and is known as the greenhouse effect. In addition, pollution causes temperature inversion, which traps cooler air under the hotter polluted air, and toxic chemicals are not released to the atmosphere.

ESSAY II

STRUCTURE OF DNA

A DNA molecule is shaped like a twisted ladder (double helix). The "sides" of the ladder are made up of sugars and phosphates while the "rungs" of the ladder are made up of the nucleotides — adenine, thymine, cytosine, and guanine.

The nucleotides form complementary base pairs. Adenine pairs with thymine and guanine pairs with cytosine. The nucleotides adenine and guanine are purines. The nucleotides thymine and cytosine are pyrimidines. The complementary base pairs are connected by hydrogen bonds. DNA is located in the nucleus. Genetic information in the DNA is in the sequence of bases. The genetic code is a four-lettered alphabet: adenine (A), thymine (T), guanine (G), and cytosine (C).

FUNCTION OF DNA

DNA contains codes for proteins that are synthesized in the cytoplasm of the cell. Condensed DNA and associated histone proteins are collectively referred to as chromosomes.

REPLICATION OF DNA

Replication or duplication of DNA takes place in the nucleus of the cell. Replication takes place before the cell divides mitotically. The strands' hydrogen bonds break and each half of the single strand acts as a template for a new complementary strand. These new strands are built from free nucleotides in the nucleus. These free nucleotides bind to the template bases according to the rules for complementary base pairing.

TRANSCRIPTION

Transcription is the creation of RNA (specifically, messenger RNA) by DNA. Certain sections along the chromosome (condensed DNA) unwind as hydrogen bonds are broken. DNA directs the formation of

RNA along these exposed sections. Free nucleotides attach to the open bases, except that uracil is substituted for thymine. RNA (called messenger RNA) forms a complementary strand, using the DNA as a template (pattern). The copy is not identical to DNA, but is complementary. The ribosome attaches to the mRNA and moves along the length of its strand, matching the codons of the mRNA to the anticodons of tRNA. The tRNA with the correct complementary sequence then goes to the correct location on the mRNA The tRNA attaches to the mRNA codon for a short time. It then leaves, but the amino acid stays behind. It becomes attached to a growing chain of amino acids that were brought to the mRNA in the ribosome by other tRNA molecules. After the mRNA has directed a certain number of amino acid sequences, it breaks apart. The chains of amino acids form proteins. Depending on the mRNA message, these proteins can be structural proteins or enzymes.

ESSAY III

Development in flowering plants begins with a seed, which consists of the embryo, stored food, and the seed coat. The seed is a period of dormancy before the plant germinates, or begins growth. Seeds are important in the dispersal of plants as they can travel long distances and do not have to germinate immediately upon soil contact.

The seed coat protects the embryo against drying out and injury. It permits it to stay in a reduced activity period until conditions are favorable for its growth. Some scientists believe there may be some hormones responsible for a role as germination inhibitors.

The endosperm contains the stored food, which is utilized in the early stages of germination. Plants that do not have endosperms, such as peas, have cotyledons. The stored food is believed to be a starch. To test this postulate, soak the seed overnight in water, then cut it in half with a sharp blade. The Lugol's Iodine test could be used. If it is a starchy material, the iodine will turn black.

The seed is a mature ovule. Its parts include some of the parent plant's genetic material, as well as some of the previous generation. The endosperm actually has 3n cells, two female and one male. The embryo, or the part that will grow, is diploid, one n from the mother and one n from the father. In the early germination stages, the seed coat splits open and falls to the ground. A root then forms to anchor the seed in the ground. It is believed that gibberellins and auxins plant hormones stimulate the growth or combat the germination inhibitors. After the plant has utilized the stored food, the root uses capillary action to obtain water and nutrients such as phosphorus, magnesium and nitrogen from the soil. When the plant breaks the soil surface, then chlorophyll can be used to make energy from the sun.

To determine if seeds contain a substance similar to antibiotics, soak the seeds in water overnight. Cut them with a sharp razor blade. Place the cut seeds onto a nutrient rich agar that has been spread with bacteria. Incubate the agar plates. For controls, use an agar plate with only bacteria and one with seeds that have not been cut. If there is an antimicrobial agent, there should be clear plaques around the seeds that were soaked and cut.

ESSAY IV

1.   FIRST LAW OF THERMODYNAMICS

The First Law of Thermodynamics states that energy cannot be created or destroyed, but instead is transformed from one form to another. The potential energy of the reactants equals the sum of the potential energy of the products and the energy that is released as a result of the reaction.

Energy that is released may not be in a form that is usable by organisms. For example, heat, a form of energy, may dissipate into the surrounding environment (to nonliving matter); therefore, it would not be available for use by biological systems. Also, living cells draw primarily on chemical energy derived from complex organic molecules rather than heat energy to do their work. Because not all energy is usable by a living system, there must be a source of usable energy outside the organism that would be available for it to use. An example is the use of the sun's energy by plants to form complex organic molecules, such as glucose, that contain potential energy for later use by the cell.

2.   THE SECOND LAW OF THERMODYNAMICS

The Second Law of Thermodynamics states that with each energy transformation that takes place, the amount of usable (free) energy at the end of a reaction is less than the amount of usable (free) energy present in the beginning. With each energy transformation, entropy (or randomness) of a system increases. An example of this law in biological systems is the tendency toward disintegration or death.

A source of energy for the system is needed (e.g., the sun). The addition of energy to a system allows the system to achieve a state of decreased entropy, or increased organization. Examples of organization in biological systems are the formation of tissues from cells, the formation of organs from tissues, the formation of systems from organs, resulting in the formation of entire organisms from systems.

3.   ANABOLISM

Anabolism is the sum of those chemical reactions in biological systems in which energy is used to synthesize or build structures within the cell or within the organism. Anabolic reactions are referred to as being endergonic reactions because they require an input of energy to start the reaction. The input of energy increases the kinetic energy of the molecules in cells, increasing the likelihood that they will collide with each other with enough force to overcome the repulsion they have for each other and to break the chemical bonds that keep the molecules in their present state. An example of an anabolic process is photosynthesis, which requires energy input from the sun to create complex organic molecules such as glucose and compounds that are synthesized from the glucose. Glucose is an example of a molecule that stores energy to be used later to do work or for growth and maintenance of the organism. ATP is the most widely used, but not the only energy currency unit.

4.   CATABOLISM

Catabolism describes those chemical reactions in which larger molecules are broken down into smaller molecules with an accompanying release of energy. This type of reaction, in which heat is released, is called an exergonic reaction. Respiration is an example of a catabolic reaction. In this reaction, energy-rich glucose molecules are broken down. The breaking-down phase allows the cell to synthesize new ATP to be used both for future anabolic reactions and for growth and maintenance of the organism.

5.   PHOSPHORYLATION OF ADP TO ATP

In phosphorylation reactions, ATP is synthesized from ADP and inorganic phosphate. Thus, a molecule of ATP, which contains much chemical potential energy, is formed. The energy needed for synthesis of ATP to occur is ultimately derived from the sun in the photosynthetic process. A non-photosynthetic organism can make ATP by using the energy from other energy-rich molecules, but not energy from the sun. Phosphorylation reactions are endergonic because energy must be added to the system when ATP is synthesized from ADP and phosphate.

412

6.   COUPLED REACTIONS

The occurrence of an endergonic reaction (e.g., photosynthesis, phosphorylation) that supplies energy to an exergonic reaction (e.g., respiration, degradation of ATP to ADP with the accompanying release of energy) is called a coupled reaction. The energy released from the exergonic process is used to "drive" the endergonic process.

Coupled reactions allow living systems to produce more complex molecules from less complex ones, despite the seeming violation of the Second Law of Thermodynamics. The Second Law is not violated during synthesis, because although free energy is needed to synthesize molecules, this anabolic process is coupled to a catabolic reaction (e.g., the hydrolysis of ATP).

THE ADVANCED PLACEMENT EXAMINATION IN

BIOLOGY

TEST VI

THE ADVANCED PLACEMENT EXAMINATION IN

BIOLOGY

ANSWER SHEET

1. Ⓐ Ⓑ Ⓒ Ⓓ Ⓔ
2. Ⓐ Ⓑ Ⓒ Ⓓ Ⓔ
3. Ⓐ Ⓑ Ⓒ Ⓓ Ⓔ
4. Ⓐ Ⓑ Ⓒ Ⓓ Ⓔ
5. Ⓐ Ⓑ Ⓒ Ⓓ Ⓔ
6. Ⓐ Ⓑ Ⓒ Ⓓ Ⓔ
7. Ⓐ Ⓑ Ⓒ Ⓓ Ⓔ
8. Ⓐ Ⓑ Ⓒ Ⓓ Ⓔ
9. Ⓐ Ⓑ Ⓒ Ⓓ Ⓔ
10. Ⓐ Ⓑ Ⓒ Ⓓ Ⓔ
11. Ⓐ Ⓑ Ⓒ Ⓓ Ⓔ
12. Ⓐ Ⓑ Ⓒ Ⓓ Ⓔ
13. Ⓐ Ⓑ Ⓒ Ⓓ Ⓔ
14. Ⓐ Ⓑ Ⓒ Ⓓ Ⓔ
15. Ⓐ Ⓑ Ⓒ Ⓓ Ⓔ
16. Ⓐ Ⓑ Ⓒ Ⓓ Ⓔ
17. Ⓐ Ⓑ Ⓒ Ⓓ Ⓔ
18. Ⓐ Ⓑ Ⓒ Ⓓ Ⓔ
19. Ⓐ Ⓑ Ⓒ Ⓓ Ⓔ
20. Ⓐ Ⓑ Ⓒ Ⓓ Ⓔ

21. Ⓐ Ⓑ Ⓒ Ⓓ Ⓔ
22. Ⓐ Ⓑ Ⓒ Ⓓ Ⓔ
23. Ⓐ Ⓑ Ⓒ Ⓓ Ⓔ
24. Ⓐ Ⓑ Ⓒ Ⓓ Ⓔ
25. Ⓐ Ⓑ Ⓒ Ⓓ Ⓔ
26. Ⓐ Ⓑ Ⓒ Ⓓ Ⓔ
27. Ⓐ Ⓑ Ⓒ Ⓓ Ⓔ
28. Ⓐ Ⓑ Ⓒ Ⓓ Ⓔ
29. Ⓐ Ⓑ Ⓒ Ⓓ Ⓔ
30. Ⓐ Ⓑ Ⓒ Ⓓ Ⓔ
31. Ⓐ Ⓑ Ⓒ Ⓓ Ⓔ
32. Ⓐ Ⓑ Ⓒ Ⓓ Ⓔ
33. Ⓐ Ⓑ Ⓒ Ⓓ Ⓔ
34. Ⓐ Ⓑ Ⓒ Ⓓ Ⓔ
35. Ⓐ Ⓑ Ⓒ Ⓓ Ⓔ
36. Ⓐ Ⓑ Ⓒ Ⓓ Ⓔ
37. Ⓐ Ⓑ Ⓒ Ⓓ Ⓔ
38. Ⓐ Ⓑ Ⓒ Ⓓ Ⓔ
39. Ⓐ Ⓑ Ⓒ Ⓓ Ⓔ
40. Ⓐ Ⓑ Ⓒ Ⓓ Ⓔ

41. Ⓐ Ⓑ Ⓒ Ⓓ Ⓔ
42. Ⓐ Ⓑ Ⓒ Ⓓ Ⓔ
43. Ⓐ Ⓑ Ⓒ Ⓓ Ⓔ
44. Ⓐ Ⓑ Ⓒ Ⓓ Ⓔ
45. Ⓐ Ⓑ Ⓒ Ⓓ Ⓔ
46. Ⓐ Ⓑ Ⓒ Ⓓ Ⓔ
47. Ⓐ Ⓑ Ⓒ Ⓓ Ⓔ
48. Ⓐ Ⓑ Ⓒ Ⓓ Ⓔ
49. Ⓐ Ⓑ Ⓒ Ⓓ Ⓔ
50. Ⓐ Ⓑ Ⓒ Ⓓ Ⓔ
51. Ⓐ Ⓑ Ⓒ Ⓓ Ⓔ
52. Ⓐ Ⓑ Ⓒ Ⓓ Ⓔ
53. Ⓐ Ⓑ Ⓒ Ⓓ Ⓔ
54. Ⓐ Ⓑ Ⓒ Ⓓ Ⓔ
55. Ⓐ Ⓑ Ⓒ Ⓓ Ⓔ
56. Ⓐ Ⓑ Ⓒ Ⓓ Ⓔ
57. Ⓐ Ⓑ Ⓒ Ⓓ Ⓔ
58. Ⓐ Ⓑ Ⓒ Ⓓ Ⓔ
59. Ⓐ Ⓑ Ⓒ Ⓓ Ⓔ
60. Ⓐ Ⓑ Ⓒ Ⓓ Ⓔ

61. Ⓐ Ⓑ Ⓒ Ⓓ Ⓔ	81. Ⓐ Ⓑ Ⓒ Ⓓ Ⓔ	101. Ⓐ Ⓑ Ⓒ Ⓓ Ⓔ
62. Ⓐ Ⓑ Ⓒ Ⓓ Ⓔ	82. Ⓐ Ⓑ Ⓒ Ⓓ Ⓔ	102. Ⓐ Ⓑ Ⓒ Ⓓ Ⓔ
63. Ⓐ Ⓑ Ⓒ Ⓓ Ⓔ	83. Ⓐ Ⓑ Ⓒ Ⓓ Ⓔ	103. Ⓐ Ⓑ Ⓒ Ⓓ Ⓔ
64. Ⓐ Ⓑ Ⓒ Ⓓ Ⓔ	84. Ⓐ Ⓑ Ⓒ Ⓓ Ⓔ	104. Ⓐ Ⓑ Ⓒ Ⓓ Ⓔ
65. Ⓐ Ⓑ Ⓒ Ⓓ Ⓔ	85. Ⓐ Ⓑ Ⓒ Ⓓ Ⓔ	105. Ⓐ Ⓑ Ⓒ Ⓓ Ⓔ
66. Ⓐ Ⓑ Ⓒ Ⓓ Ⓔ	86. Ⓐ Ⓑ Ⓒ Ⓓ Ⓔ	106. Ⓐ Ⓑ Ⓒ Ⓓ Ⓔ
67. Ⓐ Ⓑ Ⓒ Ⓓ Ⓔ	87. Ⓐ Ⓑ Ⓒ Ⓓ Ⓔ	107. Ⓐ Ⓑ Ⓒ Ⓓ Ⓔ
68. Ⓐ Ⓑ Ⓒ Ⓓ Ⓔ	88. Ⓐ Ⓑ Ⓒ Ⓓ Ⓔ	108. Ⓐ Ⓑ Ⓒ Ⓓ Ⓔ
69. Ⓐ Ⓑ Ⓒ Ⓓ Ⓔ	89. Ⓐ Ⓑ Ⓒ Ⓓ Ⓔ	109. Ⓐ Ⓑ Ⓒ Ⓓ Ⓔ
70. Ⓐ Ⓑ Ⓒ Ⓓ Ⓔ	90. Ⓐ Ⓑ Ⓒ Ⓓ Ⓔ	110. Ⓐ Ⓑ Ⓒ Ⓓ Ⓔ
71. Ⓐ Ⓑ Ⓒ Ⓓ Ⓔ	91. Ⓐ Ⓑ Ⓒ Ⓓ Ⓔ	111. Ⓐ Ⓑ Ⓒ Ⓓ Ⓔ
72. Ⓐ Ⓑ Ⓒ Ⓓ Ⓔ	92. Ⓐ Ⓑ Ⓒ Ⓓ Ⓔ	112. Ⓐ Ⓑ Ⓒ Ⓓ Ⓔ
73. Ⓐ Ⓑ Ⓒ Ⓓ Ⓔ	93. Ⓐ Ⓑ Ⓒ Ⓓ Ⓔ	113. Ⓐ Ⓑ Ⓒ Ⓓ Ⓔ
74. Ⓐ Ⓑ Ⓒ Ⓓ Ⓔ	94. Ⓐ Ⓑ Ⓒ Ⓓ Ⓔ	114. Ⓐ Ⓑ Ⓒ Ⓓ Ⓔ
75. Ⓐ Ⓑ Ⓒ Ⓓ Ⓔ	95. Ⓐ Ⓑ Ⓒ Ⓓ Ⓔ	115. Ⓐ Ⓑ Ⓒ Ⓓ Ⓔ
76. Ⓐ Ⓑ Ⓒ Ⓓ Ⓔ	96. Ⓐ Ⓑ Ⓒ Ⓓ Ⓔ	116. Ⓐ Ⓑ Ⓒ Ⓓ Ⓔ
77. Ⓐ Ⓑ Ⓒ Ⓓ Ⓔ	97. Ⓐ Ⓑ Ⓒ Ⓓ Ⓔ	117. Ⓐ Ⓑ Ⓒ Ⓓ Ⓔ
78. Ⓐ Ⓑ Ⓒ Ⓓ Ⓔ	98. Ⓐ Ⓑ Ⓒ Ⓓ Ⓔ	118. Ⓐ Ⓑ Ⓒ Ⓓ Ⓔ
79. Ⓐ Ⓑ Ⓒ Ⓓ Ⓔ	99. Ⓐ Ⓑ Ⓒ Ⓓ Ⓔ	119. Ⓐ Ⓑ Ⓒ Ⓓ Ⓔ
80. Ⓐ Ⓑ Ⓒ Ⓓ Ⓔ	100. Ⓐ Ⓑ Ⓒ Ⓓ Ⓔ	120. Ⓐ Ⓑ Ⓒ Ⓓ Ⓔ

ADVANCED PLACEMENT
BIOLOGY EXAM VI

SECTION I

120 Questions
90 minutes

DIRECTIONS: For each question, there are five possible choices. Select the best choice for each question. Blacken the correct space on the answer sheet.

1. A strand of DNA is a _____ and can generate a new _____ strand of DNA.

    (A) copy, identical

    (B) parent, identical

    (C) duplicate, duplicate

    (D) template, complementary

    (E) base, double

2. The DNA strand has a backbone of alternating:

    (A) sugar and phosphate molecules

    (B) complementary base pairs

    (C) hydrogen bonds

    (D) pyrimidines and purines

    (E) nitrogen-containing bonds

Use the following diagram with the lettered trophic levels to answer
questions 3-4:

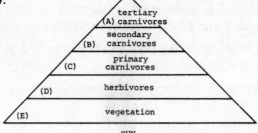

3.   If the pyramid diagram is a biomass pyramid, where would
     you find the greatest biomass of organisms?

     (A) At the top

     (B) All would be equal

     (C) At the herbivore level

     (D) Can't tell from the given information

     (E) At the bottom

4.   If the pyramid diagram indicated energy levels, what would
     be observed if the pyramid ended at trophic level "C" instead
     of "A"?

     (A) Less usable energy would be lost as you go up the food
         chain.

     (B) More usable energy would be lost as you go up the food
         chain.

     (C) No change would occur.

     (D) Usable energy would be generated as you go up the food
         chain.

     (E) It would be better to feed plants to humans rather than
         giving them to livestock.

5.   Which of the following statements about the influence of a
     predator on a prey population is not correct?

     (A) Prey organisms in poor physical condition are taken by

predators in higher numbers than healthy ones.

(B) Predation is a principal cause of determining species diversity in a particular community.

(C) Predators limit the carrying capacity of a population within a community.

(D) Food limits the carrying capacity of a population within a community.

(E) All of the above are true.

6. The description of a niche includes:

(A) predator, diet, reproductive process, use of habitat

(B) primary consumer, diet, reproductive strategy, use of habitat

(C) primary consumer, reproductive strategy, amount of rainfall

(D) predator, diet, carrying capacity

(E) it depends on the species involved.

7. Carrying capacity:

(A) is the amount of soil by volume found in a designated area.

(B) includes the maximum number of organisms the environment can support.

(C) is used to describe both aquatic and terrestrial ecosystems

(D) is limited by the resources (food, water, space) available to a species

(E) both (A) and (D)

8. Entropy:

    (A) increases in a community that is undergoing primary plant succession

    (B) increases in a community by the action of decaying organisms

    (C) requires an excess of energy to do work

    (D) both (A) and (B)

    (E) both (B) and (C)

9. Climax communities

    (A) are more diverse than pioneer communities

    (B) are less stable than pioneer communities

    (C) have greater entropy than pioneer communities

    (D) have a larger number but fewer species of plants than pioneer communities

    (E) All of the above

10. Secondary plant succession can occur after:

    (A) farmland is cultivated

    (B) forests have been burned

    (C) bare rock is exposed from a retreating glacier

    (D) sand dunes are created

    (E) None of the above

11. RNA polymerase dominates the process of:

    (A) transcription   (D) conjugation

    (B) translation   (E) transference

    (C) transduction

12. _____ carry amino acids to the ribosomes.

   (A) Messenger RNAs

   (B) Ribosomal RNAs

   (C) Transfer RNAs

   (D) DNA and messenger RNAs

   (E) RNA polymerases

13. Viruses have:

   (A) the ability to replicate their genetic material

   (B) the ability to make their own energy

   (C) their own metabolic machinery

   (D) their own enzymes

   (E) both (A) and (C)

14. _____ is/are directly associated with anticodon.

   (A) Messenger RNA

   (B) Transfer RNA

   (C) Ribosomal RNA

   (D) DNA and messenger RNA

   (E) RNA polymerase

15. Which condition is necessary for diffusion to occur?

   (A) A living cell

   (B) A permeable membrane

   (C) A differentially permeable membrane

   (D) A difference in concentration

   (E) A source of energy

16. The cytoplasm of the cell, surrounded by a semipermeable membrane, is a highly concentrated solution of various molecules. What would happen if an animal were placed in pure water?

    (A) Nothing would happen

    (B) The cell would swell and possibly burst

    (C) The cell would shrivel up

    (D) Water would diffuse into the cell

    (E) Both (B) and (D)

17. The cell membrane has the following structure(s):

    (A) one outer layer of phospholipids and one layer of proteins inside the phospholipid layer

    (B) two layers of phospholipids arranged in bilayers

    (C) protein extending through the bilayers

    (D) hydrophilic lipids

    (E) both (B) and (C)

18. In eukaryotic cells, cilia and flagella are:

    (A) a group of microtubules anchored to a point of attachment in the cell membrane called a baseplate

    (B) a 9+2 arrangement of microtubules attached to a basal body

    (C) a 9+0 arrangement of microtubules attached to a 9+2 basal body under the cilia

    (D) attached directly to the centrioles

    (E) a chain of proteins not enclosed in the cell membrane

19. Two pea plants (TT and Tt) are observed to be tall. These plants will always have the same:

(A) alleles

(D) Both (B) and (C)

(B) genotype

(E) All of the above

(C) phenotype

Question 20 refers to the following processes:

(1) $C_6 H_{12} O_6 + C_6 H_{12} O_6 \rightarrow C_{12} H_{22} O_{11} + H_2 O$

(2) monosaccharide + monosaccharide→disaccharide + water

(3) glycerol + 3 fatty acids →triglyceride + HOH

(4) Dipeptide + $H_2 O$ →amino acid and amino acid

(5) $C_{12} H_{22} O_{11} + H_2 O \rightarrow C_6 H_{12} O_6 + C_6 H_{12} O_6$

20. Example(s) of condensation reaction

(A) 1 only

(D) 4 and 5

(B) 2 only

(E) 3 and 4

(C) 1, 2 and 3

21. Proteins are formed by combining:

(A) lipids

(B) monosaccharide and disaccharides

(C) nucleic acids

(D) amino acids

(E) glycerols

22. An enzyme is a large organic molecule with a surface geometry that is composed of:

424

(A) amino acids   (D) polysaccharides

(B) monosaccharides   (E) triglycerides

(C) glycerol and fatty acids

23. Enzymes:

(A) are inactivated at high temperatures

(B) are highly sensitive to pH charges

(C) are highly specific to the reactions they catalyze

(D) work best at optimum temperatures

(E) all of the above

24. A coenzyme is

(A) a substance that makes an enzyme less effective

(B) a substance that directly reacts with the substrate

(C) a substance that some enzymes need before they can function

(D) a substance that is a catalyst

(E) All of the above

25. Thylakoids are:

(A) outfoldings of the mitochondrial membrane

(B) sites that trap light for photosynthesis

(C) sites of protein synthesis

(D) sites of transformation of ATP in the mitochondria

(E) small structures of the chlorophyll molecules

26. Oxygen is released by the:

(A) light reaction of photosynthesis

(B) dark reaction of photosynthesis

(C) formation of ATP from ADP

(D) "excited" electrons in the chlorophyll molecule

(E) splitting of the 6-carbon sugar

27. A chemical has been added to the mitochondria that causes H+ ions to be transported through the membrane. What would be the result(s)?

(A) More ATP would be formed

(B) Less ATP would be formed

(C) The electrical gradient is destroyed

(D) Both (A) and (C)

(E) Both (B) and (C)

28. The term "coupled reaction" refers to:

(A) linking endergonic and exergonic processes

(B) linking of two reactions using the same enzyme

(C) linking of ADP and inorganic phosphate group

(D) linking of reactions in a metabolic pathway

(E) the joining of a nucleotide with a five-carbon sugar

29. One set of parents are heterozygous for two gene pairs. They can produce offspring that are different from themselves. This demonstrates:

(A) the law of independent assortment

(B) the law of segregation

(C) the importance of using samples

(D) the genes are linked

(E) that recessive traits can be seen in a heterozygous individual

30. DNA differs from RNA in that:

(A) RNA has three bases while DNA has four bases

(B) RNA has a single strand and DNA has a double strand

(C) RNA has a base thymine instead of the base uracil as in DNA

(D) DNA contains nucleotides while RNA does not

(E) None of the above

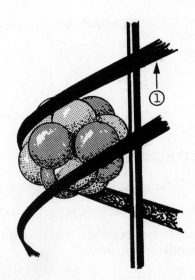

31. Which of the following structures is the one marked 1 likely to be?

(A) DNA   (D) prokaryote

(B) RNA   (E) flagellum

(C) histone protein

32. Bacteriophages are:

   (A) composed of a protein core surrounded by a coat of DNA nucleic acid

   (B) bacteria that attack viruses

   (C) bacteria that feed viruses

   (D) bacteria that "take over" the genetic machinery of viruses

   (E) viruses that attack bacteria

33. A restriction enzyme cleaves the following DNA segment between adjacent guanine and adenine.

   G \A A T T C
   C T T A A\G

   Which of the following sequences indicates that sticky ends have formed?

   (A) TTAA----------AATT

   (B) GAAT---------GAAT

   (C) G--------------G

   (D) TC-------------CT

   (E) None of the above

34. Using the previous example, what complementary base of foreign DNA could be inserted?

   (A) GGCC---------CCGG

   (B) TTCC----------AATT

   (C) TTAA----------AATT

   (D) CCGG---------GGCC

   (E) AATT----------TTAA

428

35. Which of the following DNA mutations could be deadly?

   (A) Codon substitution

   (B) A nucleotide is added or deleted

   (C) Inversion of part of nucleotide for another

   (D) Change of one nucleotide for another

   (E) Both (C) and (D)

36. The F_2 generation of a heterozygous dihybrid cross is produced by:

   (A) two distinctly different gametes

   (B) three distinctly different gametes

   (C) four distinctly different gametes

   (D) eight distinctly different gametes

   (E) None of the above

37. The F_2 generation of a heterozygous dihybrid cross has:

   (A) two distinctly different genotypes

   (B) four distinctly different genotypes

   (C) five distinctly different genotypes

   (D) eight distinctly different genotypes

   (E) nine distinctly different genotypes

38. Genes responsible for hemophilia in man are carried:

   (A) on the X chromosome not the Y chromosome

   (B) on the Y chromosome not the X chromosome

   (C) on any chromosome pair except the X and Y chromosomes

(D) on both X and Y chromosomes

(E) on either the X or Y chromosomes depending on whether the individual is male or female

39. The theory of punctuated equilibrium assumes that

(A) there are periods of time of stability in which little evolutionary change occurs

(B) speciation can occur in a very short period of time and is marked by rapid changes

(C) evolution occurs gradually within lineages

(D) Both (A) and (B)

(E) All of the above

40. The diversification of mammals that followed the extinction of dinosaurs is an example of:

(A) allopatric speciation

(B) sympatric speciation

(C) disruptive selection

(D) adaptive radiation

(E) Both (A) and (C)

41. The basis for the taxonomic and systematic classification of organisms is:

(A) binomial system of nomenclature

(B) grouping by genus

(C) grouping by morphological features

(D) grouping by species

(E) division into plants and animals

42. What traditional criterion(a) is(are) used to differentiate between prokaryotes and protists?

(A) Method of obtaining nutrition

(B) Methods of reproduction and motility

(C) Presence (or absence) of cell wall

(D) Number of cells

(E) Both (A) and (C)

43. All living organisms have a eukaryotic cell structure except:

(A) Monera   (D) Plants

(B) Protista   (E) Animals

(C) Fungi

44. Which of the following is true about the five-kingdom classification system?

(A) All one celled organisms are grouped in one kingdom.

(B) All heterotrophs are grouped in one kingdom.

(C) Organisms are divided into kingdoms based on their evolutionary history.

(D) Eukaryotes are grouped in one kingdom.

(E) All prokaryotes are grouped in a single kingdom.

45. When a biologist studies the fundamental similarities in the bones of the forelimbs of a crocodile , a bird, and a horse, he is studying the _____ of organisms.

(A) taxonomy   (D) speciation

(B) phylogeny   (E) Both (B) and (C)

(C) systematics

46. A major difference between the two major divisions of plants (lower and higher) is whether one group has

(A) protected zygotes

(B) special photosynthetic structures

(C) parasitic characteristics

(D) a protective cuticle

(E) vascular tissue

47. Directly after meiosis occurs, what structure is produced in ferns?

(A) Haploid spores

(B) Diploid spores

(C) Haploid sporophyte

(D) Haploid gametes

(E) Diploid sporophyte

48. A gametophyte is:

(A) a plant that produces gametes

(B) haploid

(C) diploid

(D) produced by the fusion of gametes

(E) Both (A) and (B)

49. Water is not required for fertilization by

(A) seed plants

(B) land plants

(C) bryophytes only

(D) ferns only

(E) angiosperms only

50. In angiosperms, the immature male gametophyte is:

(A) the megaspore

(B) the stamen

(C) the pollen tube

(D) the pistil

(E) the pollen grain

51. Double fertilization occurs only in angiosperms, and is defined as:

(A) fusion of one sperm nucleus with the egg cell, and one sperm nucleus with the polar nuclei in the ovule of the flower

(B) fusion of two sperm nuclei with the egg in the ovary of the flower

(C) fusion of one sperm nucleus with the egg cell and two sperm nuclei in the ovule of the flower.

(D) fusion of one sperm nucleus with the egg and one sperm nucleus with the synergids in the ovule of the flower.

(E) fusion of two sperm nuclei with the egg and one tube nucleus with the polar nuclei in the ovule of the flower.

52. Germination starts when:

(A) the seed coat permits water to enter the tissues of the seed

(B) the sporophyte emerges

(C) the seeds are planted

(D) the apical meristems of the root elongate

(E) None of the above

53. The length of the root (primary growth) is chiefly the result of:

   (A) differentiation of cells

   (B) elongation of cells

   (C) divisions in the apical meristem

   (D) the wearing away of the root cap

   (E) None of the above

54. Which is a characteristic of a hormone?

   (A) Large quantities are needed to produce a desired effect

   (B) They are produced in the tissue that they affect

   (C) Small quantities can produce effects

   (D) They work interdependently with other hormones

   (E) Both (C) and (D)

55. When a stem bends toward the light , it is due to:

   (A) the increased level of auxin on the light side of the shoot tip

   (B) the migration of auxin toward the dark side of the shoot tip

   (C) the migration of auxin toward the light side of the shoot tip

   (D) the elongation of cells on the light of the shoot tip

   (E) Both (B) and (D)

56. When a seedling rights itself, this is a _____ response.

   (A) gravitropic

(B) photoperiodic

(C) circadian

(D) phototropic

(E) None of the above

57. In phyla advanced beyond the level of sponges, embryonic cells aggregate into:

(A) organs

(B) organ systems

(C) gametes and somatic cells

(D) tissues

(E) amino acids

58. A nervous impulse starting at the dendrite will next pass through the:

(A) cell body

(B) axon

(C) nodes of Ranvier

(D) synaptic bouton

(E) None of the above

59. The process by which blood plasma moves out of the glomerulus capillaries and into Bowman's capsule is called:

(A) reabsorption

(B) secretion

(C) countercurrent exchange

(D) filtration

(E) multiplication

60. The subunit of a kidney that purifies blood and maintains a safe balance of solutes and water in humans is called:

   (A) glomerulus

   (B) loop of Henle

   (C) urethra

   (D) Bowman's capsule

   (E) nephron

61. Memory cells produced by B-lymphocytes help the organism to respond more quickly to an infection the second time because they:

   (A) start a cell-mediated response

   (B) have created their own antigens from the first exposure to the infection

   (C) rapidly clone antibodies picked up during the first exposure to the infection

   (D) directly attack the invaders instead of producing antibodies

   (E) are not specific to a particular antigen

62. Which of the following donors would be the best choice for a skin graft?

   (A) Identical twin

   (B) A sister with the same blood

   (C) A non-family donor and the patient will be given a drug to suppress the immune response

   (D) A parent

   (E) (A), (B) or (C) would be an acceptable donor

63. The formation of eggs and sperm are called:

    (A) metamorphosis

    (B) gastrulation

    (C) ovulation

    (D) fertilization

    (E) gametogenesis

64. Which of the following correctly shows the path of blood in the blood vessels?

    (A) arterioles - capillaries - arteries - veins - venules

    (B) arteries - arterioles - capillaries - venules - veins

    (C) capillaries - arterioles - arteries - veins - venules

    (D) venules - capillaries - veins - arteries - venules

    (E) veins - venules - arterioles - capillaries - arteries

65. The process of cleavage produces a:

    (A) zygote

    (B) blastula

    (C) gastrula

    (D) archenteron

    (E) None of the above

66. If the blood pressure of a human is 111/80

    (A) the systolic pressure is 80

    (B) the diastolic pressure is 80

    (C) the pulse rate is 80 beats per minute

    (D) the blood pressure during contraction of the heart is 80

(E)   the right atrium moved 111 milliliters of blood per beat
and the left atrium moved 80 milliliters of blood per beat

67. The heart, bones, and blood develop primarily from the:

(A)   endoderm

(B)   ectoderm

(C)   mesoderm

(D)   morula

(E)   blastula

68. The diversity of animals is chiefly a result of what characteristic?

(A)   The cells are eukaryotic

(B)   Their complex nervous system

(C)   They are multicellular

(D)   Their mode of reproduction

(E)   Their mode of nutrition

69. Animals are classified into phyla chiefly by:

(A)   number of tissue layers

(B)   basic body plan

(C)   presence or absence of true coelom

(D)   pattern of development from embryo

(E)   All of the above

70. In the animal kingdom, bilateral symmetry is related to:

(A) three germ layers

(B) two germ layers

(C) more efficient movement than radial symmetry

(D) Both (A) and (C)

(E) Both (B) and (C)

71. Which is associated with the movement from radial to bilateral symmetry?

(A) Paired organs

(B) Circulatory system

(C) A coelom

(D) Two - ended digestive system

(E) Two germ layers

72. Which of the following is <u>not</u> an assumption of the Hardy-Weinberg rule?

(A) random mating

(B) no selection

(C) no mutation

(D) large population size

(E) migration

DIRECTIONS: The following groups of questions have five lettered choices followed by a list of diagrams, numbered phrases, sentences, or words. For each numbered diagram, phrase, sentence, or word choose the heading which most directly applies. Blacken the correct space on the answer sheet. Each heading may be used once, more than once, or not at all.

Questions 73 - 75 refer to the following list of cell organelles:

    (A)   mitochondrion

    (B)   lysosome

    (C)   Golgi Bodies

    (D)   rough endoplasmic reticulum

    (E)   plasma membrane

73.   The organelle primarily responsible for intracellular digestion.

74.   The organelle responsible for cellular respiration.

75.   The site where proteins are modified, packaged and secreted from the cell.

Questions 76 - 78 refer to the following list of activities that occur in the cytoplasm of the cell.

    (A)   phagocytosis

    (B)   active transport

    (C)   endocytosis

    (D)   exocytosis

    (E)   osmosis

76. The taking in of solid material by white blood cells and unicellular organisms, such as amoebas.

77. The movement of water through a membrane from an area of high concentration of water molecules to one of lower concentration of water molecules.

78. The use of energy to move a substance through a membrane against its concentration gradient.

Questions 79 - 80 refer to the following list of terms relating to populations.

(A) Adaptive Radiation

(B) Allopatric Speciation

(C) Sympatric Speciation

(D) Directional Selection

(E) Disruptive Selection

79. This is the geographic separation of two populations accompanied by gradual divergent evolution between the two populations.

80. This process occurs when populations within the same distribution range reproduce in isolation.

Questions 81 - 82 refer to the following diagram.

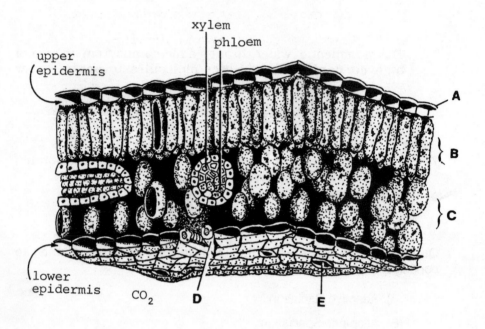

81. Used for protection against desiccation, disease, insects

82. Most photosynthesis takes place here

Question 83 refers to characteristics of:

    (A)   acoelomates

    (B)   pseudocoelomates

    (C)   coelomates

    (D)   acoelomates and pseudocoelomates

    (E)   acoelomates, pseudocoelomates, and protosome coelomates

83. Triloblasty is characteristic of which group?

Questions 84 - 85 refer to the following list of processes for functions that occur in digestion.

(A) The absorptive surface of the digestive tract is increased

(B) The food is manipulated and mixed with saliva

(C) Supportive connective tissue

(D) Used to cut food

(E) A thin layer of connective tissue covered by moist epithelium

84. Which is the function of the microvilli?

85. What is the function of the peritoneum?

Questions 86 - 87 refer to the lettered structures that are listed below.

(A) Shoot

(B) Node

(C) Rhizome

(D) Lateral buds

(E) Apical meristem

86. Leaves are produced by this structure.

87. These structures can differentiate to form branches, flowers, and shoots having a special function.

Questions 88 - 89 refer to the cluster of characteristics beside the lettered headings.

   (A) eukaryotic cells
- method of nutrition - photosynthetic only
- mostly multicellular
- can respond to some stimuli
- major groups are algae, slime molds, and protozoa

   (B) eukaryotic cells
- method of nutrition - heterotrophic or parasitic
- not mobile
- mostly multicellular
- major groups are yeasts, coenocytic organisms (many nuclei within cellular cytoplasm), or multicellular organisms

   (C) prokaryotic cells
- method of nutrition - combination of heterotrophic and photosynthetic
- mostly unicellular with a few multicellular forms
- major groups are algae, slime molds, and yeasts

   (D) eukaryotic cells
- mostly unicellular with a few multicellular forms
- can respond to stimuli
- major groups are algae, slime molds, and protozoa

   (E) either eukaryotic or prokaryotic cells

- mostly unicellular with a few multicellular forms
- method of nutrition: heterotrophic only
- major groups are algae, slime molds, and yeasts

88. Characteristics of Protists.

89. Characteristics of fungi.

Questions 90 - 91 refer to the following plant hormones.

(A) Auxins

(D) Ethylene

(B) Cytokinins

(E) Abscisic Acid

(C) Gibberellins

90. You observe that a monocot seed is shrinking. Which hormone is probably present?

91. You observe that seeds appear to be dormant, and on a dry day, the stomata of the leaves are closed. Which hormone is probably present?

Questions 92 - 94 refer to reproductive structures of a generalized flower that are identified by letters.

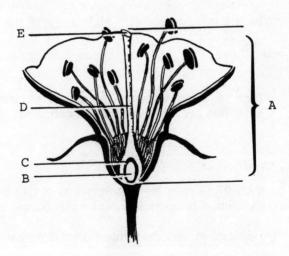

92. Structure in which the female gametophyte (megagametophyte) develops.

93. Structure in which the ovule develops and which becomes the fruit.

94. Structure which becomes the seed.

Questions 95 and 96 refer to the following enzymes of digestion and organs of digestion:

Enzymes	Organs
(A) amylase	(A) liver
(B) lipase	(B) pancreas
(C) pepsin	(C) salivary glands
(D) disaccharidase	(D) intestinal mueasa
(E) trypsin	(E) stomach

95. Which structure secretes <u>all but one</u> of the above enzymes?

96. Which of these enzymes acts first on the proteins in food after it has been ingested and swallowed?

For question 97 refer to the following lists of terms that may be in the basic body plans of acoelomates and/or pseudocoelomates.

(A) Ectoderm, endoderm, mesoderm,digestive cavity

(B) Endoderm, mesoderm, digestive cavity, pseudocoelom

(C) Ectoderm, mesoderm, endoderm, digestive cavity

(D) Mesoderm, endoderm, ectoderm, coelom

(E) Pseudocoelom, coelom, mesoderm, endoderm

97. Starting with the <u>outside</u> layer, list all structures that would be seen in a cross-section of a flatworm (platyhelminthes).

To answer questions 98 - 99 refer to the following diagram and the list of microorganisms:

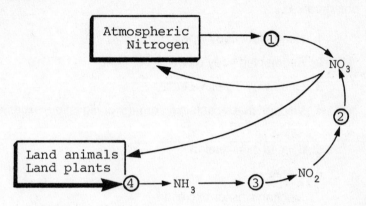

(A) nitrogen-fixing bacteria

(B) plankton

(C) nitrifying bacteria

(D) denitrifying bacteria

(E) decomposing bacteria

98. What group of bacteria belongs in circle 1?

99. What group of bacteria belongs in circle 4?

The following questions refer to experimental or laboratory situations or data. Read the description of each situation. Then choose the best answer to each question. Blacken the correct space on the answer sheet.

Questions 100 - 101 refer to the list of characteristics of mollusca, annelida, and arthropoda. Some characteristics may refer to more than one group:

1. Head - foot body type

2. Segmented body type

3. One - way digestive tract

4. Visceral mass containing organs of digestion, excretion, and reproduction.

5. Jointed exoskeleton

6. Mantle

7. Mechanisms for excretion

8. Closed circulatory system

100. The characteristics associated with mollusca include:

(A) 1,3,5,8

(B) 2,3,4,7

(C) 1,4,5,7

(D) 2,5,6,8

(E) 1,4,6,7

101. The characteristics associated with both annelida and arthropoda include:

(A) 1,4

(B) 1,3

(C) 2,5

(D) 2,6

(E) 2,7

448

Questions 102 - 103 refer to the animal characteristics listed below:

1. embryonic development has spiral cleavage pattern
2. embryonic development has radial cleavage pattern
3. mouth development near blastopore
4. anus development near blastopore
5. coelom formation by splitting of mesoderm
6. coelom formatting by out-pocketing of embryonic gut
7. presence of coelom
8. absence of coelom

102. The characteristics associated with protostomes are:

(A) 1,3,5,7 (D) 2,3,5,8
(B) 2,4,6,8 (E) 1,3,4,7
(C) 1,4,6,7

103. The characteristics associated with deuterosomes are:

(A) 1,3,5,7
(B) 2,4,6,7
(C) 1,4,6,7
(D) 2,3,5,8
(E) 1,3,4,7

Questions 104 - 105 refer to the following situation. A mother with Rh-negative blood had a baby with Rh-positive blood and two years later had another Rh-positive baby.

104. This situation can be dangerous for the second child because:

    (A) The second baby may make antibodies to a factor in the mother's blood

    (B) The mother may make antibodies to a factor in the first child's blood

    (C) The father has donated a dominant gene to the child

    (D) The father will donate antibodies to the second child

    (E) Both (B) and (D)

105. To say that the baby has Rh- positive blood means that:

    (A) The baby has an antigen on his red blood cells that the mother's red blood cells do not have.

    (B) The baby lacks an antigen on his red blood cells that the mother has

    (C) The baby has an antigen for blood group O

    (D) Rh is a multiple allele

    (E) The father has donated antibodies to the child

Questions 106 - 107 refer to the graphs below:

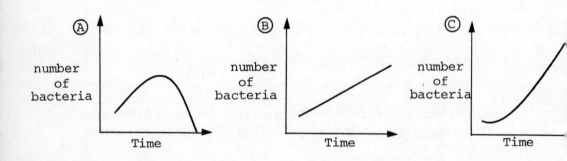

450

106. Which graph shows the long-term growth of a population of bacteria placed in a nutrient medium in a covered dish?

(A) Graph A   (D) All of the above

(B) Graph B   (E) None of the above

(C) Graph C

107. Which graph shows the long term growth of a population of bacteria placed in a nutrient medium in which there is plenty of food, space, and other natural resources?

(A) Graph A   (D) All of the above

(B) Graph B   (E) None of the above

(C) Graph C

Answer questions 108 -109 using the following information and chart.

Both the Smith family and Jones family had babies on the same day at the same time. There was a mix-up in the nursery, but the hospital had the blood types of the Smith's and the two babies. The chart below gives that information.

Mrs. Smith - Group O
Mr. Smith   - Group AB
Baby 1   - Group A
Baby 2   - Group O

108. Which of the two babies could belong to the Smiths?

(A) Baby 1 (blood group A)

(B) Baby 2 (blood group O)

(C) Either baby 1 or baby 2

(D) Neither baby

(E) Not enough information

109. If Mr. Smith had group O blood, which child would belong to the Smiths?

(A)   Baby 1 (blood group A)

(B)   Baby 2 (blood group O)

(C)   Either baby 1 or baby 2

(D)   Neither baby

(E)   Not enough information is provided

110. A large number of genetically identical mice are randomly put into two groups. The experimental group is fed mouse food mixed with 2% DTT. The control group is fed mouse food only. At the end of six months, the following data were recorded:

	Experimental Group N=40	Control N=39
% of deaths	30%	0%
% of births compared to original number	40%	82%
% diseased but not dead	20%	2%
average weight	190g	220g
food weight eaten in a day	30g	39g

The data indicates that:

(A)   The DDT fed mice appeared to be eating more frequently

(B)   The control group mice weighed less and ate less

(C)   The DDT fed mice died more frequently, reproduced less, had a higher frequency of disease

(D)   The control group mice died less frequently, reproduced more, had a lower frequency of disease, and weighed and ate more.

(E)   Both (C) and (D)

The diagrams below demonstrate three types of evolutionary trends of a population.

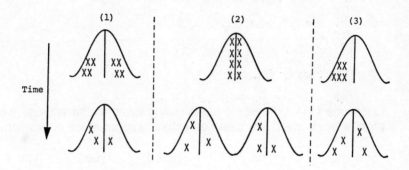

(4)   Consists of (1), (2), and (3)

(5)   Consists of (1) and (3) only

Answer questions 111 – 113 based on the diagram above.

111.   Adaptations arise from:

(A)   1   (D)   4

(B)   2   (E)   5

(C)   3

112. The optimum birth rate of babies is approximately 7 pounds. What type of natural selection is represented by this fact?

(A)   1   (D)   4

(B)   2   (E)   5

(C)   3

113. Which is an example of directional selection?

(A) 1

(D) 4

(B) 2

(E) 5

(C) 3

Questions 114 - 115 refer to the following listing of characteristics of basic muscle types which are classified either by appearance or location.

	Type 1	Type 2	Type 3
1. Found in visceral organs		+	
2. Attached to skeleton	+		
3. Voluntary	+		
4. Unstriated		+	
5. One nucleus per cell		+	+
6. Branched network of cells			+
7. Involuntary		+	+
8. Intercalated discs			+

114. Which type is classified as smooth muscle tissue?

(A) Type 1

(D) Type 1 or 3

(B) Type 2

(E) Type 2 or 3

(C) Type 3

115. Which type is classified as cardiac muscle tissue?

(A) Type 1

(D) Type 1 or 3

(B) Type 2

(E) Type 2 or 3

(C) Type 3

116. Identical plants which are sensitive to light are grown in four
     pots. Pot 1 is exposed to white light. Pot 2 is exposed to red
     light. Pot 3 is exposed to blue light, and pot 4 is kept in the
     dark. The plants growing in the four pots look different from
     each other. These differing appearances demonstrate that:

     (A) A change in environmental conditions influences the
         process of differentiation

     (B) Changes in less important proteins can cause changes in
         appearance

     (C) Organ systems work together in such a way as to change
         the activity of the genes.

     (D) The nucleus controls differentiation in the cell

     (E) The cytoplasm controls the expression of genes in cells
         that differentiate

Question 117 refers to the following experiment.

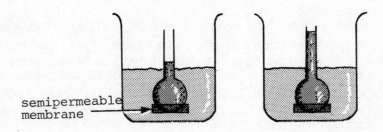

semipermeable
membrane

Experiment-

A thistle tube is filled with colored water and starch and immersed in distilled water. Distilled water has a 100% concentration of water molecules. A semipermeable membrane separates the starch solution from the water. The second diagram shows what has happened after time has elapsed.

117. In what direction does the movement of liquid occur?

(A) Down the concentration gradient

(B) Up the concentration gradient

(C) From an area of low concentration to an area of high concentration

(D) Both (B) and (C)

(E) Both (A) and (C)

Questions 118 - 120 refer to the following experiment described below.

Four test tubes are filled with H_2O. Five drops of phenol red, a solution that is pink when basic and yellow when acidic, is put in the test tubes. Carbon dioxide is added to Test Tubes 1 and 3 by blowing across them. The liquid turns yellow indicating the presence of CO_2 as carbonic acid (a weak acid). All four test tubes are put into a beaker with cold water and an Elodea leaf is put in Test Tubes 1 and 2. They are exposed to a bright light source. The results are:

Test Tube 1 (pink)	with leaf
Test Tube 2 (pink)	with leaf
Test Tube 3 (yellow)	without leaf
Test Tube 4 (pink)	without leaf

118. In which test tube(s) are gas bubbles produced?

(A) Test Tube 1   (D) Test Tube 4

(B) Test Tube 2   (E) Test Tubes 1 and 2

(C) Test Tube 3

119. The gas referred to in question 118 is:

(A) Carbon monoxide

(B) Oxygen

(C) Carbon dioxide

(D) Hydrogen oxide

(E) ozone

120. Which test tube(s) changed color upon exposure to the light source?

(A) Test Tube 1   (D) Test Tube 4

(B) Test Tube 2   (E) Test Tubes 1 and 2

(C) Test Tube 3

SECTION II

Answer each of the following four questions in essay format. Each answer should be clear, organized and well-balanced. Diagrams may be used in addition to the discussion, but a diagram alone will not suffice. Suggested writing time per essay is 22 minutes.

1. Mitosis and meiosis use much of the same cellular machinery. There are some important differences, however, between them. Explain the differences according to:

    A. Chromosomal movements and the action of the spindle apparatus (assume there are 4 chromosomes consisting of two homologous pairs)

    B. The final result of the process with regard to number of chromosomes and genetic makeup of the cell

2. Explain how the properties of water make the following statement true:

    Water is essential to the sustenance of life.

3.   The relationship of the predator and prey populations with each other represents a type of community interaction among species of organisms. In this essay you will focus on the unique interaction among predator and prey.

Discuss the dependency that predator and prey have on each other with respect to their potential numbers in a given feeding area. Include a definition of predator. Discuss what types of organisms are susceptible to predators. Relate the "carrying capacity" of the predator population to the prey population. Explain why there is a lag of time between the growth of predator numbers and prey. The parasite/host relationship is a special kind of predator/prey relationship. Explain the difference. Include the definition of "parasite".

4.   Respiration in aerobic organisms (such as man) is the intake of oxygen and the accompanying release of carbon dioxide. Discuss the respiration process as a four stage process.

Stage 1 - Discuss how bulk oxygen gets from the external environment to the lungs. Include in your explanation the path taken by the air starting at the nose, the mechanical process of inhaling and exhaling, and the control center for that process.

Stage 2 - Discuss how the oxygen gets across the membrane of the lungs and into the blood.

Stage 3 - Discuss how oxygen gets from the blood to the tissues.

Stage 4 - Discuss how carbon dioxide gets from the tissues to the blood and out the nasal passages.

ADVANCED PLACEMENT BIOLOGY EXAM VI

ANSWER KEY

1.	D	31.	A	61.	C	91.	E
2.	A	32.	E	62.	A	92.	B
3.	E	33.	A	63.	E	93.	C
4.	A	34.	E	64.	B	94.	B
5.	C	35.	B	65.	B	95.	B
6.	A	36.	C	66.	B	96.	C
7.	D	37.	E	67.	C	97.	C
8.	E	38.	A	68.	D	98.	A
9.	A	39.	D	69.	E	99.	E
10.	B	40.	D	70.	D	100.	E
11.	A	41.	D	71.	A	101.	E
12.	C	42.	E	72.	E	102.	A
13.	A	43.	A	73.	B	103.	B
14.	B	44.	E	74.	A	104.	B
15.	D	45.	E	75.	C	105.	A
16.	E	46.	E	76.	A	106.	A
17.	E	47.	A	77.	E	107.	C
18.	B	48.	E	78.	B	108.	A
19.	C	49.	A	79.	B	109.	B
20.	C	50.	E	80.	C	110.	E
21.	D	51.	A	81.	A	111.	D
22.	A	52.	A	82.	B	112.	A
23.	E	53.	B	83.	E	113.	C
24.	C	54.	E	84.	A	114.	B
25.	B	55.	B	85.	E	115.	C
26.	A	56.	A	86.	E	116.	A
27.	E	57.	D	87.	D	117.	A
28.	A	58.	A	88.	D	118.	A
29.	A	59.	D	89.	B	119.	B
30.	B	60.	E	90.	C	120.	A

ADVANCED PLACEMENT
BIOLOGY EXAM VI

DETAILED EXPLANATIONS
OF ANSWERS

SECTION I

1.   (D)
   During self-replication, the DNA unwinds at the weak hydrogen bonds which join the complementary base pairs. Free nucleotides in the environment become attached to the open bases of the parent strands if the proper catalyzing enzymes are present. The attachments follow the principle of complementary base pairing so that the strand produced is complementary, not identical, to the intact parent strand. The original strand acts as a "template" for the generation of a complementary strand.

2.   (A)
   The sugar phosphate backbones run in opposite directions so that the nucleotide bases can align with and be bonded to their complementary base in the DNA molecule. Options (B), (D), and (E) are synonyms for each other. Along with the hydrogen bonds (option C), they form the "rungs," not the backbone, of the DNA ladder.

3.   (E)
   Normally the biomass (the total dry weight of all living organisms that can be supported at each trophic level in a food chain) decreases with each succeeding trophic level.

461

4.   (A)

The shorter the food chain, the less the usable energy that is lost. Thus, if plants are fed to livestock and livestock are in turn eaten by humans (with the accompanying energy loss at each trophic level), more energy would be lost than if humans ate plants directly.

5.   (C)

Predation can affect the size, fitness and evolution in a community, but food resources are often the factor that limits the carrying capacity of a population. It has not been shown that predation limits the carrying capacity of a population.

6.   (A)

To describe the niche of a species, we must know what it eats, what eats it, what environment (weather, temperature, chemicals) it can handle, the type and size of its habitat, how it affects other species and the non-living environment, and how these factors affect it.

7.   (D)

Carrying capacity is the maximum population that a given environment can support indefinitely. No population can grow rapidly indefinitely. Eventually, some resource is used up, and then the death rate will exceed the birth rate. Eventually, the population grows again until the birth rate and the death rate are about equal. The carrying capacity is determined by environmental conditions, and if these conditions do not change, the number of individuals supported remains about the same.

8.   (E)

When a system converts energy from one form to another, there is a loss of heat energy that cannot be used for useful work. For example, when an organism dies, the ordered pattern of molecules decays (facilitated by decay organisms) into smaller molecules that are spread throughout the environment. This tendency for a system to go to disorder is called entropy. Option (A) is wrong because it represents an orderly System that begins in an area (e.g., bare rock) that has not been previously occupied by a community of living things and proceeds through a series of growth stages until a climax community is realized. The clue is "orderly" succession, the opposite of entropy.

9.   (A)

Climax communities are relatively stable with regard to ecological succession. They have a diverse array of species, so option (D) is incorrect. Entropy refers to disorder, and climax communities are relatively stable, so options (C) and (B) are not correct.

10.   (B)

Secondary plant succession occurs when an ecosystem at some stage of succession is disrupted and set back to an earlier stage. The difference between secondary succession and primary succession is that the latter occurs on virgin surfaces such as options (C) and (D). Choice (A) is unrelated.

11.   (A)

RNA polymerases are enzymes that catalyze the synthesis and assembly of nucleotide chains called RNA transcripts. These chains are formed using the unzipped area of the DNA double helix, which acts as a template for assembling the complementary base pairs. This process is called transcription.

12.   (C)

The transfer RNA with its amino acid at one end carries that amino acid to the ribosome. There, the three-nucleotide combination of the tRNA (called an anticodon) positions itself at the binding site of a complementary mRNA codon. As a second tRNA moves into a binding site on the ribosomal mRNA, the amino acids from the anticodons align with each other by the formation of a peptide bond. This is the beginning of a growing chain of amino acids that will be linked together by peptide bonds.

13.   (A)

Viruses are parasites in that they cannot multiply outside their host cell. They use the energy sources (option (B)), metabolic machinery (option (C)), and enzymes (option(D)) of the host cell. Viruses do have the ability to replicate their genetic material (DNA or RNA). They do so by inserting a copy of their genetic material and using the resources of the host cell to replicate the material and to form their own protein coats. The virus particle then escapes from the host cell and is ready to infect another host cell.

14.   (B)

Transfer RNA is transcribed from the DNA of the cell and is found in the cytoplasm. It is clover-leaf shaped with an amino acid attached at one end and a loop at the other end containing three nucleotides (one of which is usually guanine). This sequence of three nucleotides is called an anticodon. During protein synthesis, the anticodon attaches itself to a complementary mRNA codon located in the ribosome. The amino acid of the tRNA forms a peptide bond with the amino acid of a second tRNA with its particular anticodon which has attached itself to an adjacent mRNA codon in the ribosome.

15.   (D)

Diffusion occurs when there is a difference in the concentrations of substances and the goal is to have the particles distributed uniformly throughout. Diffusion can occur with or without a membrane (choices (B) and (C). Lab experiments showing diffusion can be done with inanimate objects. Since diffusion occurs down the concentration gradient, no energy is required.

16.   (E)

The concentration of water inside the cell is lower than the concentration of water outside the cell. Therefore, water would diffuse into the cell to equalize the concentration. When water enters an animal cell, there is no protective wall, so the cell would swell and probably burst. A plant cell would probably not burst because of the rigid wall.

17.   (E)
   The fluid mosaic model of the cell membrane provides a model of the structure of a cell membrane.

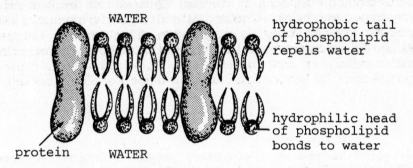

The phospholipids have a hydrophilic (water-loving) and a hydrophobic (water-fearing) tail. The lipids are hydrophobic. The proteins extending through the bilayer have hydrophobic regions where they are in contact with the hydrophobic phospholipid tails, and hydrophilic regions where they are in contact with the cytoplasm or the outer environment.

18.   (B)
   Options (A) and (E) apply to prokaryotic flagella. The microtubular arrangement in option (C) is the reverse of what it should be. The centrioles referred to in option (D) are structurally similar to the basal bodies, but they are distributed differently in the cell and may play a role in the spindle formation that occurs during mitosis. Option (B) has the correct structure. The basal body from which the cilia and flagella arise has the 9 + 0 arrangement of microtubules.

19.   (C)
   The phenotype of a plant is an observable trait. The clue word that is used to describe both pea plants is "tall." "Tall" is an observable characteristic. The pea plants have different genotypes; that is, the gene pairs are different (option (B)). The TT plant has the gene pair for "tall dominant". The Tt plant has the recessive gene "t" (short) which is masked by the "T" (tall) gene. Alleles (option (A)) are contrasting forms of a gene located at the gene locus. Examples of alleles are the genes for "tall" and "short"; these constitute an allelic pair.

20.   (C)

Condensation reactions involve the building up of organic molecules from the joining of their subunits and the removal of a water molecule at each synthesis. This is also known as a dehydration synthesis. Choices (A) and (B) are examples of the same condensation reaction expressed in different forms. Choices (D) and (E) represent hydrolysis reactions, the reverse process of condensation.

21.   (D)

Proteins are composed of carbon, oxygen, hydrogen, nitrogen, and sometimes sulfur. These elements combine to form amino acids. Each amino acid has a carboxyl group (-COOH) and an amino group ($-NH_2$) which are attached to a carbon atom. In addition, side groups (called radicals) are also attached to the carbon atom. There are 20 different kinds of amino acids each with a different side group.

22.   (A)

An enzyme is a type of protein which catalyzes biological reactions. The surface geometry plays an important part in its specificity for substance molecules.

23.   (E)

Enzymes require specific conditions of temperature, pH, and substrate under which they operate at maximum efficiency. In humans, enzymes work best at an optimum temperature of 98.6°F, 37°C, but at various pH values, dependent on the enzyme, the substrate, and the location of reaction.

24.   (C)

Coenzymes are small organic molecules that activate enzyme proteins. They are usually derived from water soluble vitamins, such as the B-complex. They are essential for the function of some enzymes. They are not proteins. They do not react directly with the substrate.

25.   (B)

Photosynthesis occurs in chloroplasts which are located in the main body of the leaf (mesophyll). These chloroplasts are composed of a double outer membrane which surrounds the stroma (a semi-fluid matrix). Inside the stroma is a series of membranes that form a series of disks stacked on one another. Each disk is called a thylakoid, and it is here that the pigments necessary for photosynthesis are found, and where "the light" reactions of photosynthesis take place.

26.   (A)

When the water molecule is split (H-OH) by the NADP, the hydrogen ions from the water molecules are picked up by the NADP which becomes $NADPH_2$ and then enters the dark reaction. The OH loses an electron which is eventually recycled back to the chlorophyll molecule. The OH molecules from the splitting of water form O_2 and H_2O. Oxygen is considered a by-product of the light reaction of photosynthesis.

27.   (E)

If the H+ ions can be transported through the membrane, no concentration gradient of H+ ions can be established. In order for ATP to be synthesized, there must be an electrochemical gradient (H+ concentration and voltage gradient) to make energy available to power the synthesis.

28.   (A)

For a coupled reaction to occur, an exergonic (energy-releasing) reaction such as respiration must occur at the same time as an endergonic (energy-requiring) reaction such as photosynthesis. The energy that results from the exergonic reaction is used to make the endergonic reaction "go". The other options are incorrect.

29.   (A)

In Mendel's law of independent assortment, the alleles assort independently into gametes producing offspring that may or may not resemble their parents. The potential for variety is very great. A dihybrid cross as described in the problem can yield nine genotypes, (multiply the number of genetic combinations by 3). For example, parents who are heterozygous for hair color (brown, blonde-Bb) and height (tall, short-Tt) form the following genetic combinations when the gametes are crossed: BBTT, BBTt, BbTT, BbTt, BBtt, Bbtt, bbtt, bbTt, bbTT. See the following Punnett square:

	BT	Bt	bT	bt
BT	BBTT	BBTt	BbTT	BbTt
Bt	BBTt	BBtt	BbTt	Bbtt
bT	BbTT	BbTt	bbTT	bbTt
bt	BbTt	Bbtt	bbTt	bbtt

30.   (B)

Both DNA and RNA have four bases. Adenine, guanine and cytosine are found on both. However, where DNA has thymine, RNA has uracil. Both DNA and RNA contain nucleotides (option (D)). Therefore, option (B) is the correct answer.

31.   (A)

The nucleosome consists of DNA as a connecting "string" that winds around histone proteins. An average gene is made up of three or four nucleosomes.

32.   (E)

Bacteriophages are viruses that attack bacteria by attaching themselves to the bacteria cell wall, injecting their genetic material into the cell and parasitizing it.

33.   (A)

When the two strands are cleaved between the G and A, the following separation results:

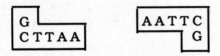

These single strands have complementary bases that protrude, and have sticky ends which are ready to receive a complementary sequence.

34.   (E)

The complementary bases for the protruding sticky end TTAA would be AATT, and the complementary bases for sticky end AATT would be TTAA.

35.   (B)

A change in a nucleotide of a codon sequence may or may not change the amino acid called for. However, an entire protein may be structurally and functionally altered upon the addition or deletion of a nucleotide.

36.   (C)

Use the following P_1 generation as an example of a dihybrid (two - trait) cross:

B = black hair (dominant)
b = blonde hair (recessive)
G = grey eyes (dominant)
g = blue eyes (recessive)

The P$_1$ gametes typical of a dihybrid cross (BG and bg) undergo fertilization producing an F$_1$ generation (BbGg) which then separates into the following gametes- BG, Bg, bG and bg. Each gamete is both male and female. Therefore, there is a total of four distinctly different gametes.

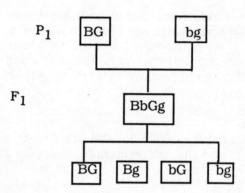

F$_1$ Gametes

37.   (E)
The nine possible genotypes using the example of a dihybrid cross given in the answer to problem 55 are: BBGG, BBGg, BbGG, BbGg, BBgg, Bbgg, bbGG, bbGg and bbgg.

	BG	Bg	bG	(Female) bg
BG	**BBGG**	**BBGg**	**BbGG**	**BbGg**
Bg	BBGg	**BBgg**	BbGg	**Bbgg**
bG	BbGG	BbGg	**bbGG**	**bbGg**
bg	BbGg	Bbgg	bbGg	**bbgg**

(Male)

38.   (A)
Hemophilia is caused by a recessive gene that is found on the X chromosome. A woman who has this gene usually has a dominant normal gene on the other X chromosome and so doesn't have hemophilia. On the other hand, a man with the gene for hemophilia has no second X chromosome with a normal gene, so he will have hemophilia.

39.   (D)
The theory of Punctuated Equilibrium can account for the sudden appearance and disappearance of fossil species. The fossil record shows periods of stability with regard to appearance and disappearance of species as well as periods of sudden change.

40.   (D)
Adaptive radiation is a pattern that occurs when a lineage (single line of descent) branches into two or more lineages, and these further branch out. This pattern can occur when a species is able to invade environments that have previously been occupied by other species. In this case, when the dinosaurs became extinct, mammals invaded their vacated ecological niches and quickly diversified to adapt to the living conditions of the niches. In addition to invading vacant ecological niches, species can undergo adaptive radiation when they partition existing environments.

41.   (D)
Classifying organisms by species is the basis for organizational systems such as the taxonomic, systematic, and evolutionary systems. "Species" is defined as a group of organisms that can interbreed and produce fertile offspring but not with members of other groups. The definition focuses on the biological aspects of organisms. Options (A), (C) and (E) are ways to classify organisms, but there are also other ways. Option (B) describes the grouping of species that are related to each other in some ways but distinct from one another in other ways. The fundamental unit is always the species.

42.   (E)

Prokaryotes are traditionally classified using absence or presence of cell wall and mode of nutrition. Option (E) is correct. Their reproductive patterns are not complex, (option (B)) and some patterns (e.g. conjugation) are found in other kingdoms (e.g. Protists). Also, both Prokaryotes and most protists are unicellular (option (D)), so this criterion cannot be used to distinguish one from the other.

43.   (A)

The kingdom Monera, (the prokaryotes) consists of two major divisions, the blue-green algae (cyanophyta) and the bacteria (schizophyta). They differ from the eukaryotes in that the DNA is in the form of a large single molecule in the cytoplasm and is not associated with histones (five basic proteins bound to the DNA in eukaryotic cells). In addition, prokaryotes do not have a membrane-bound nucleus or membrane-bound organelles. Prokaryotic cell membranes do not have cholesterol and other steroids (as eukaryotes do), and the electron transport system is located on the membrane (in eukaryotes, it is found in the mitochondria). Finally, the cell walls of prokaryotes do not contain cellulose and other complex polysaccharides (as do eukaryotes). Instead, the cell walls can contain complex polymers (peptidoglycans), in addition to lipoproteins which are not found in eukaryotic cells.

44.   (E)

Prokaryotes (Monera) constitute a separate kingdom because their differences from eukaryotic cells are sufficient to warrant this separation.

45.   (E)

Systematics is the study of relationships among organisms. One type of relationship is phylogeny which is the study of the evolutionary history of an organism. In this case, evidence would indicate that while forelimbs of the crocodile, bird, and horse may have different functions and appearances, the basic structure is the same.

472

46.   **(E)**

The two major divisions of plants are the Bryophytes (lower plants), which lack vascular (conducting) tissue, and the Tracheophytes (higher plants), which have vascular tissue throughout the plant. Option (C) is incorrect because all plants carry out photosynthesis and can therefore make their own food. Options (A), (B), and (D) are characteristics of all plants.

47.   **(A)**

In the life cycle of fern, meiosis occurs in the sporangia (the bodies that produce haploid spores). These sporangia are located on the underside of the fronds of the diploid sporophyte. The single celled, haploid spores which result from meiosis undergo many cell divisions to produce a multicellular plant known as the gametophyte. The gametophyte produces gametes. When fertilization occurs, a diploid zygote results. The zygote is a new sporophyte. It undergoes cell division to produce, ultimately, a new adult sporophyte plant.

48.   **(E)**

In the alternation of generations life cycle, the gametophyte grows from the haploid spores released by the mature sporophyte plant. Mature gametophytes produce sperm cells and egg cells which are located in protective structures on the plant. Fertilization occurs when a sperm cell fuses with an egg cell, and the resulting zygote divides mitotically and gives rise to a mature diploid sporophyte.

49.   **(A)**

Seed plants (gymnosperms and angiosperms) do not need water for the sperm to swim to the egg. The life cycles of the seed plants can make use of dispersing agents such as the wind to distribute pollen grains containig the microspores. The development of the pollen tube conducts the sperm to the egg without the need for water as a conduit for the sperm cells. In contrast, ferns and bryophytes need water for the sperm to swim to the egg.

50.   (E)

As plants evolved, the male gametophyte stage that is part of the life cycle of alternation of generations shrank in size. In angiosperms and gymnosperms, the pollen grain is the male gametophyte. The megaspore is the result of meiosis in the female structures. It divides several times and the multicellular structure is the megagametophyte. The remaining choices refer to reproductive structures of a flowering plant.

51.   (A)

52.   (A)

The seed coat has protected the embryo from water and carbon dioxide until the conditions are favorable for growth and survival. Germination is initiated when water enters the seed coat which has been worn away or weakened by the elements. The seed coat then breaks and the embryo begins to grow into a young sporophyte.

53.   (B)

Growth is first started by cell divisions in the apical meristem. At the point above which cell division is reduced, the cells elongate causing the root to grow in length. Thus, while growth depends on cell divisions in the apical meristem (choice (C)) and those cells eventually become elongated, the principal cause of root growth is the elongation of the cells.

54.   (E)

Choices (A) and (B) are incorrect. Hormones of animals are produced in certain tissues but are transported to and affect other tissues. They can exert specific influences using very small quantities.

55.　　　(B)

This response to a light stimulus is called phototropism. The hormone auxin, in response to the light, migrates from the light to the dark side of the shoot tip. The cells on the dark side now contain more auxin which causes the cells on that side to elongate more rapidly than the cells on the light side. The result is that the plant bends toward the light.

56.　　　(A)

A gravitropic response refers to the tendency of shoots to grow upward and of roots to grow downward even if the seedling is placed in an unfavorable position. Gravitropism is an important survival response for young plants. Photoperiodicity is a response to the duration and timing of light and dark conditions. Circadian refers to changes in response over a given time period, usually 24 hours. Phototropism is a directional response toward light.

57.　　　(D)

Cells which are similar in structure and function unite into tissues.

58.　　　(A)

The impulse received by the dendrite is then passed to the cell body of the nerve cell which then conducts it to the axon which carries the nerve impulse away from the cell body to other cells and organs. Neurons may vary in size, but they have a basic structure of dendrite, cell body, and axon. The nodes of Ranvier are found at breaks of myelin along the axon. They allow for impulse transmission. The synaptic bouton is found at the end of the axon, and releases neurotransmitters which stimulate the dendrites of the next nerve cell.

59.　　　(D)

Within the nephron and associated capillaries of the kidneys, three processes take place – filtration, secretion choice (B), and reabsorption choice (A). After filtration takes place, the filtrate passes into the nephron tubule where most of the water and solute that entered the nephron are returned to the bloodstream. This process is called reabsorption. Some substances such as metabolic wastes are also actively transported from the peritubular capillaries into the nephron tubule and they become part of the urine. This process is called secretion. Countercurrent multiplication (choice (E)) and countercurrent exchange (choice (C)) are processes that help to keep the osmotic gradient that is necessary for maintaining consistent solute concentrations in the urine. It takes place across the loop of Henle.

60.   (E)
The glomerulus (choice (A)), the loop of Henle (choice (B)) and Bowman's capsule (choice (D)) are part of the nephron. The urethra (choice (C)) is the structure used to remove urine from the body, but it is not part of the nephron.

61.   (C)
The memory cells are part of the secondary immune response. When the B-cells and T-cells initially clone to fight the first infection, some of the clones are not used. Instead they can be activated to clone when they later come in contact with the same antigen. The antigens may then be destroyed before they cause the disease.

62.   (A)
The graft will be least likely to be rejected if it comes from the person who approximates the patient's body chemistry as much as possible. An identical twin is the best choice. Drugs can be given to suppress the immune response, but as infection is a potential problem it is best not to suppress the general immune response.

63.   (E)
The sperm and the egg form within the parental reproductive system. The sperm grows a tail that will help it move to the egg cell, and the egg cytoplasm gains nutrient.

64.   (B)

Arteries carry blood away from the heart. They can expand and recoil thus forcing blood into the arterioles. These structures function in controlling blood flow distribution in the body by contraction and expansion of their diameters. The blood flows from the arterioles to the thin walled capillaries which have a large surface area for materials-exchange between blood and interstitial fluid. Capillaries merge into venules which then become veins that return blood to the heart.

65.   (B)

Cleavage occurs after fertilization. It is a time of cell division without a growth in size. As cleavage proceeds, a blastula forms. The blastula is a hollow sphere with a cavity called the blastocoel, resulting in an embryo with three germinal layers. The archenteron is the primitive digestive cavity of the gastrula.

66.   (B)

Blood from the veins fills the atrial chambers then both atria are at rest. As blood fills the atrial chambers, pressure rises and the blood is forced into the ventricles by the contraction of the arteries when the pressure inside the atria becomes greater than the pressure inside the ventricles. The contraction of the atria is called "systole" and the relaxation of the atria is called diastole. When a measure of blood pressure is taken, both the systole and diastole readings are taken at large arteries. The systolic measurement is that of the highest pressure in the artery. The diastolic measurement is taken at the lowest pressure of the artery. Diastolic pressure is usually about 80 millimeters of Hg (Mercury).

67.   (C)

The endoderm (choice (A)) gives rise to the inner lining of the digestive and respiratory tract, as well as the liver and pancreas. The ectoderm (choice (B)) gives rise to the epidermis and nervous system. The morula and blastula are early zygotic stages of division.

68.   (D)

Animal diversity is primarily due to their mode of reproduction: sexual fertilization and meiosis. These patterns increase the chance of diversity in the offspring by the incorporation of crossing over and independent assortment of the chromosome complement of the gametes.

69.   (E)

Each of these criteria is used in classifying animals.

70.   (D)

Bilaterally symmetrical animals have three germ layers (ectoderm, mesoderm, and endoderm) and can move more efficiently than animals whose body structures are radially symmetrical. Bilateral symmetry is characteristic of higher level animals starting with the phylum platyhelminthes.

71.   (A)

Bilateral symmetry means that the body is organized longitudinally having the left and right halves as approximate mirror images of the other. The existence of paired organs fit this criterion. The other options are not associated with bilateral symmetry. As bilateral symmetry evolved, animals developed anterior - posterior ends, dorsal-ventral orientation, and three germ layers-ectoderm, mesoderm and endoderm.

72.   (E)

The Hardy-Weinberg rule predicts that in the absence of agents of change (i.e., mutation, migration, natural selection), the frequencies of different alleles and genotypes in a population will remain stable.

73.   (B)
Lysosomes are located in the cytoplasm. They contain digestive enzymes which break down molecules and malfunctioning cell parts.

74.   (A)
The mitochondria, located in the cytoplasm, are the place where cellular respiration, which produces energy for activities in the cell, occurs.

75.   (C)
Some proteins pass through the endoplasmic reticulum and are enclosed in stacks of membrane which separate from the endoplasmic reticulum. This stacked membrane system is called the Golgi Bodies. Proteins may be changed and stored there until they are needed by the organism. The rough endoplasmic reticulum is the site of protein synthesis. The plasma membrane is the outer membrane of a cell.

76.   (A)
Phagocytosis or "cell-eating" refers to the engulfment of solid particles by amoebas and white blood cells of vertebrates. Endocytosis is the uptake of large particles across the cell membrane. It occurs by the invagination of the cell membrane until a membrane enclosed vesicle is pinched off within the cytoplasm. In exocytosis cellular excretions are enclosed in a membrane-bound vesicle which merges with the membrane and releases its contents to the extracellular environment.

77.   (E)
For osmosis to occur, there must be both a selectively permeable membrane and a concentration differential between the water solutions on either side of the membrane. The movement is from an area of higher concentration of water molecules (lower concentration of solute) to an area of lower concentration of water molecules.

78.   (B)
Active transport in a cell occurs when materials are moved across a membrane from an area of lower concentration (fewer water molecules) to an area of higher concentration (more water molecules). Since this movement is against a concentration gradient, energy is required.

79.   (B)
Allopatric speciation can occur when physical barriers between sections of a population prevent interbreeding among offspring.

80.   (C)
Sympatric speciation is a type of speciation that occurs without geographic isolation. Adaptive radiation results from the competition between two closely related species which diverge into two distinct and different species niches. Directional selection acts against one extreme characteristic, resulting in a population shift in one direction. Disruptive selection acts against individuals in the mid-part of a distribution, favoring both extremes. This results in a split into two subpopulations.

81.   (A)
The cuticle is on the epidermis, and is especially thick on the part closest to the sun.

82.   (B)
Most chloroplasts are located in the palisade layer. The chloroplasts are located closer to the leaf surface which puts them closer to the sun. The spongy layer (C) stores starch. The guard cells (D) control the degree to which the stoma (E) are open.

83.   (E)
Acoelomates, pseudocoelomates, and protostomes are all triloblastic, which means they have three tissue layers—ectoderm,

endoderm, and mesoderm which is located between the ectoderm and the endoderm. The basic body plan of three-tissue layers, the characteristic of the acoelomates, becomes more complex as we shift from acoelomates to pseudocoelomates to protostomes.

84.   (A)
The microvillus is a thin extension of the surface of the animal cell that increases cellular absorption and secretion. They are found in great abundance in the small intestine of the human digestive tract.

85.   (E)
The peritonium is composed of a thin layer of connective tissue covered by moist epithelial cells which line the abdominal cavity containing the stomach, intestines, pancreas, and liver.

88.   (E)
Just as root growth is initiated by cell divisions in the apical meristem of the root. leaves are produced by the primordium apical meristem of the shoot.

87.   (D)
Lateral buds can also remain dormant during one stage of growth. They are located in the axil where the leaf joins the stem. The shoot is the stem with leaves. flowers. and other appendages. The node is the point on the stem where a leaf or bud was/is attached. The rhizome is an underground stem.

88.   (D)

89.   (B)

90.   (C)
   In monocot seeds such as grasses, the hormone gibberellin stimulates the production of enzymes that convert the stored food in the
endosperm into sugars and amino acids. These converted foods can
be used by the seedling, causing the seed to shrink.

91.   (E)
   Abscisic acid is a growth-inhibiting hormone. It also affects stomatal closures and plays a role in dormancy. Of the remaining choices,
auxins promote cell elongation, cytokinins promote cell division, and
ethylene stimulates fruit ripening, a function of the aging process.

92.   (B)
   The ovule encloses and protects the female gametophyte which
has developed from the megaspore.

93.   (C)
   The ovary contains the ovules and will become the fruit of the
angiosperm.

94.   (B)
   The ovule located in the ovary is the structure from which the
female gametophyte develops and which develops into the seed once
fertilization occurs.

95.   (B)
   The pancreas secretes all the enyymes except pepsin which is secreted by the stomach. The pancreas secretes a large number of enzymes which digest a wide range of foods – carbohydrates (amylase
and disaecharidase), fats (lipase). proteins (trypsin); and nucleic acid.

96.   (C)

The digestion of proteins begins in the stomach which secretes the digestive enzyme pepsin. This enzyme breaks down the peptide bonds holding the proteins together. Further digestion of proteins is accomplished by the enzymes trypsin, chymotrypsin, and carboxypeptidase which are secreted by the pancreas.

97.   (C)

Flatworms (platyhelminthes) have three-layered bodies with no body cavity (coelom) other than the digestive cavity.

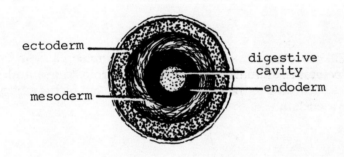

98.   (A)

The microorganisms in circle 1 are nitrogen-fixing bacteria. They live in the soil and convert (or fix) the gaseous nitrogen into organic nitrogen-containing compounds (called nitrates -NO_3)which dissolve in soil and water and are taken in by plant roots.

99.   (E)

When plants and animals die, decomposing bacteria and fungi break down the larger organic nitrogen molecules in the dead bodies into forms of ammonia (NH_3). This process is called ammonification. The bacteria in circle 3 are nitrifyng bacteria which oxidize the forms of ammonia to nitrates (NO_2), a compound that is toxic to plants. The bacteria in circle 2 oxidize the nitrates which then can be used by plants.

100. (E)

The mantle secretes the shell.

101. (E)

Annelids are characterized by a one-way digestive tract, paired nephridia (excretory organs), a closed circulatory system, and a well-defined nervous system. Arthropods, the largest phylum in the animal kingdom, are characterized by a hard exoskeleton made of chitin and a complete digestive and nerve cord with paired ganglia. Characteristics that are common to both annelids and arthropods include a segmented body type and excretory mechanisms (nephridia in annelida and malpighian ducts in arthropods).

102. (A)

Coelomate animals are divided into protostomes and deuterostomes according to the pattern of development by the embryo, but all have a coelom.

103. (B)

104. (B)

When an Rh-negative woman has an Rh-positive baby, the woman's blood will usually form antibodies at the time of delivery, when the infant's blood enters the mother's bloodstream. In subsequent pregnancies, these antibodies can be passed into the fetal bloodstream during the last month of pregnancy and cause destruction of the red blood cells of an Rh-positive fetus. This is called erythroblastosis fetalis.

105. (A)

An Rh-negative women can bear an Rh-positive baby if the father is Rh positive. "Rh positive" describes a genetically determined antigen (antigen D) found on the surface of red blood cells. A person with Rh-negative blood has no antigen D and will produce antibodies when exposed to an Rh-positive fetus.

106.   (A)

Without subsequent addition of nutrients, the population will accelerate toward maximum growth. As the resources are depleted, fewer numbers can be supported and there is a decrease in the population.

107.   (C)

The graph shows an ever-increasing number of individuals over a unit of time under ideal conditions. This is called 'exponential' or geometric growth. Graph (B) shows arithmetic growth which is non-characteristic of bacteria.

108.   (A)

The I^A and I^B and ii are forms or alleles of the gene responsible for human ABO blood groups. The I^A and I^B are co-dominant when they are paired together. The "i" allele is recessive when paired with I^A or I^B. Mr. Smith has blood type AB. Therefore, he has the genotype $I^A I^B$ so he cannot be the father of baby 2 who has blood type O. This blood type has the genotype ii. Therefore the father must have had at least one "i" gene to pass on. Baby 1 has blood group A and could have either I^{Ai} or $I^A I^A$ genotype. Mr. Smith with blood type AB would contribute the I^A allele to Baby 1, while Mrs. Smith with blood type O would contribute the "i" allele resulting in Baby 1 having an $I^A i$ genotype.

109.   (B)

Both Mr. and Mrs. Smith have blood group O which has the "ii" genotype. That "ii" form is both recessive to both I^A and I^B and would not come out unless both the mother and father contributed an "i" gene.

110.   (E)

The difference is dramatic. The DDT affects the rats' ability to gain weight, reproduce, and resist disease.

111.   (D)

Adaptations arise primarily from natural selection. The diagrams (A), (B), and (C) represent three types of natural selection—stabilizing selection, disruptive selection, and directional selection, respectively.

112.   (A)

Choice (A) represents a type of natural selection called "stabilizing selection." In this type, extreme phenotypes (e.g., low birth weight) are eliminated, and the most frequently occurring phenotypes are maintained in the population.

113.   (C)

In directional selection, a species adjusts gradually to changes in its environment with the result that a phenotypic character of a population shifts as a whole in a consistent direction. Choice (B) is an example of disruptive selection. It favors extreme phenotypic characters. Such a shift results in two distinct subpopulations.

114.   (B)

Smooth muscle tissue is composed of small individual cells each with a single nucleus. The tissue is unstriated (no stripes when seen under the microscope) and is found in the walls of internal organs such as digestive organs. Smooth muscle is categorized as involuntary because contraction is not under the control of the organism. Type 1 represents a skeletal muscle tissue and Type 3 represents cardiac muscle tissue.

115.   (C)

Cardiac muscle cells are branched and shorter than skeletal muscle tissue. Cardiac muscle cells are fused end to end at locations called intercalated disks so that when one cell receives a message to contract, the other cells also contract. Both cardiac and smooth muscle tissue (Type 2) are unstriated and involuntary.

116.   (A)

Environmental conditions such as declining food supply, light and temperature serve as signals which influence those genes that are necessary for cell differentiation. Options (B), (D) and (E) can affect gene differentiation, but under different situations.

117.   (A)

The movement occurs from an area of high concentration to an area of low concentration. The area of high concentration is the distilled water (100% water molecules) and the area of lower concentration is the starch solution (with a lower percentage of water molecules).

118.   (A)

For photosynthesis to occur, carbon dioxide and water are needed as raw materials along with sunlight, an energy source. Carbon dioxide is present in the exhaled air (blown over the test tubes). One end product of photosynthesis is oxygen. The gas bubbles of oxygen were produced in Test Tube 1 because all the raw materials and the energy source were present.

119.   (B)

In the photosynthesis reaction, oxygen is produced as a by-product as carbon dioxide is used up.

120.   (A)
Test Tube 1 changed color because oxygen was produced as carbon dioxide was used up. The yellow color changed back to pink reflecting the decreased level of carbon dioxide.

SECTION II

ESSAY I

- In mitosis, chromosomes duplicate themselves before mitosis begins. When a chromosome is duplicated, it becomes in time, two chromatids joined by a centromere.

- In meiosis, chromosomes duplicate themselves before meiosis begins. When a chromosome is duplicated, it becomes, for a time, two chromatids joined by a centromere.

PROPHASE

- In mitosis chromosomes become visible, the nuclear envelope and nucleolus disappear and the mitotic spindle is formed.

- In mitosis chromosomes have doubled, forming 2 identical sister chromatids attached at the centromere.

- In meiosis each chromosome pair condenses and lines up with its homologue (One chromosome of a pair containing one of two genes that constitutes a gene pair). This lining up produces four chromosome equivalents apiece.

- In meiosis, crossing over occurs between nonsister chromatids of each homologous chromosome.

METAPHASE

- In mitosis, homologous chromosomes line up on the mitotic spindle on opposite sides of the midplane of the spindle.

- In meiosis four chromosomes, each made up of two identical sister chromatids, attached at the centromere line up on the midplane. The chromosomes line up without pairing.

ANAPHASE

- In mitosis the centromeres divide and each identical chromosome moves to the poles of the cell on the mitotic spindle. There is now a complete set of four chromosomes at each pole identical to the nucleus.

- In meiosis the homologous chromosomes separate and move to opposite poles. The chromosomes have formed two groups, each made up of 2 sister chromatids. Therefore the genetic material is not the same at the poles of the cell.

TELOPHASE

- In mitosis the chromosomes uncoil forming a mass of DNA

- In mitosis the nuclear membrane forms around each set of chromosomes.

- In mitosis the cytoplasm divides when the cell membrane is constricted down the middle and pinches apart. This process is called cytokinesis.

- In mitosis each daughter cell now has chromosomes identical to those in the parent nucleus.

- In meiosis, one of each pair of homologous chromosomes is clustered around each of the two spindle poles. The chromatids are still attached to the centromere. The cell starts to divide and the nuclear membrane reforms. The chromosome number will be haploid (one of each kind of chromosomes) but the centromeres remain. Prophase II begins.

PROPHASE II

- In each daughter cell the chromosomes prepare for the next division.

- The genetic material is not replicated before prophase II

METAPHASE II

- The spindles reform and the replicated chromosome aligns at the midplane of the spindle. The centromere divides and each replicated chromosome becomes two separate chromosomes.

ANAPHASE II

- The separate chromosomes move toward opposite ends of the spindle.

-   The cytoplasm divides and the nuclear envelopes form.

TELOPHASE II

-   Four daughter cells each with a haploid nucleus are formed. Each of the nuclei has one member of each pair of chromosomes from the original nucleus. From the 4 original chromosomes, each nucleus has two chromosomes.

-   All haploid nuclei contain different genetic material due to crossing over and independent assortment.

ESSAY II

Living organisms are dependent on water for their existence. It is the most abundant biomolecule, comprising most of the mass of living cells. The chemical and physical properties of water, which make it so important, are the polarity, specific heat, heat of vaporization, heat of fusion, and the density of the liquid state.

Consisting of an oxygen covalently bonded to two hydrogens, there is polarity, or uneven distribution of charge, throughout the molecule of water. This allows it to form intermolecular bonds, giving water its unusual properties. The oxygen is more electronegative than the hydrogens, thus it tends to attract positive hydrogen atoms from other nearby water molecules, forming hydrogen bonds. As the bonds between water molecules form, a lattice pattern due to the repulsion of like charges is observed.

Polarity also affects water's adhesion to other surfaces, and gives rise to capillary action, where water is passively drawn upward. This is most important in plant roots, which use capillary action to obtain water from the soil. It makes water a good solvent for ionic compounds, such as salt (NaCl), giving rise to the term "hydrophilic" — water loving. Water's polarity also causes nonpolar substances to become "hydrophobic" — water fearing — which makes them tend to group together, minimizing surface contact. If a substance is amphipathic, or has both hydrophilic and hydrophobic tendencies, in water the substance would form a micelle, with the hydrophobic tails inside and hydrophilic heads outside.

This is important in cellular membrane formation.

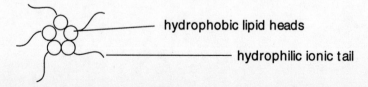

hydrophobic lipid heads

hydrophilic ionic tail

As expected, the cumulative effects of many hydrogen bonds give rise to its unusual properties. Water's specific heat, or the amount of heat required to raise the temperature of one gram of liquid by one degree Celsius, is greater than that of most other liquids because more heat is required to raise the kinetic energy. Since most cellular mass is water, this minimizes temperature fluctuations in the cell. This is important as most biologically significant reaction rates are temperature sensitive.

The heat of vaporization is also greater than that of most liquids. Water requires more heat to evaporate as the intermolecular hydrogen bonds must be broken to permit disassociation of molecules. Because it is an endergonic, or heat-absorbing reaction, this makes perspiration decrease body temperature.

About 80 calories of heat per gram is needed for heat of fusion, which causes a substance to change or go from the solid to the liquid state. This is also due to the number of hydrogen bonds. This moderates body temperatures for living organisms during external temperature fluctuations.

The density of most substances increases upon freezing; however, due to the lattice structure of the hydrogen bonding, this is not so with water. The maximum density of water is at four degrees Celsius. Below this, it expands, which causes it to be less dense than water. If it were more dense than water, it would sink to the bottom, killing the organisms in it. Since it floats on top of water, it sustains life in water bodies by keeping the water temperature below the ice layer relatively constant.

ESSAY III

- A predator is an organism that captures and eats organisms in order to get nutrients for growth and energy. The size of the predator population is dependent on the carrying capacity of the environment of the prey population. If the resources of the environment cannot support a critical mass of prey, the prey population may start to decrease, and eventually the predator population may also be adversely affected

- The prey numbers in a given feeding area may not be as adversely affected if the predator is territorial, and if the numbers of prey are larger than the predator can handle. Fish traveling in schools are an example of strength in numbers, since a predator cannot eat enough of them to decrease their numbers appreciably.

- Concentrations of predators on one part of a population (old, sick, young) can affect the prey population. In this case, the fittest may Survive to perpetuate the special genetic make-up that allowed them to survive to subsequent generations.

- Predators can also affect the actual numbers of a prey population by eating enough of them to reduce their number. This decrease could affect whether the prey can survive as a species or die out.

- There is often a lag time between the growth of a predator population in response to the growth of a prey population. One reason is that it takes time for predators to move into an area where the prey population is larger.

Another reason is that the size of the predator population is dependent on such natural factors as

- Fertility
- Natural catastrophes that may decimate a population
- The activities of man
- The activities of the "predators of the predators" in the food web

Some mechanisms that prey use to escape from predators include:

- mimicry — Some organisms escape from predators by looking or behaving like those prey that may have an offensive taste or be-have aggressively enough toward predators that they might be frightened away.

- camouflage — Using protective coloring or behaving like a harmless plant or inanimate object (e.g. grass, rock), a prey organism can actually "hide" in the open.

- display behavior — Some organisms, when cornered, will behave in such a way as to startle or intimidate an attacker.

- Some organisms produce many young in a short period of time so that a predator cannot possibly eat all of them.

- Chemical defenses such as obnoxious odors can be sprayed toward attackers to drive them away.

- Some organisms hibernate, so that they "wake-up" a generation later and hopefully avoid their predators. The 17-year cicada is an example of an organism that hibernates.

- Some organisms burrow into the ground or take to higher ground to escape predators.

- The parasite/host relationship, while it may be a form of a predator/prey relationship, is primarily different in that the parasite gets nourishment from the host but usually does not kill the host outright. The host may die from weakness or infection, but it is not usually eaten by the parasite.

- Another difference is that the parasite is often smaller than the host, while a predator is usually bigger than the prey.

ESSAY IV

STAGE 1

- Air enters through the nose which is lined with cilia and hairs to trap dust.

- The nasal cavity is lined with epithelial cells which secrete mucus that keeps the air moist and warm as it travels to the lungs.

- The air goes to the pharynx or throat cavity.

- The air then goes to the larynx which contains the vocal chords.

- The air then goes to the trachea which branches into two bronchi, which in turn, subdivide into bronchioles that are located in the lungs.

- The exchange of oxygen for carbon dioxide occurs in the alveoli which are grape like clusters located at the ends of the smallest bronchioles.

- Changes in the pressure gradients between the alveoli and the atmosphere causes air to flow into and out of the chest cavity.

- These changes are the result of changes in the volume of the chest cavity which are caused by the contraction and relaxation of the diaphragm and the intercostal muscles.

- During exhalation the diaphragm contracts, moving downward and flattens while the intercostal muscles act to pull the rib cage up and out. Thus, the column of the chest cavity enlarges.

- Air pressure is then lower in the chest cavity than in the outside atmosphere. The pressure differential causes the air to move into the lungs resulting in the lung tissue enlarging.

- During exhalation, the muscles relax, the diaphragm returns to its original position, the volume of the chest cavity decreases, the air is forced out from the lungs, into the air tubes and out into the atmosphere.

- The respiratory center in the brainstem is called the medulla. It controls the rate of breathing and the amount of air that is taken in.

- The brainstem receives signals from parts of the body such as the lungs about the level of substances such as oxygen and carbon dioxide in the lungs and blood vessels. The respiratory center then sends signals to the diaphragm and intercostal muscles and they respond by contracting or relaxing.

STAGE 2

- The air coming into the alveoli has been warmed and made moist to prepare it for diffusion, a process that requires a thin membrane and a concentration differential.

- Oxygen from the air diffuses from the alveoli into the interstitial fluid (a connective tissue separating the alveoli from the capillaries) and into the capillaries.

- The alveoli have a large surface area which brings them into contact with many capillaries.

STAGE 3

- Oxygen transport is increased by the protein hemoglobin in the blood to which oxygen molecules can bind. The amount of oxygen that can bind to hemoglobin is directly proportional to the partial pressure of the oxygen.

- The hemoglobin gives up the oxygen more readily to a tissue which is undergoing increased metabolic activity and producing increased levels of carbon dioxide.

STAGE 4

- Blood flowing into the capillaries has a lower pressure of carbon dioxide compared to the level of carbon dioxide in the tissues. Therefore, the carbon dioxide diffuses from the tissue into the capillaries.

- Most carbon dioxide in the capillaries is carried in the blood in the form of a bicarbonate ion. This ion is formed when carbon dioxide and water form carbonic acid which then can dissociate into bicarbonate ions and hydrogen ions. This process is also reversible.

- The lower concentration of carbon dioxide in the alveoli (compared to that of the capillaries) causes the carbonic acid to dissociate into water and carbon dioxide. The carbon dioxide then diffuses into the alveoli, then into air passageways, and out into the atmosphere.

REA's **Problem Solvers**

The "PROBLEM SOLVERS" are comprehensive supplemental text-books designed to save time in finding solutions to problems. Each "PROBLEM SOLVER" is the first of its kind ever produced in its field. It is the product of a massive effort to illustrate almost any imaginable problem in exceptional depth, detail, and clarity. Each problem is worked out in detail with a step-by-step solution, and the problems are arranged in order of complexity from elementary to advanced. Each book is fully indexed for locating problems rapidly.

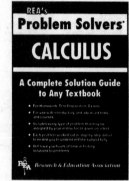

ACCOUNTING
ADVANCED CALCULUS
ALGEBRA & TRIGONOMETRY
AUTOMATIC CONTROL
   SYSTEMS/ROBOTICS
BIOLOGY
BUSINESS, ACCOUNTING, & FINANCE
CALCULUS
CHEMISTRY
COMPLEX VARIABLES
DIFFERENTIAL EQUATIONS
ECONOMICS
ELECTRICAL MACHINES
ELECTRIC CIRCUITS
ELECTROMAGNETICS
ELECTRONIC COMMUNICATIONS
ELECTRONICS
FINITE & DISCRETE MATH
FLUID MECHANICS/DYNAMICS
GENETICS
GEOMETRY
HEAT TRANSFER

LINEAR ALGEBRA
MACHINE DESIGN
MATHEMATICS for ENGINEERS
MECHANICS
NUMERICAL ANALYSIS
OPERATIONS RESEARCH
OPTICS
ORGANIC CHEMISTRY
PHYSICAL CHEMISTRY
PHYSICS
PRE-CALCULUS
PROBABILITY
PSYCHOLOGY
STATISTICS
STRENGTH OF MATERIALS &
   MECHANICS OF SOLIDS
TECHNICAL DESIGN GRAPHICS
THERMODYNAMICS
TOPOLOGY
TRANSPORT PHENOMENA
VECTOR ANALYSIS

If you would like more information about any of these books,
complete the coupon below and return it to us or visit your local bookstore.

RESEARCH & EDUCATION ASSOCIATION
61 Ethel Road W. • Piscataway, New Jersey 08854
Phone: (732) 819-8880   **website: www.rea.com**

Please send me more information about your Problem Solver books

Name _____

Address _____

City _____ State _____ Zip _____

REA's Test Preps
The Best in Test Preparation

REA's Test Prep Books Are The Best!
(a sample of the <u>hundreds of letters</u> REA receives each year)

" I am writing to congratulate you on preparing an exceptional study guide. In five years of teaching this course I have never encountered a more thorough, comprehensive, concise and realistic preparation for this examination. "
Teacher, Davie, FL

" I have found your publications, *The Best Test Preparation...*, to be exactly that. "
Teacher, Aptos, CA

" I used your *CLEP Introductory Sociology* book and rank it 99% – thank you! "
Student, Jerusalem, Israel

" Your GMAT book greatly helped me on the test. Thank you. "
Student, Oxford, OH

" I recently got the French SAT II Exam book from REA. I congratulate you on first-rate French practice tests."
Instructor, Los Angeles, CA

" Your AP English Literature and Composition book is most impressive."
Student, Montgomery, AL

" The REA LSAT Test Preparation guide is a winner! "
Instructor, Spartanburg, SC

(more on front page)